Lionel Rothschild

The avifauna of Laysan and the neighbouring islands

With a complete history to date of the birds of the Hawaiian possessions

Lionel Rothschild

The avifauna of Laysan and the neighbouring islands
With a complete history to date of the birds of the Hawaiian possessions

ISBN/EAN: 9783337274481

Printed in Europe, USA, Canada, Australia, Japan

Cover: Foto ©berggeist007 / pixelio.de

More available books at **www.hansebooks.com**

THE

AVIFAUNA OF LAYSAN

AND THE

NEIGHBOURING ISLANDS:

WITH A COMPLETE HISTORY TO DATE OF THE

BIRDS OF THE HAWAIIAN POSSESSIONS.

BY

THE HON. WALTER ROTHSCHILD, PH.D.

ILLUSTRATED WITH COLOURED AND BLACK PLATES
BY MESSRS KEULEMANS AND FROHAWK;
AND PLATES FROM PHOTOGRAPHS, SHOWING BIRD-LIFE AND SCENERY.

LONDON:
R. H. PORTER, 7 PRINCES STREET, CAVENDISH SQUARE, W.
1893-1900.

CONTENTS.

a 2

Part I. (Pages i–xiv, 1–58) appeared August 1893.

Part II. (Pages 59–126) appeared November 1893.

Part III. (Pages i–xx, (Di.) 1–(Di.) 21, 127–320) appeared December 1900.

PREFACE.

I was induced to take a great interest in the fauna of the Sandwich or Hawaiian Islands when I was an undergraduate at Cambridge, and Professor Newton showed me some of the wonderful species of birds discovered on those islands by Mr. Scott Wilson. At that time I had just engaged Mr. Henry Palmer to make a collection of birds, &c., for me on the Chatham Islands, south of New Zealand, and I determined that if the first trip turned out a success I would send him to the Sandwich Islands. On receipt of a most interesting collection from the Chatham Islands (which included the new Pigeon, *Hemiphaga chathamensis*), I immediately instructed Palmer to start for Honolulu, where he arrived in December 1890, and stayed on the Sandwich Islands till August 1893. During this period he collected 1832 birds on the following islands:—Oahu, Hawaii, Kauai, Maui, Lanai, Molokai, Niihau, Laysan, French Frigate Shoals, Lisiansky, and Midway or Brooks. Although, considering the long time spent on this group of islands, the collection was by no means large, it was most interesting, for Palmer procured all the known resident land-birds, with the exception of *Moho apicalis*, *Chætoptila angustipluma*, *Hemignathus ellisianus*, *Heterorhynchus lucidus*, *Psittirostra olivacea*, *Peunula ccandata*, and *Peunula sandwichensis*, which are all undoubtedly extinct, and most of the sea-birds. In addition to these Palmer discovered fifteen species entirely new to science, and several birds new to the islands. The proof of the exhaustiveness of Palmer's work is that Mr. Perkins, though an experienced zoologist and a thoroughly trained collector, found only one new bird after Palmer left, and failed to get several found by the latter, viz., *Drepanis pacifica*, *Ciridops anna*, *Hemignathus lanaiensis*, *Rhodacanthis flaviceps*, and *Loxops rufa*.

In conclusion, I can only hope that my readers will not find this final part has suffered by the long lapse of seven years between the appearance of the second and third parts of the book. I have much pleasure in thanking all those who have very kindly helped me during the progress of this work, and more particularly Mr. Scott Wilson, Mr. Perkins, Dr. G. Hartlaub, Professor Studer, Professor Brandt in Kiel, Dr. von Lorenz of Vienna, Professor Reichenow, Count von Berlepsch, Dr. H. O. Forbes, Professor Schauinsland,

and also Messrs. F. Spencer, Gay, Robinson, Damon, the Hon. C. Bishop, Mr. Meyer, Brother Matthias, Mr. Johnston, and many other residents on the Hawaiian Islands, who assisted Palmer most generously and courteously.

I also wish to thank Mr. Ernst Hartert for his assistance. He has taken great interest in the work from its commencement, and has helped me a great deal, especially with the synonymy and introductory chapters.

It will be seen that I have only used trinomials in a very few cases, where they are generally adopted by ornithologists. I have not in this work used them for the forms representing each other in the various Hawaiian Islands. As there are not, as a rule, many very similar species in the same islands, the question did not seem to be so highly important in this case, and as I did not begin to use trinomials in the first parts, I thought it better, for the sake of uniformity, not to introduce them more than necessary into the last part, although in future some of the forms (of *Chlorodrepanis* and *Loxops* especially) will probably with more right be degraded to the rank of subspecies.

Descriptions and full synonymies are, as a rule, only given of those birds which are resident and breed in the Hawaiian Possessions.

THE PRINCIPAL LITERATURE

BIRDS OF THE HAWAIIAN ISLANDS.

1780. FORSTER, G. Göttinger Mag. Wissenschaften, i. (6) p. 346.

Description of " *Certhia coccinea* " (our present *Vestiaria coccinea*) from specimens brought to Cassel by Barthold Lohmann, who had sailed with Cook's last expedition.

1781-1785. LATHAM, JOHN. Gen. Synopsis of Birds.

A number of species are described for the first time, apparently all, or nearly all, from Cook's Voyages. These birds received afterwards scientific names from Gmelin (Syst. Nat.), and from Latham himself, in the 'Index Ornithologicus' (1787). (Cf. *Chlorodrepanis virens, Drepanis pacifica, Himatione sanguinea, Psittirostra psittacea, Phæornis obscura, Loxops coccinea, Chasiempis sandwichensis, Hemignathus obscurus, Pennula sandwichensis.*)

[See also the Supplement to the 'General Synopsis,' and the same author's much later 'General History of Birds' (1821-1824).]

1782. ELLIS, WILL. Narrat. Voy. Capt. Cook & Clerke, vol. ii. p. 156.

Mr. William Ellis describes himself in this " irregularly published narrative " as " Assistant-Surgeon to both vessels." A number of poor drawings of birds from his hand—among them that of the extinct *Pennula sandwichensis*—are in the British Museum. His narrative is not important to ornithologists.

1784. KING, JAMES. Voyage to the Pacific Ocean, vol. iii. pp. 119, 120.

A few short notes of little importance. The *Drepanis pacifica, Vestiaria coccinea, Corvus tropicus, Himatione sanguinea, Psittirostra psittacea,* and *Moho nobilis* can be recognized with absolute certainty. Cook's ships landed only on Hawaii, Kauai, and Niihau. Some of the skins of this voyage are still preserved in the Liverpool, Cambridge, and Vienna Museums, and also probably in Göttingen.

1784. MERREM, B. Beyträge bes. Gesch. d. Vögel, fasc. i. p. 8, pl. ii. p. 14, pl. iv.

First scientific notice of our present *Moho nobilis* (sub nomine *Gracula nobilis*) published. Also figure and description of *Vestiaria coccinea,* sub nomine *Mellisuga coccinea.*

[1786. Merrem : Avium rariorum et minus cognitarum, Icones et Descriptiones—the same in Latin !]

1789. DIXON, G. A Voyage round the World . . . . . . in the King George' and 'Queen Charlotte.'

On pages 55-61 the stay in the Sandwich Islands, principally Oahu, is described. In the Appendix, on plate 8 is distinctly figured, under the name of " Yellow Tufted Bee Eater," the extinct *Moho apicalis* ! Latham's description of the " Yellow Tufted Bee Eater " is copied, but the author " takes the liberty to add, that the specimen from which the annexed engraving was made differed from that described by Latham in having all the tail-feathers spotted with white at the ends " !

1826. BLOXAM, A.   Voy. II.M.S. 'Blonde.'   Appendix.

This Appendix is, to use Professor Newton's words, "a disgrace to all concerned, since, so far from advancing the knowledge of the subject, it introduced so much confusion as to mislead many subsequent writers"[1]. The blame, however, does not rest on Bloxam, as his work was edited by a lady, Mrs. Maria Graham, who had only a few of Bloxam's notes to guide her. Bloxam had, indeed, made a fairly good collection, having obtained nine species of land-birds, and among them an otherwise unknown and now evidently extinct species of *Phæornis* on Oahu![2] All the specimens were placed, properly labelled, at the disposal of the Lords of the Admiralty, but most of them have long ago disappeared!

It will, perhaps, be interesting to my readers to peruse the following letter and extracts from letters of Bloxam to Swainson, which belong to the Swainson correspondence recently purchased by the Linnean Society, and which I reprint by the kind permission of the President and Council of the Linnean Society :—

Copy of Letter from BLOXAM *to* W. SWAINSON.

Valparaiso, Sept. 18, 1825.

DEAR SIR,—

    I have just received your kind letter dated the 27th of last November, and much regret that I did not receive it before. I have been very unsuccessful in my collection—having besides the Sandwich Islands only touched at two other small islands in our voyage hither, and at both we were not on shore more than two hours; one was tenanted by nothing but sea birds; from the other I procured a beautiful small dove, a kingfisher, and starling. I have well examined into the nature of those birds peculiar to the Sandwich Islands, I mean those with curved bills, and can confidently affirm that they bear no relation to any other species of bird that I am acquainted with. I hope on my return, which I expect will be in March or April, to furnish you with all my notes relative to them. I am sorry to say that I can hear nothing of the package of books which were sent from Liverpool, for which however I still feel greatly obliged, tho' I have all the books you have been kind enough to send me, by me. The following comprise my natural history books :—Graves's 'Naturalist's Guide,' Bullock's 'Directions Taxidermy,' Turton's 'Linnæi Systems,' and Mawe's 'Collector's Pilot' and his 'Linnæan Conchology.' I regret to say that I was very unsuccessful in procuring shells at the Sandwich Islands, not having a single mitre and but few cones which are well known as the *C. ebræus.* I have procured however a good selection of the beautiful little land-shells, amounting to about ten varieties, some with reversed mouths, they apparently belong to the volutes of Mawe's Linnæan System. Insects there are scarcely any ; I have found but one butterfly and a beautiful sphinx moth, no coleopterous insects or any in the water ponds did I observe—these islands are certainly very barren in this respect. With regard to birds I only met with one bird of prey, a brown owl which I was unable to procure. I have preserved all the tongues of the different birds I obtained. Everything that I collect belongs to the Government : but my Brother I hope will give you duplicates of most of the Sandwich Island birds, also the land-shells. On the coast of Chili particularly in this bay I have met with a great variety of the Chiton, different species of the Patella comprising the *crepidula, fissurella & infundibulum*; also many turbo's and trochus and two species of land-shells; some of the chitons grow to an immense size. I have measured some upon the rocks which measure more than 6 inches ; among these the *C. spinosus* is common. there are eight or nine varieties of them. The *Buccinum Concholepas* also is very common here, and I hope my Brother will present you with a series of them. We are going this week to Concepcion, where I expect to find some new varieties—the opportunities I have had for collecting have been so bad and few that I can hardly call what I have procured a collection, I have not more than 100 birds, and few or scarcely any rare shells ; the only place where I had some chance of adding to my collection was the small island of Mauti about 400 miles S.W. of Otaheite ; our stay here however was only two hours, we were unable to touch at Otaheite on account of the wind being very adverse the whole time. I met with no gulls at the Sandwich Islands. At the Galapagos, where we were for two days under a most burning sun and where I had a narrow escape of my life in consequence of the great heat, I procured only one dove ; several birds that I had reserved for stuffing one evening were so bad in the morning that they were obliged to be thrown overboard. The heat prevented our doing anything there and we were glad to leave them, the Sun at the time crossing the line. Sea birds comprising different species of the tern, petrel and pelecanus were very numerous ; I met only with one species of the penguin there, which was small and nothing remarkable about it. I have procured about a dozen different varieties of land birds from here,

---

[1] Introduction to Wilson's 'Aves Hawaiienses,' p. xiii.   [2] *Phæornis oahensis,* p. 309.

but the country is in such a state that it is not safe to go about by oneself, and it is not often that I can find persons to accompany me. I am not allowed even from the ship a boy or sailor to accompany me to carry any thing for me, and in these hot countries I find it sufficient to carry only my gun, powder and shot. I have also other difficulties to encounter which have precluded me making the best use of my time. The observations I have made on my return will I hope be not uninteresting to you. I met with no swallows or any species of *Caprimulgus* at the Sandwich Islands; I was quite astonished at the paucity of birds there; one species of bird from which the yellow feathers are procured is so scarce that the whole time I was on the several islands I did not observe one and with great difficulty procured a single tolerable specimen from one of the natives; he asked a high price for it. These birds have only a few yellow feathers under the wings which are paid as tribute to the chiefs, and they are so scarce that a dollar is frequently paid by the natives themselves for a pair; they are nearly exterminated. The red birds are more common tho' scarce. I was obliged to trust to a native to procure these as in all my excursions I did not shoot more than three of them. I was frequently out the whole day without killing any thing. I have however described accurately and minutely everything respecting the Sea birds which I have procured, and all my notes will be at your service when I arrive home.

I remain, Dear Sir, with many thanks for your kind and interesting letter,
Your obliged & faithful Servant,
(signed) ANDREW BLOXAM.

Extract from a Letter by BLOXAM *to* W. SWAINSON, *dated " Rugby, March 1826."*

" But a friend of mine on board ship intends sending me a few specimens which I hope you will accept of, though without mentioning my name or from whom they come. They will be, I believe, three different species of the birds belonging to the Sandwich Islands, and I have enumerated four in my journal which I have written for the Admiralty. From their peculiar habits I have termed them the *Nectarina* class. It is from these birds that the feathers for ornamenting the chiefs are procured, and one species, the *Nectarina nigra*[1], is so scarce that during the whole time I was in the islands I could only procure one specimen with the tail- and other feathers perfect, and this was procured me by a native. This of course I sent to the Admiralty; but there will be the head and legs of this bird, the former with the tongue well preserved, which I hope to be enabled to send you. It was procured by a friend of mine, who did not think the rest of the bird worth preserving as it was bare of feathers. I send you with this my papers where I originally described the birds, and from these I wrote out a fair copy for the Admiralty. I am uncertain yet what will be done respecting it, and should therefore wish nothing to be published from my notes, as without doubt there are many inaccuracies in it. I procured a tolerable collection of birds from the coast of Chili, but upon the whole have not succeeded in bringing home more than one hundred. . . . . Of the Sandwich Islands I have described accurately the few birds which inhabit them at Woahoo. I was out with my gun day after day without scarcely seeing or killing one bird. I found them, however, more numerous in Owhyhee. . . . . The islands are very barren in affording objects of natural history. I observed only one species of *Papilio* which I hope to send you. There are no beetles or Coleopterous insects, and nothing remarkable in entomology. The owl is the only bird of prey. Few sea birds are found on the coast, but at the Galapagos I was very much struck at the vast quantity of different species of them."

1839. LICHTENSTEIN, H. Abh. kön. Akad. Wissenschaften, Berlin, 1838, p. 440, pl. v.

Genus *Hemignathus* established, *H. lucidus* described.

1839. DE LAFRESNAYE, in Magasin de Zoologie, deuxième série, première année, Oiseaux, pl. 10 and text.

Description of *Heterorhynchus* (nov. subgen.) *olivaceus* = *H. lucidus*.

[See also 'Voyage autour du Monde sur la frégate 'La Venus,' Atlas de Zoologie,' Oiseaux, pl. i. (1846), text, by Prévost and Des Murs, pp. 183-193 (*Heterorhynchus lucidus*).]

[1] *Drepanis pacifica* or *Moho nobilis*.—W. R.

1848. PEALE, TITIAN.   U.S. Expl. Exp. (Zool.), Birds.

Only a few copies of this work exist, nearly the whole edition having been destroyed soon after its issue. Nevertheless of great importance for the Ornithology of the Pacific Ocean. Described for the first time from the Sandwich Islands :—*Buteo solitarius, Corvus hawaiiensis, Chætoptila angustipluma*; many others mentioned.

1850. CABANIS, JEAN.   Mus. Hein. vol. i. pp. 99, 100.

Himatione chloris and *Himatione maculata = Oreomyza maculata* described.

1852. HARTLAUB, G., in Archiv für Naturgeschichte, Heft i. pp. 93–138.

Review of the ornithological results of the U.S. Exploring Expedition.

1854. HARTLAUB, in Journ. f. Orn. p. 170.

List of 30 species supposed to be known from the Sandwich Islands. First list of Hawaiian Birds !

1858. CASSIN, JOHN.   U.S. Exploring Expedition: Mammalogy and Ornithology.   With Atlas.

This work may be called a new edition of Titian Peale's work of 1848. While many mistakes are corrected, some alterations are most unhappy and erroneous.

In the Atlas is figured : *Buteo solitarius* (sub nomine gen. " *Pandion* ! "), *Corvus hawaiiensis, Phæornis obscurus* (s. n. gen. *Tænioptera), Chætoptila angustipluma* (s. n. gen. *Mohoa), Fulica alai*.

1860. GOULD, J.   Proc. Zool. Soc. Lond. p. 381.

Moho apicalis, sp. nov.

1869. DOLE, SANFORD.   Synopsis of Birds hitherto described from the Sandwich Islands. (Proceed. Boston Soc. Nat. Hist. vol. xii. pp. 294–309.)

Forty-eight species enumerated. Most valuable, though full of errors.

1871. SCLATER, P. L.   Remarks on the Avifauna of the Sandwich Islands. (Ibis, vol. i. 3rd series, pp. 356–362.)

Critical and other remarks on Dole's work of 1869.

1872. PELZELN, A. von.   Ueber Geschlechtsdifferenzen bei den Meliphagiden der Sandwich-Inseln. (Journ. f. Orn. 1872, pp. 24–30.)

This article contains some mistakes ; for example " *Himatione flava* Bloxam " is considered to be the adult male of *H. chloris* Cab. ! We are told that the Vienna Museum possesses a pair of *Drepanis pacifica* from the Leverian Museum, and some of Deppe's specimens from Oahu.

1877. OUSTALET, M. E.   Bull. Soc. Philom. Paris, p. 99.

Loxioides bailleui described.

1877. STREETS, TH.   Description of a new Moorhen from the Hawaiian Islands. (Ibis, pp. 25–27.)

Figure : frontal shield of *Gallinula sandwichensis*.

1878. SCLATER, P. L. Proc. Zool. Soc. Lond. pp. 340-351.

List of 13 species collected during the voyage of H.M.S. 'Challenger' on Hawaii. *Anas wyvilliana* described for the first time.

[See also Report Scient. Res. Challenger, Zool. vol. ii. pp. 93-99, which is a reprint of above with a few additions and plates of *Buteo solitarius* and *Anas wyvilliana* (1881).]

1879. DOLE, SANFORD. List of Birds of the Hawaiian Islands. Corrected for the Hawaiian Annual, with valuable additions.

This is usually quoted as 'Hawaiian Alman.,' 1879. It is considered as a new edition of the work of 1869. 53 species enumerated. Some errors are corrected, and four species described as new, of which however, two only belong to really undescribed forms.

[In 'The Ibis' for 1880 (pp. 240-242) the descriptions of the four supposed new species are reprinted, which is most useful, as it is very difficult to get a copy of the article in the 'Hawaiian Annual and Almanac.']

1880. FINSCH, O. Letter from the Pacific. (Ibis, pp. 77-81.)

Dr. Finsch made a short stay on the Sandwich Islands, and gave some interesting notes. The several weeks which this ornithologist stayed at Maui were, however, not very fruitful.

1885. SHARPE, R. B. Cat. Birds Brit. Mus. vol. x. pp. 3-9, 49-51 (*Drepanidæ*).

1885. SCLATER, P. L. On the Muscicapine Genus *Chasiempis*. (Ibis, pp. 17-19, pl. i.)

1887. STEJNEGER, L. Birds of Kauai Island, collected by Mr. Waldemar Knudsen. (Proc. U.S. Nat. Mus. vol. x. pp. 75-102.)

Very valuable, especially for Kauai. *Himantopus knudseni*, *Phæornis myiadestina*, *Himatione parva*, and *Oreomyza bairdi* described for the first time.

1888. RIDGWAY, R. Note on *Æstrelata sandwichensis*. (Proc. U.S. Nat. Mus. vol. xi. p. 104.)

1888. STEJNEGER, L. Further Contributions to Hawaiian Avifauna. (Proc. U.S. Nat. Mus. vol. xi. pp. 93-103.)

Second collection from Kauai.

1888. WILSON, SCOTT B. On *Chloridops*, a new Generic Form of *Fringillidæ* from the Island of Hawaii. (Proc. Zool. Soc. Lond. p. 218.)

1889. STEJNEGER, L. Notes on a third Collection of Birds made in Kauai. (Proc. U.S. Nat. Mus. vol. xii. pp. 377-386.)

1889. WILSON, SCOTT B. On three undescribed Species of the Genus *Hemignathus*. (Ann. & Mag. Nat. Hist. ser. 6, vol. iv. p. 400.) (Appeared in November.)

1889. CABANIS, JEAN. Journ. f. Orn. p. 331.

Description of *Hemignathus procerus* from "Kauai" (*sic* instead of Kauai). The author, learning that Mr. Wilson made inquiries about the forms of the genus *Hemignathus*, of which he was going to describe a new species, hurried to describe a specimen sent to the Berlin Museum by a Mr. Wentscher; but the description would have been behind that of Wilson's *H. stejnegeri*, which appeared in November, had it not appeared in the 'Vossische Zeitung' of September 14th !! (see editorial note, J. f. O. 1889, p. 331). The October number of the J. f. O. did *not* appear in October.

1889. WILSON, SCOTT B. Description of some new Species of Sandwich-Island Birds. (Proc. Zool. Soc. Lond. pp. 445-447.)

Chrysomitridops, Loxops, Himatione.

1890. BERLEPSCH & LEVERKÜHN. Ornis, vol. vi. pp. 2-4.

About *Chasiempis*. The authors having only material from Oahu (Kiel Museum, Behn coll.), erroneously united the forms from the various islands.

1890. WILSON, S. B. On some of the Birds of the Sandwich Islands. (Ibis, pp. 170-196, pl. vi.)

Hemignathus hanapepe and *H. stejnegeri.*

1890. WILSON, S. B. On a new Finch from Midway Island, North Pacific. (Ibis, p. 339, pl. ix.)

Telespiza cantans, erroneously said to have come from Midway Island.

1891. WILSON, SCOTT B. On the Muscicapine Genus *Chasiempis*, with a Description of a new Species. (Proc. Zool. Soc. Lond. p. 164.)

1891. WILSON, SCOTT B. Description of a new Species of the Genus *Himatione*. (Ibid. p. 166.)

1891. WILSON, SCOTT B. Descriptions of two new Species of Sandwich-Island Birds. (Ann. & Mag. Nat. Hist. ser. 6, vol. vii. p. 460.)

Himatione mana and *Phæornis lanayensis.*

1892. WALLACE, A. R. Chapter XV. "The Sandwich Islands," in 'Island-Life,' 2nd edition, pp. 310-329.

On pp. 314-315 a list of 37 *Passeres* then known is given. That this is incomplete and not without errors is only natural, considering how much we have learnt since 1892, but it is a valuable list. The impression of the author is (p. 316) that the affinities of the Sandwich-Island Birds "are chiefly with Australia and the Pacific Islands; but they exhibit in the buzzard, one of the owls[1], and perhaps in some of the *Drepanididæ*, slight indications of very rare or very remote communication with America." In my opinion, however, the American element is more recent.

[See also Wallace's 'Island Life,' 1st edition, 1880-81.]

1892. ROTHSCHILD, W. Descriptions of seven new Species of Birds from the Sandwich Islands. (Ann. & Mag. Nat. Hist. ser. 6, vol. x. pp. 108-112.)

Two new genera—*Rhodacanthis* and *Viridonia*—established; seven new species, of which two—*Bernicla munroii* and *Telespiza flavissima*—are not valid.

---

[1] This is based on the erroneous statement of Cassin, that the Short-eared Owl of the Sandwich Islands is the Galapagos Island form.

1892-93. ROTHSCHILD, W. In Bull. B. O. Club, vol. i. p. xvii (*Anas laysanensis*, sp. n.); p. xxiv (*Hemignathus lanaiensis*, sp. n.) ; pp. xxxv, xxxvi (*Pseudonestor xanthophrys*, gen. & sp. n.) ; pp. xli, xlii (*Acrulocercus bishopi*, *Himatione newtoni* and *wilsoni*, spp. nn.) ; p. xlviii (*Diomedea immutabilis*, sp. n.) ; p. lvi (*Loxops wolstenholmei*, sp. n. (=*rufa*)) ; p. lvii (*Anous hawaiiensis*, sp. n.).

1893. PERKINS, R. C. L. Notes on Collecting in Kona, Hawaii. (Ibis, pp. 101–112.)

Most interesting observations, and it is only to be regretted that similar notes from the experience of the author are not published about other islands.

1893. NEWTON, A. Proc. Zool. Soc. Lond. p. 690.

Description of *Drepanis funerea*.

1893. ROTHSCHILD, W. Description of three new Birds from the Sandwich Islands. (Ibis, pp. 112–114.)

Heterorhynchus affinis, *Loxops ochracea*, and *Palmeria mirabilis* = *P. dolei*.

1893-94. ROTHSCHILD, W. In Bull. B. O. Club, vol. iii. p. ix (*Palmeria mirabilis* = *Himatione dolei*); p. xxv (*Palmeria* a genus of *Drepanidæ*) ; p. xlii (*Chætoptila* exhibited).

1894. ROTHSCHILD, W. In Novitates Zoologicæ, vol. i. p. 692.

Pseudonestor xanthophrys a *Drepanine* bird.

1895. ROTHSCHILD, W. In Nov. Zool. vol. ii. p. 54.

On the *Loxops* of Oahu.

1895. HARTLAUB, G. Beitrag z. Geschichte der ausgestorb. Vögel der Neuzeit, sowie derjenigen deren Fortbestehen bedroht erscheint. In Abhandlung. d. naturwissenschaftl. Vereins zu Bremen, xiv. 1 Heft.

A number of Hawaiian birds spoken of at length.

[See also the same, 2nd edition, "Als MS. gedruckt," 1896.]

1895. BEDDARD, F. Text-book of Zoogeography, p. 203.

General statements about Hawaiian fauna compiled from other authors.

1897. NEWTON, A. Proc. Zool. Soc. Lond. pp. 892, 893 (plate li. figs. 6, 7).

Description and figures of eggs of *Himatione virens*.

1898. FINSCH, O. Notes Leyden Museum, vol. xx. p. 77.

Description of "*Pennula wilsoni*" from the only—and perhaps typical—specimen of *Pennula sandwichensis* Gm.

1890-99. WILSON, SCOTT B., & EVANS, A. H. Aves Hawaiienses : the Birds of the Sandwich Islands. (In 8 parts, xxvi and 257 pages, 72 plates.)

This is the most important and (except the present work) the only illustrated monograph of the Hawaiian birds. Special charm and importance is given to it by the fact that the author himself has made two journeys to the islands. Dr. Gadow has contributed "Remarks on the Structure of certain Hawaiian Birds, with reference to their systematic position ;" and Professor Newton an account of the discovery of the islands and of their visitors down to the year 1891, partly reprinted from 'Nature,' vol. xiv. pp. 465 *et seqq.* (March 1892).

1899. ROTHSCHILD, W. In Bull. B. O. Club, vol. viii. p. lvi.

Telespiza flavissima = cantans.

1899. SCHAUINSLAND, H. Drei Monate auf einer Kornlleninsel (Laysan). (Bremen, Verlag von Max Nössler.)

A most fascinating little book, containing many notes on birds. On p. 101 a list of the birds of Laysan. On p. 100 a more detailed article about the birds in the Journ. f. Orn. is promised, but has not appeared till now, September 1900.

1899. SCLATER, W. L. & P. L. Geography of Mammals, pp. 21 & 49.

A "Hawaiian Subregion" recognized.

1900. HENSHAW, H. W. Occurrence of *Larus glaucescens* and other American Birds in Hawaii (Auk, vol. xvii. pp. 201–206); Description of a new Shearwater (*Puffinus newelli*) from the island of Ulani (ibidem, p. 246).

1900. SCHAUINSLAND, H. Ein Besuch auf Molokai. Separat-Abdruck aus Abh. nat. Ver. Bremen, Bd. xvi. Heft 3.

Most interesting account of the Leper island, with a few short notes on birds on pages 12, 13, and 14.

ON THE ORIGIN AND DISTRIBUTION

OF THE

HAWAIIAN AVIFAUNA.

———————

THERE is in the Sandwich Islands evidently an older and a more recent avian element. We may safely assume that the most modified forms which are restricted in their distribution to the Hawaiian Group, and form now the so-called "*Drepanidæ*" or "*Drepanididæ*," are the descendants of the oldest bird-inhabitants. The next in age seem to be the *Meliphagidæ*, *i. e.* the genera *Moho* and *Chætoptila*, and the Flycatchers (*Chasiempis*); while those species which are identical with or nearest and still very little removed from American forms are the most recent arrivals. As such close allies of American forms we have to regard *Himantopus knudseni, Fulica alai, Gallinula sandwichensis*, and apparently also the *Buteo*, while *Nycticorax nycticorax nævius*, though resident, cannot be separated from the American form, and a constant immigration from America is now going on every autumn (see, for example, Henshaw's remarks under *Larus glaucescens*, p. 286).

The limits of the family *Drepanidæ* have been very uncertain, and their origin is by no means clear. Dr. Gadow, in 1891, allowed only the genera *Drepanis, Vestiaria, Himatione, Chlorodrepanis, Loxops, Oreomyza, Hemignathus*, and *Heterorhynchus* to belong to that family. *Psittirostra* (which had been recognized as *Drepanine* by Cabanis and Sclater) and *Loxioides* he rejected to the *Fringillidæ*, their tongue being "truly *Fringilline*," as well as their alimentary canal and the rest of their internal and external characters. Recently (1899) Dr. Gadow believes that all the Finch-like birds of the Hawaiian Islands, *i. e.*, not only *Psittirostra* and *Loxioides*, but also *Chloridops, Rhodacanthis* (and I may add *Telespiza*), and *Pseudonestor*, are *Drepanidæ*. It is thus evident that his former assertions of the Fringilline structure were less valuable than the mere external knowledge of the systematical skin-ornithologist, who, by the wonderful transition (through *Pseudonestor*!) from the bill of *Hemignathus* to that of *Psittirostra* is bound to conclude that these forms belong to the same "family." It is, on the other hand, not necessary for the ordinary skin-ornithologist to connect *Rhodacanthis* and *Chloridops* with these forms, nor does Dr. Gadow give any reasons for his belief. On the contrary, he admits that we cannot define either *Drepanidæ* or *Fringillidæ*, *Cærebidæ* or *Tanagridæ*. It is true that Dr. Gadow says Mr. Perkins has arrived at the notion that all these birds are *Drepanidæ* from the study of the habits, the voice, and the peculiarly strong and disagreeable scent of these birds. There is, however, scarcely enough peculiarity in the habits and voice of these birds to define their systematic position, and the peculiar smell they have is equally strong or stronger in the genus *Moho*, which undoubtedly belongs to the *Meliphagidæ*[1]. This is even now noticeable in dried-up skins. It is thus really the external morphology, and not the anatomy, that has altered Dr. Gadow's view. The view, nevertheless, is apparently quite correct.

On the other hand, I cannot at all understand the force of Dr. Gadow's contention that the *Drepanidæ* are derived from forms of the south-east, which are offshoots of the Colombian fauna. If we, as Dr. Gadow himself admits, do not know the origin of the *Drepanidæ*, and if he "hinted" merely at the *Cærebidæ*, and if his hint is only "probable," we cannot from that conclude that the *Drepanidæ* are of "Colombian" origin. We must therefore consider the oldest indigenous avifauna of the Sandwich Islands as being of uncertain origin.

To sum up. There are three differently-aged stocks of bird-population :—

1. A very old original one, the origin of which is uncertain.
2. A distinctly Polynesian branch, which is also rather old, but probably less so than No. 1.
3. An American stock, which is the most recent.

With regard to the affinities between the birds of the various islands, it is quite evident that the more distant islands have a more modified avifauna. Thus the differences between Kauai, Oahu, Hawaii, and the central group of islands are most marked and about equally apparent, while the islands of the central group (Molokai, Lanai, and Maui) have many forms in common. Thus Lanai and Molokai have the same form of *Phæornis*, the genus *Chasiempis* is absent from all the three islands, the *Chlorodrepanis* of all the three is the same, *Palmeria* is found on Maui and Molokai; while *Loxops*, *Hemignathus*, *Drepanorhamphus* are each confined to a single island, but not represented by allied forms on the other islands of the central group.

The avifauna of Hawaii is by far the richest. There alone are the wonderful thick-billed forms, such as *Chloridops*, *Rhodacanthis*, *Loxioides*; there live the Raven and the Buzzard, the *Ciridops* and the Goose; and Hawaii was also the only known home of the extinct *Chætoptila* and *Pennula millsi*—all forms without representative allies anywhere in the group.

---

[1] It must also be noticed that in Mr. Perkins's own publications in the 'Ibis,' 1895, p. 118 (and foll.), the *Pseudonestor* as well as the *Chloridops* are most distinctly spoken of as *Fringillidæ*.

The ornis of the far-outlying island of Laysan is quite different, though the presence of the *Himatione*, which is closely allied to *H. sanguinea*, and the *Telespiza*, which is not far from *Rhodacanthis*, show clearly the affinity with the Sandwich Islands.

As far as our knowledge goes, Oahu has lost more species than the other islands. The *Psittirostra*, *Hemignathus ellisianus*, *Heterorhynchus lucidus*, *Moho apicalis*, *Phæornis oahuensis* are evidently extinct, and the *Loxops* is on the verge of extinction. On Hawaii at least one (or two?) species of *Pennula* and the *Chætoptila* are gone, while *Ciridops*, *Drepanis*, and perhaps others are doomed.

It is, however, to be feared that on most of the islands species have disappeared of which we have no trace or knowledge left. This may be inferred from the isolated occurrence of *Hemignathus lanaiensis* and *Drepanorhamphus funereus*, and the very imperfect knowledge of the Hawaiian fauna before Wilson's, Palmer's, and Perkins's explorations of the islands.

A peculiar feature is the absence in the Hawaiian Archipelago of some families of birds which are otherwise numerous and characteristic for the islands in the Pacific Ocean. Thus the *Pigeons* and *Kingfishers* are entirely absent, and an *Acrocephalus* is only found on Laysan.

LIST OF PLATES.

[1] On Plate " *Himatione*."

d

[NOTE BY E. HARTERT.—The figures 44 to 47 and 27 to 34 show the great similarity between the bills of *Hemignathus* (*Heterorhynchus*), *Pseudonestor*, and *Psittirostra*. Many years ago Cabanis and Sclater had already noticed the fact that *Psittirostra* is a *Drepanine* bird, but the *Pseudonestor* shows the most wonderful transition. The tongue of *Pseudonestor* (54 and 54 *a*) is much frayed out, and shows the position of the bird at a glance. Nevertheless, Mr. Rothschild was, when first describing this form, perfectly right in comparing it with *Psittirostra*, only making the mistake that at the time he accepted Gadow's view that the latter was a *Fringilline* bird. Dr. Gadow has since not only admitted that *Psittirostra* is *Drepanine*, but he places also all the other thick-billed birds of the Sandwich Islands among the *Drepanidæ*. The tongue of *Psittirostra* is very peculiar, and differs much from the tongues of true *Fringilline* birds, such as *Passer*, *Fringilla*, *Chloris*, and *Coccothraustes*, which have a much thicker, more compact, and less frayed-out tongue. The figure given by Dr. Gadow of the tongue of *Psittirostra*, in his celebrated chapter on Hawaiian birds in the 'Aves Hawaiienses,' is evidently totally schematical, and does not at all look like the *Psittirostra*-tongue before me, which are distinctly frayed out at the tip and hollowed out along the upper surface as if laterally contractile into a sort of tube, or, at least, developed from a tubularly-contractile form of a tongue. Dr. Gadow seemed formerly to attach value to the form of the crop; but not only Finches have a big and pouch-like crop, but also *Panurus biarmicus*, which is, in my opinion, a Tit, and certainly not a Finch. It is to me evident that the crop is much sooner altered by the food than the external characters used for classification by the "ordinary cabinet-ornithologist." Though, however, seeing much in *Pseudonestor* and *Psittirostra* that differs from the *Fringillidæ*, it seems to me that *Telespiza*, *Rhodacanthis*, and *Chloridops* cannot by the form of their bills or their tongues be separated from the Finches. Their bills differ much from that of *Psittirostra* (see the figures !), and their tongues are much thicker and more compact at the tips, as if quite unmovable and in no way laterally contractile. Why, then, are they not *Fringillidæ*? It is quite possible that Dr. Gadow's recent view that they are *Drepanidæ* in disguise is correct, a view which is also accepted by Mr. Rothschild in this work. It is, however, a bad sign for our present state of systematic knowledge of the *Passeres* that morphologists like Dr. Gadow are in uncertainty about a bird being a Finch or a *Drepanide*. His view that all (or, better, most of) our so-called families of *Passeres* do not deserve that rank is certainly correct, but, nevertheless, there are subdivisions, and it is hardly advisable to take geographical distribution as a character to define a family, as Dr. Gadow (Aves Hawaiienses, Part vii.) thinks we might have to do in future. With little time and osteological experience, and not enough material at hand, I cannot say much about the palatine bones of these birds, but it seems to me, from comparing those of *Hemignathus*, *Pseudonestor*, and *Psittirostra*, that they alter surprisingly in connection with the form of the bill, and that slight differences in the roof of the mouth of *Psittirostra* and *Telespiza* mean therefore nothing. Nevertheless, there is for me no proof that ALL Hawaiian birds are *Drepanidæ*, while *Psittirostra* doubtless *is* a *Drepanine* bird. More investigations about the anatomy, biology, and nidology of these birds are certainly needed.]

THE

AVIFAUNA OF LAYSAN

AND THE

NEIGHBOURING ISLANDS;

WITH A COMPLETE HISTORY TO DATE OF THE

BIRDS OF THE HAWAIIAN POSSESSIONS.

BY

THE HON. WALTER ROTHSCHILD.

*Illustrated with Coloured and Black Plates by Messrs. Keulemans and
Frohawk, and Collotype Photographs.*

LONDON:

R. H. PORTER, 18 PRINCES STREET, CAVENDISH SQUARE, W.

1893.

(To be completed in **Three Parts**, price **£3 3s.** each, Net.)

I. THE ISLAND OF LAYSAN.

INTRODUCTORY NOTES.

SCATTERED about in a north-westerly direction from the Sandwich Islands are a number of small islands, rocks, and reefs—namely, Necker Isle, French Frigate Shoals, Gardner Island, Maro Reef, Lisiansky, Laysan or Möller Island, and some others. Although ornithologists have had their eyes on these islands for a long time, they had not been trodden by the foot of any ornithologist until I sent my collector, Henry Palmer, to explore them.

From the results of his collecting-tour it was evident that none of these islands are inhabited by land-birds, except Laysan or Möller Island. This island is therefore by far the most interesting one to ornithologists.

The island was named Möller by Capt. Stanikowitch, in 1828, after his vessel; but it had been discovered before by an American ship and received the name of Laysan, under which it is best known at the present time. (See Findlay, 'North Pacific Ocean and Japan Directory,' p. 1113.)

Captain Brooks, in 1859, describes it as follows:—" Laysan Island is in lat. 25° 46′ N., long. 171° 49′ W., is 3 miles long and 2¼ miles broad, and covered with a luxuriant growth of shrubs. It is surrounded by a reef about half a mile from the land. Outside of this reef there is a bank 5 miles wide, on which I found from 14 to 19 fathoms water. There is a boat-passage inside the reef nearly the whole way round the island, the only obstruction being on the south and S.E. sides. Good landing can be found anywhere, excepting on the south and S.E. sides; good anchorage anywhere on the west side; the best, however, is about half a mile from the S.W. point, in from 8 to 12 fathoms water. It can be approached from any point of the compass, no dangers existing within half a mile of the reef.

" There is a lagoon on the island, about 1 mile long and half a mile wide, with 5 fathoms

a

water in the centre, and coral bottom. On the shores of this lagoon I found salt of good quality.

"There are (1859) five palm-trees on the island, 15 feet high, and I collected twenty-five varieties of plants, some of them splendid flowering shrubs, very fragrant, resembling plants I have seen in gardens in Honolulu. I saw on the beach trunks of immense trees, probably drifted from the N.W. coast of America. The island contains about 50 acres of good soil. It is covered with a variety of land- and sea-birds; some of the land-varieties are small and of beautiful plumage. Birds' eggs were abundant.

"There is a very small deposit of guano on the island, but not of sufficient quantity to warrant any attempts to get it. Dug a well and found very good water. The reefs here abound in fish and turtle."

But a long time before, in 1834, the well-known German traveller and ornithologist, F. H. von Kittlitz gave, in the 'Museum Senckenbergianum,' vol. i. pp. 117 *et seq.*, a highly interesting account of Laysan and some of the other islands, and a list of all the birds that were observed.

Unfortunately Herr von Kittlitz was on board the 'Senjawin' and not on the 'Moller,' which visited the islands in March 1828, "at the commencement of the breeding-season of the innumerable birds which live on these lonely spots, and are especially plentiful on the small flat island that was then discovered for the first time, and was named Moller, after the ship."

"Unfortunately," Herr von Kittlitz proceeds to say, "nobody on board the 'Moller' was able to do anything of importance for natural history; but the ship's surgeon, Herr C. Isenbeck, did his best to bear all he saw in mind, and to prepare and keep as many of the birds, which were mostly caught by hand, as the very unfavourable circumstances allowed him to do. The following notes are written down from his and his companions' reports given to me when we met later in Kamtschatka."

I here translate the most important information given by Kittlitz, *l. c.*:—

"The crew of the 'Moller' did not land on the rocky island of Necker, but they saw enormous quantities of different birds around it. They were too far off to recognize the species.

"Herr Isenbeck landed on Gardner, although the landing was very difficult and dangerous, and a very small part of the island only was accessible. Most of the birds kept to the inaccessible high part, and therefore very few eggs were found.

"On March 12 (24) Herr Isenbeck landed on Moller (Laysan), which was originally a coral-island with a long reef round it. It seems that it was raised higher and became a real island from the accumulations of the birds' excrements. It is covered with a strong bushy kind of grass and partly with low shrubs, between which a few pigmy palms had grown up. Although there was no fresh water on the island, there were not only sea-birds but also several land-birds, as the following list will show. Most of the larger birds were already breeding, or had paired at least.

"On the 22nd of March (April 3), when on Lisiansky Island, they again found all the larger birds as on Laysan, and mostly breeding; but none were found that they had not seen on Laysan.

"The following is the list I was able to compile from the information I had received:—

"1. *Diomedea* (an *exulans* ?)[1]. Albatross: white, with flesh-coloured bill, varying with white; grey and black wings. Plentiful on Gardner, where they seemed to live on the highest parts, Moller and Lisiansky, where they lived in a like manner to no. 2. Its voice resembles somewhat that of most of the Gulls, but is not so strong, and more like a cackle with a howling sound in it.

"2. *Diomedea* (an *fuliginosa* ?)[2]. Albatross: chestnut-brown, with black bill and feet; bill with a white line along it; size about the same as no. 1. Common on Laysan and Lisiansky, living on the flat ground: extremely foolish and fearless; can be caught with the hands; must run a good distance before being able to get up, and stands still if coming across anything. If two meet they bow to each other, uttering a low cackling. When Herr Isenbeck met one he used to bow to it, and the Albatrosses were polite enough to answer, bowing and cackling. This could easily be regarded as a fairy tale; but considering that these birds, which did not even fly away when approached, had no reason to change their customs, it seems quite natural. The voice is similar to that of no. 1. Its nest is a lump of earth, with a hole in the middle, in which the single egg is placed with the point downwards[3]. The breeding-season was over on Laysan; only bad eggs were found, but a good many young ones, which were covered with grey down and fed with fish by both parents.

"3. *Tachypetes aquilus.* Obviously the same as Buffon's figure of the Great Frigate-bird, which is the real *Pelecanus aquilus* of Linné. The specimens found here, of which Herr Isenbeck has preserved several of both sexes, are especially beautiful, with large bright red crop in the male and rich metallic gloss on the long narrow feathers on the back, which are much less conspicuous in the female. The latter is also a little smaller, and has the throat covered with white feathers, instead of having the bare crop of the male. The male blows out the crop like a ball when flying in the air, and this is said to be a very peculiar sight. A single such bird only was seen over Gardner, but very many on Moller, where they sat in pairs on the nests, and were so little shy that they were often caught with the hands. No eggs were as yet found on Laysan, but many on Lisiansky. A single egg only is in each nest, of the size of a Goose's egg, and not very pointed; white in colour. The nests were loose structures of twigs placed on the bushes.

"The sailors further remarked that, like the Swifts, they firmly believed the Frigate-birds were unable to fly up off flat ground, but always throw themselves off the higher rocks so as to be able to use their huge wings. This gives one good reason to suppose that they are also unable to swim, or at all events do so very reluctantly. [Mr. Hartert, however, who has seen these birds in numbers in the West Indies, saw them fly up easily from the bushes and scrub, and a bird he wounded swam very well.—W. R.]

"4. A species of *Frigate-bird* that has not yet been distinguished from the *Tachypetes aquilus*; but although similar to the common species in size and proportions, differs from it in

---

[1] *Diomedea immutabilis*, Rothsch.—W. R.

[2] *Diomedea chinensis*, Temm.—W. R.

[3] Palmer gives no record of this, which is so contrary to all other observations, that I firmly believe he could not have overlooked it.—W. R.

having the abdomen mostly white, the feathers of the back rounded, the head sometimes pure white, sometimes overspread with rusty yellow, the bill red and lacking the great gular pouch. I would consider this, without doubt, an immature plumage of the former species, if there were not so many varieties that one appears to be able to distinguish the old and young of this form, and none of them have a trace of the long and pointed glossy feathers on the back. Since the officers of the 'Moller' clearly recognized the two different species, one is induced to believe that they are better distinguished in life and by their habits than they seem to be in museum cabinets. It must be admitted that this does not make the difference quite certain, but renders it very probable; since both forms were always found on the same islands, but not in company, the white-headed variety can certainly not be the winter plumage.

"This variety was common on Gardner Island, while the other was rare; it flew generally very high, together with the *Phaëton*, and did not cry. On Moller and Lisiansky, on the other hand, where the other was common, this form was scarce, not in pairs, and not nesting; it was often seen sitting or running on the rocks, but mostly flying very high.

"5. *Phaëton* (an *candidus*?), with white, somewhat broad tail-feathers; seen singly near Gardner, only flying very high. The voice was similar to that of *Ph. phœnicurus*, which has some resemblance to that of our *Larus ridibundus*. The visitors to Gardner mentioned as a very remarkable sight the persistent flight of one of these birds with a Frigate-bird, that tried in vain to steal a fish from it that it had just caught.

"Here, too, the *Phaëton* was not seen sitting or swimming. This is probably the most persistent flier of all the birds in existence, although its wings have not the enormous appearance of those of the Albatrosses, for example, which are so often seen resting on the water, but which can by no means be compared with the *Phaëton* in rapidity and lightness of flight. I have seen innumerable *Phaëtons*, especially of the red-tailed species, which never showed the slightest intention to swim, and they frequent chiefly those parts of the sea where not a single rock is to be found, so that they cannot possibly rest on dry land at night-time. I have also never seen or heard that a *Phaëton* rested on a ship at night, as other birds, for example the species of *Sula*, so often do. Very likely the *Phaëton* sleeps whilst swimming, and spends the rest of its life, except the breeding-season, on the wing. At least this applies to *Phaëton phœnicurus*, which I observed so often, and which, in the Pacific Ocean, is common in both the Tropics, but especially in the northern Tropic, north of which, contrary to its general habits, it ranges up to 40 degrees in the Pacific.

"6. A species of *Carbo*?[1], of about the size of *Pelecanus piscator*; chestnut-brown. Not very numerous on Moller, where it nests on the above-mentioned palm-trees. Three nests, built of twigs, were placed on one tree; the birds were sitting on the nests in pairs and showed little fear. They were just then sitting on two eggs each, about the same size as those of the Lapwing, grey in colour and spotted. This bird was also seen on Lisiansky.

"7. *Pelecanus piscator*, L. White, with blackish wings; bill bluish, with red gular skin, with

[1] *Sula sula* (Linn.).—W. R.

a black triangle in the female; feet red. Rather common on Moller and Lisiansky: breeds in colonies on the beach; lays one egg only without nest, which is fiercely defended by the parents; these eggs were white in colour and of an excellent flavour. At the time of the visit they were fortunately not yet set.

"8. *Larus*, sp., perhaps *argentatus*, L. A large Gull, with flesh-coloured beak. Seen in great numbers flying round the top of Gardner Island; later on Moller and Lisiansky too.

"9. *Sterna* (?)[1], about 9 inches long; white, with greyish wings, back, and crown; tail with two long lateral rectrices. This bird was termed a Petrel by Herr Isenbeck and his companions; but this did not correspond with the fact that they would recognize it on seeing my *Sterna kamtschatica*, for then it ought to have a black crown, which, however, may be absent in winter, as is the case with several Terns. The eggs which the travellers believed to belong to this species were found together in numbers, without nest, in caves and fissures of the rock; they were about as large as the eggs of domestic Pigeons, much compressed at the blunt end, greyish, and spotted with brown.

"10. A *Petrel*[2] about 9 inches long; all over deep chestnut-brown, with blackish bill and feet, and a cuneate tail. Only seen on Gardner.

"11. Another *Petrel*[3], a little larger; breast, abdomen, and neck white; upper surface mixed white and brown; the forked tail only moderately emarginate. Seen on all three islands. On Moller it was sitting under the scrub in pairs, and was so little prepared for the intrusion that several were trodden upon. Eggs were not found.

"12. A species of *Duck*[4], with no conspicuous plumage, living in small flocks on Moller and Lisiansky, but not breeding.

"13. A species of "*Snipe*"[5] (most probably a *Totanus*), which the observers were unable to describe. In flocks on Moller and Lisiansky, and by no means shy.

"14. A species of *Sandpiper* (perhaps *Tringa minuta*); also in flocks on Moller and Lisiansky. [Perhaps *Calidris*?—W. R.]

"15. A kind of *Fowl*[6], about as large as a Ptarmigan; mixed grey and brown; running on the ground, singly, but at the same time rather numerous, on Moller and Lisiansky; very rapid and rather shy. Eggs were not found.

"16. A small *Sparrow-like bird*[7], brownish grey above, yellowish green below. Only found on Moller, where it was often seen running on the ground under the grass very quickly. One, however, was knocked down with a cap.

"17. A small *red bird*[8] with black wings. On the same island, where it is not very common; flying round the bushes.

[1] *Haliplana lunata*, Peale.—W. R.
[2] *Puffinus nativitatis*, Streets.—W. R.
[3] *Puffinus cuneatus*, Salvin.—W. R.
[4] *Anas laysanensis*, Rothsch.—W. R.
[5] *Totanus incanus* (Gm.),—W. R.
[6] Although the description is very different, nothing else can be meant but *Porzanula palmeri*, Froh.; but at the present time it does not exist on Lisiansky Island.—W. R.
[7] *Telespiza*.—W. R.
[8] *Himatione freethi*, Rothsch.—W. R.

"18. **A small bird**[1], somewhat resembling a Humming-bird; like it hovering in the air. Brownish, glossy greenish from beneath. Also seen only on Moller. It would be very curious if this should prove to be a Humming-bird.

" Birds observed on each of these Islands.

" Gardner Island :

    1. *Diomedea* (no. 1). Very common.
    2. *Tachypetes* (no. 3). Single.
    3. *Tachypetes* (no. 4). Common.
    4. *Phaëton* (no. 5). Single.
    5. *Larus* (no. 8). Very common.
    6. *Sterna* (no. 9). Numerous.
    7. *Procellaria* (no. 10). Numerous.
    8. *Procellaria* (no. 11). Numerous.

" Besides these an enormous number of larger and smaller birds were seen, which appeared to be Gulls and Petrels.

" Moller Island (Laysan) :

    1. *Diomedea* (no. 1). Common.
    2. *Diomedea* (no. 2). Common.
    3. *Tachypetes* (no. 3). Common.
    4. *Tachypetes* (no. 4). Not numerous.
    5. *Carbo* (no. 6). Not numerous.
    6. *Pelecanus* (no. 7). Common.
    7. *Larus* (no. 8). Common.
    8. *Procellaria* (no. 11). Common.
    9. *Anas* (no. 12). Rather common.
    10. *Totanus ?* (no. 13). Rather common.
    11. *Tringa ?* (no. 14). Rather common.
    12. Fowl-like bird (no. 15). Rather numerous.
    13. Sparrow-like bird (no. 16). Rather common.
    14. Small red bird (no. 17). Single.
    15. Humming-bird ?? (no. 18). Single.

" *Note.*—On the beach several small Seals and very large Sea-Turtle were found. [Seal = *Otaria ursina*.]

" Lisiansky :

" All the same birds as on Moller to no. 12 inclusive. The three last-named small birds were not seen. It seems as if they had come through an extraordinary accident to that island."

---

[1] There is no such bird on the island.—W. R.

DIARY OF HENRY PALMER.

From May 5th to August 18th, 1891.

May 5 & 6 (Honolulu).—These two days were spent in looking for a vessel for my expedition and obtaining permission from the Laysan Guano Company to collect on their islands.

May 9.—Arranged with Capt. Walker for the trip and put my things on board. I am very nervous, as he lost his last vessel among these same islands; but I must go with him, as this is the only chance.

May 10 & 11.—I spent these days in preparing for my voyage and in hunting up a Captain Rosskill, who had brought three living specimens of the Laysan-Island Rail to Honolulu.

May 12.—I went to see the Rails and bought two of the three specimens. As there was no steamer leaving for Europe, I was obliged to kill and skin them. On the journey from Laysan they had been fed on canary-seed and potatoes.

May 21.—It is now a little more than a week since I last wrote in my diary, but I was too disgusted at the innumerable delays and crotchets of Capt. Walker. Yesterday only did we get the schooner alongside the wharf, so many alterations and improvements had to be made ; but now everything is settled, and if the weather is favourable we start to-morrow.

May 23.—At last everything is ready and I am once again on the water, bound for the islands I have to explore. This evening we have anchored 30 miles from Honolulu to get everything straight before going right out to sea.

May 24.—At sea. Nothing to note.

May 25.—Weather fine, but heavy sea running. I expect to-morrow to reach Niihau, or Bird Island.

May 26.—Reached Bird Island at 2 o'clock, but the swell was too heavy to land. We sailed all round it before leaving. There seemed to be any quantity of sea-birds on it. The island itself is only a large rock with very scanty vegetation; all I could see were two small groups of some kind of palms. While sailing round the island some Gannets came and settled on our bowsprit, and I caught three of them, all of one kind, although I saw another species flying round. I here mention the various species of sea-birds we saw on Niihau as far as I could identify them :—

2 species of Albatross. [Probably *Diomedea immutabilis*, Rothsch., and *Diomedea chinensis*, Temm.—W. R.]

2 species of Gannet. [*Sula sula* and *S. piscator.*—W. R.]

3 species of Petrel. [These are probably the same as those of Laysan.—W. R.]

6 or 7 species of Tern. [Probably *Anous stolidus*, *A. hawaiiensis*, *Haliplana fuliginosa* ad. and juv., *A. cinereus*?, *Gygis candida*, and *H. lunata.*—W. R.]

A small flock of Akekeke came flying off the island when we were opposite the landing-place, took a circle round, and disappeared. [*Strepsilas interpres*, L.—W. R.]

In the evening I saw another flock of these birds flying the same way as the vessel was steering and as though they were migrating to another island. I also noticed some Red-tailed Tropic-birds [*Phaëton rubricauda.*—W. R.] flying round the island, and their love-making in the air was accompanied by some peculiar evolutions. The male, on approaching the female, swung his tail from side to side and up and down, almost doubling it up under him, and this it continued for a long time.

May 30.—On account of the very bad weather we have had since Wednesday I have been unable to write my log, but I take notes as I go along, which I here record. On the 27th at sea all day and nothing to note.

May 28.—Passed and sailed round Necker Island. A strong wind and a heavy sea made landing impossible. Necker Island is a large rock with no vegetation whatever upon it; nearly all round its sides appeared perpendicular, except one small place on the S.W. side, where in calm weather I believe I could have landed, but I did not see the use of staying round this island for an indefinite period to await a calm sea: we therefore made straight for French Frigate Islands, where we have just anchored (May 30), but cannot land yet on account of the surf. All the way from Niihau we have seen plenty of the same species of sea-birds I have mentioned before, but yesterday, for the first time, I saw a Storm-Petrel [*Thalassidroma* ?—W. R.].

French Frigate Islands are simply large low sandbanks with little or no vegetation on them. There is also one rock a hundred and twenty-five feet high, which it is perfectly impossible to land on, but it is covered with sea-birds.

May 31.—This evening I have camped out on one of the sandbanks, the ship being three miles off. This bank is literally covered with birds, chiefly two species of Tern. Most of their young are nearly ready to fly, some even are flying. Besides these Terns there are a few Gannets and Albatross. My assistant and I have to-day carefully examined all the birds on this shoal, inspecting eggs, young, &c. The White-breasted Tern (*Haliplana fuliginosa*) lays only one egg and deposits it anywhere on the sand; their young are of a dark brown colour with spots, and are the same bird which I mistook for a different species on Niihau (cf. *antea,* p. vii).

The other Tern mentioned above also lays one egg only, but seems to prefer the centre of small growing bunches of grass to deposit it on. A few of the Gannets are still incubating, although some young birds are almost ready to fly. They lay their eggs on the sand, two in number, and the egg is extremely small for the size of the bird. We also found what I believe to be Albatross eggs, but as we could not identify the parents I did not collect any. There are also a few "Ulili" (*Totanus incanus*) on the island, and I also saw a few "Akekeke" (*Strepsilas interpres*) when we landed. There were also a lot of turtle on the islands; but the photographs will give a better idea of the immense number of sea-birds on these shoals than anything I can write.

June 1.—All to-day I have been busy skinning and preserving specimens. Capt. Walker came off from the ship to us and talks of remaining till the 5th; but as our water has almost all leaked out, it depends on what provisions he sends to-morrow whether I can stay on the sandbank or not. To-day we obtained specimens of a Petrel which breeds here, and so I believe I have obtained every species of bird that breeds on this island.

June 2.—Capt. Walker sent a boat this morning to say he intended at once to change his anchorage to another island, so I went on board again about mid-day.

June 3.—We left our anchorage this morning, sailed round the rock, and cast anchor again in the evening off two small sandbanks.   As I saw no birds on them I did not go on shore.   Captain Walker's sons, however, went and found a few turtle and Boobies (*Sula sula*). We went quite close to the previously-mentioned rock, and, on firing a rifle, among the birds that got up I saw a pair of another species of Tern I had not got and also a third species of Gannet.

I see further ahead an island apparently full of birds, which I hope to reach to-morrow, as the weather is now nice and fine.

June 4.—Weighed anchor and sailed up to the island I have just mentioned.   It is only a rather larger sandbank than the one we had just left.   There were plenty of birds on it, but chiefly the same as I had got before, though the young of the Black Albatross were much more plentiful.   I also got another species of Petrel which was sitting on the ground, but only obtained one out of the pair I saw.   Here I also found the White Gannet [*Sula piscator.* —W. R.] nesting.   This bird builds a nest among, and made of, some species of vine growing on the island, and the nest is not unlike those of the Slugs (*Phalacrocorax*).   All the nests I examined were very similar and built in the centre of the vine, round and flat, and about six inches above the ground.   Each nest had only one egg.   The young of the Pied Tern (*Anous fuliginosus*) were all much younger than on the island I had been before, very few being able to fly.   The large Petrel, of which there were a few stragglers on the former island, were here very plentiful, sitting on the sand in pairs.   There were also a few Frigate-birds round about, but I did not trouble to shoot any, as, according to report, they are very plentiful on Laysan Island.

June 5.—At 6 in the morning I went on to the island again, as we were to leave at 11 o'clock.   While turning up some turtle-shells that had been heaped up by a shipwrecked crew, I discovered a third species of Petrel sitting on its egg; of this we obtained ten or eleven besides six eggs.   I also obtained some more of the Petrel I got one specimen of yesterday.   Of the large Petrel I saw several males in the act of copulation, so I have no doubt they will soon commence to lay.   I saw several of them in their burrows in the sand.

June 6.—We continued our journey last night and are now bound for Gardner Island. The weather is very calm.   Unfortunately my assistant (Munroe) is ill with "La Grippe," and it is so terribly hot that I hardly know what to do with my skins.   The young birds in down I cannot do anything with, which is very discouraging, but in future I shall put them in spirit.

June 7.—At sea.   Weather very calm.   Hope to reach Gardner Island to-morrow.

From June 7th to June 13th we were occupied in beating round for Gardner Island, having experienced bad weather and for some time had lost our course.   Owing to the rough sea there was no possibility of landing on Gardner Island, and the only birds I saw were the White Tern (*Gygis*), the Grey-backed Tern [*Haliplana lunata.*—W. R.], the Tropic-bird, and the Frigate-bird.   On the 13th we lay in a dead calm, and I fear we may be a long time before we reach Laysan, and all this delay is owing to bad management on board.

June 16.—At last we reached Laysan, having sighted it just after daylight.   From what I can see of the island I am in hopes that I shall be able to make up a little of the lost time, for, to my great delight, the island appears quite alive with birds, especially the Rail, which, although I had been told was tolerably plentiful, to my intense astonishment literally covers the island. Everywhere one walks this little creature hurries out of your way. [This Rail is *Porzanula palmeri*, Froh.—W. R.]

Before going into particulars about the birds I think it best to give a description of Laysan Island, as I found it.

The island is a simple atoll with a lagoon in the centre, which Mr. Freeth (the governor of the island and manager of the Guano Company) tells me has comparatively little water in it at the present time, although I consider there are quite 100 acres under water [1]. With the exception of the lagoon the island is covered with vegetation, consisting chiefly of a coarse kind of grass about three feet high and small thick scrub between four and five feet high. Here and there are bare patches entirely covered with young Albatrosses, Terns, and Petrels, and under the grass and scrub it is just the same, almost at every step you sink into the burrow of some bird. Mr. Freeth took me to the guano-field on the tram-line he has built, and he had to send a man on ahead to clear the track of the young Albatrosses. I have seen so many birds and have been so excited that I must leave my description of them till I have seen more at my leisure.

June 17.—Have spent the day catching and skinning birds; I will not try to give a description of their habits until I have seen more of the birds; there are so many that it makes one quite confused. A most touching thing occurred: I caught a little red Honey-eater [*Himatione freethi.*—W. R.] in the net, and when I took it out the little thing began to sing in my hand. I answered it with a whistle, which it returned and continued to do so for some minutes, not being in the least frightened.

June 18.—I and my assistant went about the island and collected some eggs and birds. I shot four Ducks, which, besides the Curlew, are the only birds which cannot be caught with a hand-net. Mr. Freeth has just told me his little boy caught one of the Finches this morning and then got an egg and offered it to it; the bird broke and ate the egg while being held in the boy's hand. This would give you an idea how tame all the birds are here.

June 19.—Again the day was spent in collecting and preparing birds. While out this morning both my assistant and I saw a little Rail break and eat an egg. We had disturbed from its nest a Noddy (*Anous*), immediately the Rail ran up and began to strike at the egg-shell with its bill, but the egg being large and hard he was quite a long time before making a hole. The Rail would jump high into the air and come down with all its force on the egg, until it accomplished the task, which once done the egg was soon emptied. By this time the Tern came back and gave chase, but in vain. At the beginning I ran after the Rails with the net, but soon found that it was much better to put the net edgeways on the ground, when the very inquisitive Rails would run up to look at it, and then were easily caught.

June 20.—To-day I packed up two boxes of birds to go by the 'Mary Foster' (guano-ship) to Honolulu.

June 21.—While walking about the island I turned some of the Frigate-birds which had young off their nests. Scarcely had I pushed one off when another Frigate-bird would rush up, seize the young one, fly off and eat it. Sometimes the parent bird would give chase, but it always ended in one or the other eating the young bird. I could scarcely believe my own

[1] I am told that on analysis the water in this lagoon proves to be three times as salt as ordinary sea-water.

eyes, so I tried several, but they would even take young birds out of the nest which were almost fully feathered.

June 23.—I have spent half the day hunting for the eggs of the Rail, but although we found several nests which I believe were Rails' nests, there were no eggs in any of them. My assistant saw one of the red Honey-eaters holding a moth with its claw, which it pulled to pieces and ate.

June 24.—My assistant and I went out again this morning searching for eggs of the land-birds, and I am pleased to state that we found two nests of the Rail. One nest contained three eggs, the other an egg and a small chick. Since bringing the eggs home, the one found with the chick has hatched into a young Rail, which I have put in spirits. I now have Rails of every size from the egg upwards. I also found a Finch's nest with two young in it, which I have put in spirit, also an egg with them. We also found two nests of the " Flycatcher" [*Acrocephalus familiaris.*—W. R.].

June 26.—To-day I spent in catching live Rails to take on board with me, and if all is well we leave to-morrow. I have collected specimens of all birds found on the island, with the exception of those previously obtained on Kauai. Mr. Freeth has done everything he possibly could for me, and I have a nice lot of specimens. A small fly on the island has given me a great deal of trouble with my skins. Almost before the birds were dead they were fly-blown. It is the smallest blowfly I have ever seen, but does quite as much damage as the larger ones. A small beetle also caused me a great deal of anxiety, laying its eggs in the skins. I am now trying some insect-powder Mr. Freeth gave me.

June 27.—We are leaving to-day for Lisiansky Island.

June 29.—I have just arrived and am in camp on Lisiansky Island. As yet I have not been round the island, but from what I can see round the tent there is vegetation on the island, but nothing in comparison to that on Laysan. As we before found on all the islands, the Pied Tern (*Anous fuliginosus*) is in great abundance here, but the young are very small. There are also young White-breasted Albatross [*Diomedea immutabilis.*—W. R.], large Gannet [*Sula cyanops.*—W. R.], Grey-capped Tern [*Anous hawaiiensis.*—W. R.], and the Petrels. These are all the birds I can see close round the tent.

June 30.—I have made a thorough inspection of the island to-day, but the only additional species of birds I saw was the White and Brown Gannet (*Sula sula*), which I saw when passing Bird Island. This, to distinguish it when writing my diary, I will call the " Brown Gannet." I perceived them flying about on the weather side of the island, and shot all I saw They seem to be rather shy, but that may be because they are not breeding, for I found the other two species of Gannets much more shy when not breeding. This " Brown Gannet " is much the scarcest of the three species on all the islands I have visited hitherto. Another to me strange fact is the large preponderance of males, all those I shot being of this sex. The White-breasted Albatross are here in thousands all over the island, but the dark one [*Diomedea chinensis*, Temm.—W. R.] is very scarce. There are only a few on the weather side of the island, but, as before mentioned, the Pied Tern is by far the most numerous of any bird here. On Laysan Island they call it the " Wideawake Tern," which is a very good name for them, for night and day they keep up a continual cry, and some are always to be seen flying about. The large Gannets (*Sula cyanops*) are sitting with their young all along the beach, some of the latter being just able to fly. The small Gannets (*Sula piscator*) are sitting on their nests, which are built on some small scrub that is growing round what I believe at one time must have been a lake, on the south side of the island. This scrub, although it does not grow so high, is, I think, the same as that found on Laysan, and which, with the exception of two or three small spots only, grows round this dried-up lake. The Frigate-birds are here also in large numbers. They have their nests on the scrub round the lake. It is very interesting to watch them getting their food : in the daytime they soar about all over the island, and every now and then one of them picks up a young Tern, then a number of others chase him and keep taking the prey one from the other till at last it is eaten or drops to the ground ; but they generally make for the sea with their prey, as it is easier for them to pick up when dropped than on land. When soaring they hardly seem to move a wing, and sometimes I have seen them cleaning and picking their feathers as they floated along in the air. In the evening, just before sunset, they hover close round the island waiting for the Petrels and other birds to come home with food, when they give chase and do not leave the unfortunate bird until it has disgorged some if not all of its food. I have seen Petrels when thus chased drop on the water from sheer exhaustion, but even then the Frigate-bird would not leave it till it has disgorged. In addition to these birds there are a few "Noio" and Grey-backed Terns on the island, also a few Curlew in poor plumage, some "Akekeke" (*Strepsilas interpres*), Kolea (*Charadrius fulcus*), "Ulili" (*Totanus incanus*), and three species of Petrels, but there is no sign of any land-birds.

July 1.—This morning, on searching for some more of the "Brown Gannet," I was fortunate enough to obtain two females and two young, but I could find no eggs, and these were the only young ones on the island. When skinning the Brown Gannets I found they had much larger brains than *Sula cyanops*, which is almost twice as large. The female of the Brown Gannet is also larger than the male, and has different coloured soft parts. While out to-day we came across some of the small Gannet (*Sula piscator*) asleep on some bushes, with their heads hanging straight down. The first one I saw I thought was dead. The young Gannets also when sleeping on the sand stretch their necks out straight.

July 2.—My assistant to-day killed a seal, which I am trying to prepare for a specimen. I had shot two before, but their skins were too much injured. (The skin of the seal unfortunately was lost.) On returning to the tent we captured two young albino Albatrosses. This seems very singular, as we had examined during the voyage thousands upon thousands of young Albatrosses and seen no variation among them. Now the only bird seen on my voyage of which I have no specimens is the little Grey Tern [*Anous cinereus*, very likely.— W. R.].

July 4.—Left Lisiansky for Pearl and Hermes Reef.

July 6.—With a nice breeze all the way from Lisiansky we made these islands this morning. We have spent all day trying to get into the entrance of a sort of lagoon formed by the coral-walls round the central island, which is a mere sandbank with no vegetation.

July 8.—Left Pearl and Hermes for Midway Island, having spent the last two days at anchor in a perfectly calm sea, three or four miles away from the island, but Capt. Walker, in spite of all I could do, refused to put me ashore. I know these waters are very dangerous, on account of the coral-rocks, but the water was so calm that I consider he had no excuse for not landing me in the small boat.

July 11.—It has taken us till to-day to reach Midway Island, although it is only 60 miles from Pearl and Hermes. Soon after we anchored in the bay, I came ashore. What I am now on is what is known as Sand Island. This island is almost bare and has hardly any birds: the two or three I saw were of the same species as I had seen all through the voyage. Although this island is comparatively large, it is the most desolate place I ever was on. There is hardly any vegetation except a few tufts of grass on the south end, and in rough weather most of the island is under water. This is the place where Captain Walker was wrecked in the 'Wandering Minstrel,' and was fourteen months on the island before a passing ship relieved him. There is a house, originally built by an American Surveying Expedition, but it has been much altered and rebuilt by various shipwrecked crews. Wreckage is strewn about all over the island; in one place there is almost the whole of a schooner lying on the beach, which Captain Walker tells me was the 'General Seigel': an account of this wreck I find is hung up in a bottle in the house, together with directions for obtaining food and water, for the use of any future shipwrecked crew.

About 100 yards from the house is a graveyard, with crosses marked with the names of the sailors buried there. Altogether I cannot imagine a more melancholy picture than this island presents. As there are no specimens of any kind to be got here, I shall go on board to-night and arrange to go to the other island, which, together with this one, forms the so-called Midway Island, though really the proper name for these two shoals is Brooks Island and Lower Brooks Island. We sighted a schooner to-day, but cannot yet make out anything further.

July 12.—I came back on board, and the schooner we sighted has put in alongside of us. She is the 'Charles G. Wilson,' of San Francisco, just come from the Caroline Islands.

July 13.—Have just landed on Brooks Island, which is 4 miles from the ship. The scrub is thicker here than on any of the islands yet visited, and about five feet high, also the grass is much coarser than on Laysan. The island is about 1½ mile long by ¾ of a mile wide. I have wandered all over it, but have seen no species of bird different from what I have collected on the other islands. However, the Red-tailed Tropic-bird is more plentiful on this island, but Albatrosses are very scarce; also the little White Tern [*Gygis.*—W. R.] was very abundant here. My assistant found a rookery of the Brown Gannet and was fortunate enough to get two eggs. There are three or four houses here chiefly built of grass, but in a very dilapidated condition. Captain Walker's son liberated on this island a pair of Laysan Island Rail and a Finch.

July 14.—To-day was spent in collecting a few specimens. Both my assistant and I saw an adult White-breasted Albatross feeding a young of the black species [*Diomedea chinensis.*—W. R.]. I am quite sure of this, as we were close to the birds at the time, although from having seen old birds drive away all other young from the one they were feeding, I still believe they know their own young. The White Terns (*Gygis*) I notice do not sit on their eggs like most birds, but stand up and cover the egg between their legs by drawing the breast-feathers over it, which has made some people believe that it puts its egg in a pouch to incubate it.

July 15.—To-day I spent in hunting for eggs and young birds. I found one nest, with two eggs in, of the Brown Gannet, and a number of young; but to my surprise I found the old birds belonging to these just as shy as those without eggs or young, so, in spite of my former statements, I must come to the conclusion that this is a shy species. The nests were built of grass on the ground and each nest contained two eggs, but I never found more than one young one in a nest.   To-day I saw for the first time some of the Frigate-birds chasing the Red-tailed Tropic-birds.

July 16.—I have nothing of interest to report to-day. except that I watched the small White Tern (*Gygis*) feeding its young.   The adult bird brings small fish to the young in its bill, and the young one takes it from its mouth.   One I saw had no less than four in its beak at once : I cannot understand how it can catch fish while it already has some in its bill; I could never see them in the act of fishing, in order to find out.

July 17.—I am once more on board.   Owing to losing my tape and rule I was unable to measure any of the birds from the two shoals of Midway Island.

July 18.—Started for Muro Reef.

July 20.—Captain Walker has changed his mind and is going straight back to Honolulu.

From July 21st to the 18th of August.—We spent all this time in going back to Honolulu from Midway Island.   We saw a few birds of well-known sorts round the vessel all the way, and I caught several specimens.   The only bird I saw fresh was a large brown Gull (probably a Skua).   I had a terrible time with the skins, nothing seemed to kill the beetles (*Dermestes*), for unfortunately the whole vessel swarmed with them.   I believe I saw also another species of Petrel as we neared Kauai; it was large with white breast and brown back A Kolea (Golden Plover, *Charadrius fulcus*) flew also round the ship, and considerably astonished me by settling on the water several times to rest.

August 18.—At last we have reached Honolulu after a month's dreadful journey from Midway Island.   I never was more pleased in my life than when I found myself on shore in Honolulu once more.

So ends three months' suffering under the famous Captain F. D. Walker.

NOTE.—I have given this Diary as fully as possible, only cutting out such parts that I considered of no interest to anyone but myself, such as details of expense, quarrels with the captain, &c.   Of course I had to condense the matter to some extent.—W. R.

LIST AND DESCRIPTION

SPECIES OF BIRDS

FROM

LAYSAN AND THE NEIGHBOURING ISLANDS.

-

1. ACROCEPHALUS FAMILIARIS (*Rothsch.*).

MILLER-BIRD.

Tatare familiaris, Rothsch. Ann. & Mag. Nat. Hist. 1892, x. p. 109.

Adult male. Upper parts greyish brown, rather warmer in tinge on the rump, much paler on the sides of the head ; wings and tail dark brown, the feathers externally margined with buffy brown ; tail apparently slightly rounded ; wing somewhat rounded, the first primary rudimentary, only about 0·3 inch in length, second primary 0·4 inch shorter than the third and in length between the eighth and ninth, the ninth and following quills much shorter than the former. Underparts, together with the chin and throat, buffy white. Total length about 4½ inches, exposed culmen 0·6 to 0·65, wing 2·15 to 2·55, tail 2·4, tarsus 0·8 to 0·9.

Adult female. Closely resembles the male.

" Iris light brown ; bill blackish ; lower mandible paler and more bluish grey ; legs greyish brown with a bluish tinge ; soles of feet lighter."

Hab. Island of Laysan.

D

I QUITE agree with Canon Tristram, who wrote on the subject (in 'Ibis,' 1883, pp. 38–46), that the members of the so-called genus *Tatare* cannot be separated generically from *Acrocephalus*. Mr. Hartert is also of the same opinion.

The nest is a deep and cup-shaped soft structure, chiefly built of down and feathers of Albatrosses and other sea-birds, interwoven with some fine roots and twigs; the outer parts consist chiefly of the down of young Albatrosses, while the inner lining is made up of fine roots and white feathers.

The eggs are remarkably large for the size of the bird, and measure 0·8 by 0·54, and one even 0·85 by 0·65 inch. In colour they are quite Acrocephaline, closely resembling those of other Reed-Warblers. The ground-colour is a very pale bluish. Large blotches of olive-brown are distributed pretty evenly over the surface and are relieved by small spots of the same colour, while the underlying blotches show bluish grey through the ground-colour. In another specimen the dark spots are more blackish, much more numerous and smaller, and the underlying pale blotches are not so distinct, while in the largest one the colour is somewhat abnormally restricted to the broad end. The measurements will show that the eggs are not very much smaller than those of *Acrocephalus turdoides*, while the bird is so much smaller. The Plate shows the colour of the three varieties represented in my collection.

The "Miller-bird" is plentiful on Laysan and very tame. All specimens sent were caught with a hand-net. It is very active in its habits, hopping and creeping among the grass and scrub in search of insects. It takes its name from its fondness for some large white moths, called on the island "Millers." For the size of the bird it has a harsh deep note, much resembling that of a Thrush. It lays three eggs, and the nest is placed in a tussock of grass.

2. HIMATIONE FREETHI, *Rothsch.*

LAYSAN HONEY-EATER.

Kleiner rother Vogel, Kittlitz, Museum Senckenbergianum, i. p. 125, no. 17 (1834).
Himatione fraithii, Rothsch. Ann. & Mag. Nat. Hist. 1892, x. p. 109.

Adult male. Head, throat, breast, and upper abdomen of a bright scarlet-vermilion, with a faint tinge of golden orange, which is more pronounced in freshly moulted and living birds; lower back and upper tail-coverts similar in colour to the head, rest of upper parts orange-scarlet; the feathers of these parts are greyish brown at base, with a pale shaft-stripe and broadly tipped with orange-scarlet, those of the head and back of neck with distinct whitish spots before the red tips; on the hind neck all the basal part is quite concealed and only the red tips visible, but on the back of the neck the sub-terminal colour becomes somewhat conspicuous; lower abdomen dull ashy brown, fading into whity brown on the under tail-coverts; under wing-coverts dull ashy brown; quills dusky brown, with blackish shafts, paler at the tips, where the shafts, too, are pale; primaries externally narrowly edged with whitish, inside with buff; secondaries with pale scarlet outer edges. Bill and feet black; iris reddish yellow. Total length about 5½ inches, wing 2·55 to 2·7 (average 2·65), tail 2·4, culmen 0·55, tarsus 0·9.

Adult female. Similar to the male, but the red somewhat paler in tinge.

Young. General colour dull brown on the upper surface and light ashy brown on the under-parts, many of the feathers margined with rich buffy brown, and those of the head and hind neck with a blackish terminal and a buffy subterminal spot, thus producing a somewhat mottled appearance; wings and tail dark brown, the primaries narrowly and the secondaries broadly margined with rich brownish buff; chin and upper throat orange-buff; lower abdomen and under tail-coverts white tinged with buff.

" Iris brown ; bill blackish, brownish in the young, base of lower mandible orange in young birds; feet black."

The present species resembles *Himatione sanguinea*, from the Sandwich Islands, but is easily distinguished by its shorter bill and its scarlet-vermilion colour instead of the blood-red darker colour of *Himatione sanguinea*.

Hab. The island of Laysan, where Palmer discovered it in June, 1891.

PALMER says of this bird :—" This is by far the rarest of the Laysan-Island birds, though I have observed a fair number, generally in pairs. They are very active in their movements, flitting about in the scrub. It feeds on very small insects as a rule, but I have also observed it sucking honey from the flowers. It has a low sweet song, consisting of several notes, but very seldom gives utterance to any sound, except during the breeding-season. Mr. Freeth tells me it is in full song in January and February, when the plumage also has a beautiful golden gloss over the red. So no doubt the breeding-season lays between that time and June, when I met with full-grown young ones."

(NOTE.—*Himatione sanguinea* (Gm.) is figured on the same Plate for comparison.)

3. TELESPIZA CANTANS, *Wils.*

LAYSAN FINCH.

Telespyza cantans, Wils. Ibis, 1890, p. 341, pl. ix. (" Midway Island ").

Adult male. Feathers of the head deep brown, margined with greenish yellow; feathers of the rest of the upper surface deep brown, broadly margined with pale brown, this latter colour more or less strongly tinged with yellowish green, apparently more so in fresh plumage; rump and upper tail-coverts nearly or quite uniform brown; wing-coverts like the feathers of the back; primaries deep blackish brown, with narrow yellow outer margins; secondaries of the same colour, but with broad pale brown outer margins; tail-feathers deep brown, margined with greenish yellow; throat and breast yellow, with deep brown shaft-stripes; flanks and sides of abdomen and thighs similar but more brownish; abdomen and lower tail-coverts brownish white. Under wing-coverts white, with a yellow tinge.

The *female* does not differ from the male, but some specimens (apparently older ones) have the breast almost uniform, like that figured by Wilson in 'The Ibis' and the one shown in the foreground of my Plate, and if in abraded plumage the upper surface does not show much of the yellow tinge, but looks almost sandy brown, broadly streaked with deep brown.

Colour of the soft parts as in *T. flavissima,* Rothsch.

Total length about 6 to 6·5 inches, wing 3·1 to 3·26, tail 2·4, culmen 0·55 to 0·69, height of bill at base 0·49, tarsus 1.

Hab. Wilson (*l. c.*) states that this species comes from Midway Island, but Palmer found it common on Laysan Island, and did not see it on Midway Island, where he collected several days and where he says there are no land-birds whatever.

Of the two Finches the following notes occur in the collector's log-book :—
" Common all over the island of Laysan. They are quite omnivorous, eating insects and other birds' eggs, and seem fond of bathing in water. Mr. Freeth says he has seen a small flock feeding on a dead Albatross. The nests we obtained were built of grass and small twigs and were placed in the scrub or in the tussocks of grass. I found only one and two eggs, but I came across a nest containing two young and one egg, and one of Mr. Freeth's men

found four in a nest; I am therefore unable to say what the usual number of eggs might be. They begin breeding early in May."

The nest Mr. Palmer sent me is an artless and somewhat flat structure of grass and small twigs, lined more softly inside; it measures about $2\frac{1}{4}$ inches across inside. The eggs are white with very little gloss, marked with russet-brown blotches and narrow spots of the same colour and larger and smaller underlying blotches of a light bluish grey ; the spots are more numerous at the thick end. The eggs are short, oval, and measure 0·77 by 0·63 inch.

One egg out of another nest is more thickly spotted and is much larger, measuring 0·88 by 0·65 inch, and I think that it belongs to the somewhat larger *T. flavissima*.

Obs.—One of my specimens of *T. cantans*, Wils., is in an entirely abraded plumage, and others are almost similar, showing that they cannot be young birds. I expressly state this, as it has been surmised that *T. flavissima* might be only the full-plumaged bird of *T. cantans*, which certainly is not the case.

4. TELESPIZA FLAVISSIMA, *Rothsch.*

YELLOW LAYSAN FINCH.

Telespyza flavissima, Rothsch. Ann. & Mag. Nat. Hist. 1892, x. p. 110 (Laysan).

Adult male. Head all round, neck, and underparts down to the lower abdomen bright yellow, brightest on the head. Upper parts ashy brown, without striations; interscapular region and back strongly overspread with yellow; rump uniform ashy brown; upper tail-coverts washed with yellow; wing-coverts deep brown, broadly margined with bright yellow; quills deep brown; the primaries externally narrowly and the secondaries broadly margined with light apple-yellow; lower abdomen and under tail-coverts dirty white, more or less indistinctly washed with yellow; tail-feathers dark brown, the lateral ones narrowly, the central ones more broadly margined with yellow. Bill blackish horn-colour. Under wing-coverts pale yellow. Iris greyish brown; legs and feet deep brown. Total length about 6½ inches, culmen 0·7 to 0·79, height of bill at base 0·5 to 0·54, wing 3·3 to 3·4, tail 2·6, tarsus 1·05 (given 2·05 by mistake in the original description, *l. c.*).

The *female* does not differ from the male.

The Plate represents a male and a female.

Hab. Palmer found this bird very common on Laysan Island, but nowhere else. He sent me a fair series, but no specimens in immature plumage.

THIS is the second species of the genus "*Telespyza*" established by Scott B. Wilson, 'Ibis,' 1890, p. 341, for his *Telespyza cantans.* The name, however, should be spelt *Telespiza,* as it is derived from σπίζω.

It is very curious that two species, *Telespiza cantans* and *T. flavissima,* should occur on one small island like Laysan. But although one is at first inclined to consider them to be young and old, or male and female, they are two distinct species, as I have males and females of both, shot at the same time, and no intermediate-coloured specimens. Besides, *T. flavissima* has a stronger bill and is a little larger.

5. PORZANULA PALMERI, *Froh.*

LAYSAN CRAKE.

Porzanula palmeri, Frohawk, Ann. & Mag. Nat. Hist. 1892, ix. p. 247.
Pennula palmeri, Hartl. Abh. naturw. Ver. Brem. xii. p. 390 (1892).

Adult male. Top of the head and entire upper surface with tail pale sandy brown; feathers slate-coloured and whitish along the shaft at base, and with large deep blackish-brown central patches on the exposed parts; the feathers of the back are mostly mottled with white, but in some specimens the white is scarcely perceptible; a line over the eye, sides of the head and neck, chin, throat, breast, and abdomen pale slaty grey to ashy; feathers on the sides of the body of the colour of the upper surface, each with several white spots, faintly outlined with black. Wings very small and rounded, first primary about a third of an inch shorter than the second, second, third, and fourth about equal and longest; outer webs sandy brown with very pale margins, inner webs darker and more smoky brown. Bill light olive-green, darker and inclining to purplish on the tip and culmen; iris ruby-red; eyelid pale grey-green; tarsus and feet light olive-grey-green. Total length about 6 inches (but the bird is capable of extending its neck considerably, adding much to its entire length), culmen 0·65 to 0·7, wing 2·2 to 2·3, tarsus 0·9 to 1, middle toe with claw 1·2 to 1·3, hind toe with claw 0·4. The tail is very inconspicuous and only 0·8 to 0·9 inch in length, the tail-feathers only a little stiffer than the coverts and so much like them that it is almost impossible to make out their exact number.

The sexes are similar, only that the females seem to be a little paler.

Young birds have the entire under surface pale buff.

The *nestlings* have a yellow bill and are covered with black down.

Hab. Laysan Island only.

THE nest of the Laysan Crake is built under the thickest bunches of grass and there is a cover placed over it with a hole on the side for the bird to enter. It consists of grass woven together with very fine shreds of grass, fibres, and a little down, with here and there a feather intermixed, the materials softer inside. It is loose, but has a cup about 2 inches deep and measuring about 3 inches across.

c

The eggs are longish oval and measure on an average 1·15 by 0·86 inch. The colour is pale creamy buff, flecked with pale reddish brown and pale purplish grey, both colours being somewhat indistinct and covering the egg equally over the whole surface. Some are rather more rufous and more distinctly spotted with fewer and larger spots.

The nest described above was found on June 24th, 1891.

Palmer's notes on this bird are as follows:—"It is very plentiful all over the island, is diurnal in its habits, very active, fearless, and extremely inquisitive : I could always catch them by placing my net edgeways on an open space of ground, for they would immediately run up to see what it was. Their food consists of all kinds of insects and other birds' eggs, but they are especially fond of the maggots off dead birds. It cannot fly."

The following notes were taken by Mr. F. W. Frohawk from the living birds that were in my possession :—

" It is incapable of flight, very active and swift on foot, tame, and fearless. They never attempted to make use of their feeble wings, and they only opened them when springing up to perch.

" During the day they kept up an incessant chirping, consisting of from one to three soft, short, and clear notes ; but soon after dusk they all, as if by one given signal, struck up a most peculiar chorus, which lasted but a few seconds, and then all was silent. I can only compare the sound to a handful or two of marbles being thrown on a glass roof and then descending in a succession of bounds, striking and restriking the glass at each ricochet.

" The tail is at times held drooping, sometimes elevated, and frequently jerked up and down."

Although the size was given wrong by the sailors, there cannot be any doubt that no. 15 in Kittlitz's List refers to this Rail, and that it was also this Rail that Captain Walter Edw. Wood saw on Laysan Island in 1872, describing it to Dr. Otto Finsch as a "small wingless bird, a kind of Woodhen" (cf. Hartl. Abh. naturw. Verein Bremen, xii. p. 400, 1892).

6. CHARADRIUS FULVUS, *Gm.*

KOLEA (Sandwich Islands).

Fulvous Plover, Lath. Gen. Syn. iii. 1, p. 211 (1785).
Charadrius fulvus, Gm. Syst. Nat. i. p. 687 (1788) (Tahiti) ; Dole, Proc. Bost. Soc. Nat. Hist. xii. p. 304
(1869) ; id. Hawaii. Almanac, 1879, p. 50 ; Seeb. Geogr. Distr. Charadr. p. 99 (1887) ; Wils. Av. Hawaii.
pt. iii. (1892).
Charadrius dominicus fulvus, Ridgw. Proc. U.S. Nat. Mus. 1880, p. 108 ; Stejn. Proc. U.S. Nat. Mus. 1887,
pp. 80 & 126 (Kauai).

THE Pacific Golden Plover has been known to frequent the shores of the Sandwich Islands
almost since their discovery, and it was also frequently observed by Mr. Palmer on most of
the small islands visited on his trip to Laysan and on Laysan Island itself.

It is very interesting that the Asiatic form and not the American one (*Charadrius
dominicus,* Müll.) is found on these islands and also on the Hawaiian Islands.

7. STREPSILAS INTERPRES (*L.*).

AKEKEKE (Sandwich Islands).

Tringa interpres, Linn. Syst. Nat. ed. x. i. p. 148 (1758).
Strepsilas interpres, Ill. Prodr. p. 263 (1811); Dole, Proc. Bost. Soc. Nat. Hist. xii. p. 304 (1869); id. Hawaii. Almanac, 1879, p. 51; Wils. Aves Hawaii. pt. iii. (1892).
Tringa oahuensis, Bloxham, Voy. Blonde, p. 251 (1826).
Arenaria interpres, Stejn. Proc. U.S. Nat. Mus. xii. p. 380 (1889).

THE Turnstone is known to be not uncommon on the shores of the Sandwich Islands, and Palmer found it also common on all the smaller islands visited during his trip, and also on Laysan. Several specimens in different plumages were sent.

8. TOTANUS INCANUS (Gm.).

WANDERING TATTLER. (*Ulili*, Sandwich Islands.)

— .. —

Ash-coloured Snipe, Lath. Gen. Syn. iii. 1, p. 154 (1785) (Eimeo and Palmerston Islands).

Scolopax incana, Gm. Syst. Nat. i. p. 658 (1788) (ex Latham).

Totanus incanus, Vieill. Nouv. Dict. vi. p. 400 (1816); Seeb. Geogr. Distr. Charadr. p. 360 (1887); Wilson, Aves Hawaii. pt. iii. text & pl. (1892f.

Heteroscelus brevipes, Baird, Expl. & Surv. R. R. Route Pacific, ix. pt. ii. pp. 728 & 734 (1858); id. B. N. Amer. pl. 88 (1860).

Actitis incanus, Dole, Proc. Bost. Soc. Nat. Hist. xii. p. 303 (1869); id. Hawaiian Almanac, 1879, p. 52 (Sandwich Is.).

Heteractitis incanus, Stejn. Auk, 1884, p. 236; id. Bull. U.S. Nat. Mus. no. 29, p. 132 (1885); id. Proc. U.S. Nat. Mus. 1887, p. 83 (Hawaiian Is.); Ridgw. Man. N.-Amer. B. p. 168 (1887). (Ridgway and Stejneger very clearly give the distinctive characters of *T. incanus* (Gm.) and its near ally *Totanus brevipes*, Vieill.)

A FEW were met with by my collector on Laysan, Lisiansky, Midway, French Frigate, and other small islands. The species has long been known to occur on the Sandwich Islands.

9. NUMENIUS TAHITIENSIS (*Gm.*).

BRISTLE-THIGHED CURLEW.

Otaheite Curlew, Lath. Gen. Syn. iii. 1, p. 122 (1785).
Scolopax tahitiensis, Gm. Syst. Nat. i. p. 656 (1788) (ex Latham).
Numenius tahitiensis, Lath. Ind. Orn. ii. p. 711 (1790); Vieill. Nouv. Dict. viii. p. 308 (1817); id. Enc.
   Méth. p. 1157 (1823); Gray, Gen. B. iii. p. 569 (1847); Ridgw. Proc. U.S. Nat. Mus. 1880, p. 201;
   Baird, Brew., & Ridgw. Water-B. N. Amer. i. p. 324 (Fort Kenai, Alaska—Paumotou Group); Ridgw.
   Man. N.-Amer. B. p. 171 (1887); Seeb. Geogr. Distr. Charadr. p. 333 (1887); Wiglesw. Aves Polynes.
   (in Abh. u. Mitth. Mus. Dresd.) p. 66 (1891); Lister, P. Z. S. 1891, p. 299; Wilson, Av. Hawaii. pt. iii.
   (1892).
Numenius femoralis, Peale, U.S. Expl. Exp., B. p. 233, pl. 64 (1848) (Vincennes Is., Paumotou Group);
   Hartl. Arch. f. Naturg. 1852, i. p. 120; id. J. f. O. 1854, p. 170 (Paumotou); Cass. U.S. Expl. Exp., Orn.
   p. 316, pl. xxxvii. (1858); Finsch & Hartl. Beitr. Fauna Polynes. p. 175 (1867); Ridgw. Am. Naturalist,
   1874, p. 436 (occurrence in N. America); Finsch, Ibis, 1880, p. 220 (Taluit, Marshall Is.); id.
   tom. cit. p. 432 (Tarowa, Gilbert Is., where it is stated to be a winter visitor only!); Tristr. Ibis,
   1881, p. 251 (Marquesas); id. op. cit. 1883, p. 47 (Fanning Is.); Stejn. Proc. U.S. Nat. Mus. 1887,
   p. 83 (Kauai).
Numenius phæopus, partim!, Schleg. Mus. P.-B., _Scolopaces_, p. 93 (1864).
Numenius australis, Dole, Proc. Bost. Soc. N. H. xii. p. 303, and Hawai. Almanac, 1879, p. 51 (nec
   N. australis, Gould!).
Numenius taitensis, Coues, Check-list N.-A. B. 2nd ed. p. 105 (1882); id. Key N.-Am. B. 2nd ed. p. 616
   (1884).

Stejneger (*l. c.*) thinks the name *tahitiensis* cannot stand, but, although Latham did not
mention the bristly thigh-feathers, his description seems to suit the present species so
well, and the name given by Gmelin, who never saw the bird and merely based his
description on Latham's, has been so generally accepted, that I do not think it
necessary to reject it.

This rather rare species can be easily recognized by its thigh-feathers having their shafts
elongated into long, glossy, hair-like bristles; often even longer than in the figure in
Seebohm's 'Charadriidæ,' on p. 333. Top of head dark sooty brown, limited by broad
superciliary stripes and divided by a mesial line of buff; some dusky streaks in front and
behind the eye; chin and throat, neck, and all the lower parts pale buff; the cheeks, neck,
and jugulum streaked with brown; the sides of the body irregularly barred with the
same colour; axillaries pale rufous buff, with a dark brown shaft and broad bars of dark
brown. Upper parts sooty brown, all the feathers spotted or irregularly varied with buff;

D

feathers of the lower back broadly margined with bright buff, and the upper tail-coverts nearly or quite uniform ochraceous buff; tail ochraceous buff, barred with dark brown. Primaries dark sooty brown, inner webs banded with buff, shafts white; secondaries spotted or banded with buff on both webs. Under wing-coverts spotted brown and buff.

The sexes are similar.

" Iris dark hazel ; bill blackish brown ; base of lower mandible dull flesh-colour, tip darkest ; legs bluish." (*H. C. P.*)

Total length about 17 to 18 inches, wing 9 to 10, tail 4, culmen 3·50, 3·60 to 4·1 (this latter an adult female from Laysan Island), tarsus 2·1 to 2·3, middle toe (without claw) 1·35 to 1·50.

THE Bristle-thighed Curlew was for a long time known from the Pacific Ocean only, but has more recently been found on the coast of Alaska. From these later discoveries it had been supposed that it was merely a wanderer in the Pacific Ocean and that its breeding-grounds were the coasts of Alaska. This statement has also been made by Seebohm (*l. c.*), but whether it is true or not must remain doubtful until we have more exact information about it, and until the eggs have actually been found. Finsch (*l. c.*) says that it is only a winter visitor on the Gilbert Islands; but this can only be a surmise, as he has not been there during all seasons of the year to prove it.

Wilson (*l. c.* p. 3, in his article on *Numenius tahitiensis*) did not find it breeding, but some natives assured him that it *did* breed in the Sandwich Islands.

Now Palmer not only observed it on several small islands during his trip to Laysan, but sent me specimens from Laysan and Brooks or Midway Islands, and he says that Mr. Freeth —then living on Laysan, and who carefully protected the birds of that now lonely and unprotected bird-paradise—did not allow him to kill more than the one specimen he sent me, *because they were then* (in the latter half of June) *in the height of the breeding-season.* It is a thousand pities he did not send unmistakable proofs to confirm the statement, and did not discover the yet unknown eggs of this bird. All he says further is that Mr. Freeth told him they were—like most of the Laysan birds—very fond of sucking other birds' eggs. Palmer's statement, however, is so positive, that some weight must certainly be attached to it.

10. ANAS LAYSANENSIS, *Rothsch.*

LAYSAN TEAL.

— · —

"*Eine Art Ente*," Kittlitz, Mus. Senck. i. p. 124 (1834).
Anas laysanensis, Rothsch. Bull. B. O. C. no. iv. p. xvii (1893) ; id. Ibis, 1893, p. 250.

Male. (Whether this is in full nuptial plumage or whether there is a different plumage in winter and summer I do not yet know.) Forehead deep shining brown, almost black, the black tips and margins of the feathers being quite abraded ; top of head and neck similar, but showing more brown on the sides ; the chin shows some white feathers, and there is a somewhat irregular ring of white feathers round the eye. Feathers of the upper surface deep blackish brown, with more or less irregular U-shaped or rather "obomegoid" (Ridgw. Nomencl. Col. pl. xv. fig. 8) markings of a light rusty brown ; scapulars and greater wing-coverts with several cross-markings, and all the feathers with more or less abraded borders of the same colour ; rump and upper tail-coverts more shining black ; a large speculum of deep green, more velvety black behind and on the upper margin, bordered below by a broad white line. Primaries pale brown, with pale almost whitish edges. Feathers of the lower surface very pale rusty brown, with irregular darker bars and spots, somewhat whitish near the shaft and base. Under wing-coverts brown and rusty with a good deal of white, axillaries white with pale brown spots.

Adult female. Differs from the male in having more white on the chin and upper throat, in having much bolder and more patch-like rusty markings on the upper surface, and in the speculum being only indicated.

"Iris brown ; upper mandible blackish, green towards the edges ; lower mandible bluish brown, with a dull orange mark at the base ; tarsus and toes dull orange, webs brown with a bluish tinge."

Total length about 15 to 17 inches, wing 7·5 to 7·7, tail 3·5, culmen 1·6, tarsus 1·4 to 1·7, middle toe with claw 2, hind toe with claw 0·45.

THIS Duck was found by my collector on Laysan Island only. The captain who told Dr. Finsch that there was a flightless Duck on Laysan Island no doubt referred to this species, but on account of its extraordinary tameness he was led to believe that it could not fly. It flies, however, well enough. Kittlitz, in his article, says that it was observed on Laysan and Lisiansky, but Palmer did not notice it on the latter island.

D 2

According to Palmer's observations, this Duck was generally seen in pairs, but sometimes a dozen or more would sit round a freshwater well. They are very tame, often walking up to the front of the house. They were never observed in the water, though often on the beach. They frequented the scrub all over the island, but were not very plentiful. June seems to be their breeding-season.

The only near ally of *Anas laysanensis* is *Anas wyvilliana*, Scl., from the Sandwich Islands, which, however, is one third larger, and the adult male of which is very different in colour.

11. FREGATA AQUILA (*Linn.*).

FRIGATE-BIRD.

Pelecanus aquilus, Linn. Syst. Nat. ed. x. p. 133 (1758) (I. Adscensionis aliisque pelagicis) ; id. ed. xii. i.
p. 216 (1766) ; Gm. Syst. Nat. ii. p. 572 (♂ ad.) (1788) ; Lath. Ind. Orn. ii. p. 885 (1790) ; Cuv. Règn.
Anim. i. p. 525 (1817) ; id. ed. ii. i. p. 561 (1829) ; Vieill. Enc. Méth. p. 45 (1823).

La Frégate, Briss. Orn. vi. p. 506, pl. xliii. (1760).

La Frégate, Buff. Hist. Nat. Ois. viii. p. 381 (1781) ; id. ed. fol. ix. p. 246 (1784).

La grand Frégate, de Cayenne, Daubent. Pl. Enl. 961.

White-headed Frigate, Lath. Gen. Syn. iii. p. 591.

Palmerston Frigate, Lath. Gen. Syn. iii. p. 593.

Pelecanus leucocephalus, Gm. Syst. Nat. ii. p. 592 (1788) (Ins. Ascensionis) (av. jun.) ; Lath. Ind. Orn. ii.
p. 886 (1790) ; Rafll. Trans. Linn. Soc. xiii. p. 330 (1822).

Pelecanus palmerstoni, Gm. Syst. Nat. ii. p. 573 (1788) (Ins. Palmerston, S. Pacific) (♀ ad.).

Italiaeus aquilus, Ill. Prodr. p. 279 (1811).

Tachypetes aquila, Vieill. Nouv. Dict. xii. p. 143 (1817) ; id. Gal. Ois. ii. p. 187, pl. 274 (1825) ; Spix, Av.
Bras. ii. p. 82, tab. 105 (1825) ; Less. Trait. d'Orn. p. 609 (1831) ; Wied, Beitr. iv. p. 885 (1832) ; Aud.
Orn. Biogr. iii. p. 495 (1835) ; id. B. Am. vii. pl. 421 (1844) ; Schomb. Reis. Brit. Guiana, iii. p. 763 ;
Cass. Proc. Ac. Philad. 1856, p. 255 ; Sel. P. Z. S. 1856, p. 144 (Ascension) ; Sel. & Salv. Ibis, 1859,
p. 233 (Centr. Amer.) ; Taylor, Ibis, 1862, p. 201 (Florida) ; Gould, Handb. B. Austral. ii. p. 499 (1865) ;
Frantz. J. f. O. 1869, p. 379 (Costa Rica) ; Dohrn, J. f. O. 1871, p. 9 (Cape Verde Is.) ; Layard, P. Z. S.
1875, pp. 29, 411 (Fiji) ; Masters, Proc. Linn. Soc. N. S. W. i. p. 64 (1876) (Torres Str.) ; Finsch,
P. Z. S. 1877, p. 770 ; Cab. J. f. O. 1878, p. 366 (captivity) ; Rams. Proc. Linn. Soc. N. S. W. iii. p. 303
(1878) (Kerapoona) ; id. op. cit. iv. p. 102 (1879) (Pt. Moresby, S. Cape, Louisiades, Duke of York Is.,
Solomon Is.) ; Sharpe, P. Z. S. 1879, p. 354 (Borneo) ; Finsch, P. Z. S. 1880, p. 577 (Ruk, Carolines) ;
Simroth, Archiv f. Naturg. 1888, in "Kenntniss zur Azorenfauna" (Azores, accidental) ; Sharpe, Ibis,
1891, p. 141 (Fernando Noronha).

Fregata aquilus, Steph. Gen. Zool. xiii. p. 120 (1826) ; Less. Man. d'Orn. ii. p. 376 (1828) ; Gosse, B. Jamaica,
p. 422 (1847) ; Garrod, P. Z. S. 1873, p. 467 (anatomy) ; Hume, Stray Feath. viii. p. 116 (1879).

Fregata leucocephalus and *palmerstoni*, Steph. Gen. Zool. pp. 122, 123 (1826).

Tachypetes aquilus, Vig. P. Z. S. 1831, p. 62 (habits) ; Kittl. Kupfertaf. p. 15, tab. 20. fig. 1 (1833) ; id.
Mus. Senckenb. i. p. 121, no. 3 (p. 122, no. 4, juv.) (1834) (Laysan I.) ; Swains. Class. B. ii. p. 371
(1837) ; Peale, U.S. Expl. Exp., B. p. 179 (1848) ; Hartl. J. f. O. 1854, p. 160 (Ceylon) ; id. t. c. p. 170
(Paumotu Is.) ; Gloger, J. f. O. 1855, p. 21 ; Bp. Consp. ii. p. 166 (1855) ; Tsch. J. f. O. 1856, p. 146
(habits) ; Burm. Syst. Ueb. Th. Bras. iii. p. 459 (1856) ; Thienem. Abb. Vogeleier, tab. xciii. (1856) ;
Gosse, Jamaica, p. 422 (1857) ; Gundl. J. f. O. 1857, p. 239 (Cuba) ; Cass. U.S. Expl. Exped., B. p. 358
(1858) ; Baird, B. N. Am. p. 873 (1858) ; A. & E. Newton, Ibis, 1859, p. 369 (St. Croix) ; Blas. & Bald.
Naum. Vög. Deutschl. xiii. 2, p. 287 (1860) (Weser near Hann. Muenden, according to Beehst. Naturg.
Deutschl. ed. 2, iii. p. 756) ; Bryant, Proc. Bost. Soc. N. H. vii. p. 126 (1860) (Bahamas, excellent notes

on nidification, eggs, and habits); Albrecht, J. f. O. 1861, p. 57 (Bahamas); Gundl. J. f. O. 1861,
pp. 348, 349, 1862, p. 96, 1875, p. 406, 1878, pp. 163, 191, 193 (Cuba, Porto Rico); Dress. Ibis, 1866,
p. 45 (S. Texas); Hartl. & Finsch, Orn. Centralpolyn. p. 265 (1867); Roscub. Reis naar Zuidoostereil.
pp. 54, 82 (1867) (Aru, Kei, Ceram); Marie, Act. Soc. Linn. Bord. xxvii. (1870) (Nova Caledonia);
Melliss, Ibis, 1870, p. 105 (St. Helena); Finsch, J. f. O. 1870, p. 377; Sundev. P. Z. S. 1871, p. 125
(Galapagos); Finsch, J. f. O. 1872, pp. 33, 58, 260 (Samoa, New Zealand); Roscub. Reist. naar Geel-
vinkb. p. 9 (Ternate); Layard, Ibis, 1876, p. 393 (Fiji); id. P. Z. S. 1876, pp. 490 (Navigator's Is.), 504;
Cab. & Reichen. J. f. O. 1876, p. 329 (Kerguelen); Bocage, J. f. O. 1876, p. 434 (S. Vincente); Pen-
rose, Ibis, 1879, pp. 275, 276 (Ascension); Layard, Ibis, 1880, p. 233 (Loyalty Is.); Finsch, t. c. p. 333
(Marshall Is.); id. J. f. O. 1880, p. 296 (Ponapé); Cory, B. Bahamas, i. p. 209 (1880); Schmeltz, Ethn.
Abth. Mus. Godeffr. 1881, pp. 239, 295, 353; Finsch, J. f. O. 1880, p. 310 (Kushai); id. Ibis, 1881, p. 109
(Kushai); id. t. c. p. 115 (Ponapé); id. t. c. p. 248 (Nawodo, Pleasant I.); id. t. c. p. 540 (New Britain);
Cory, B. Haiti & S. Domingo, p. 173 (1885).
Tachypetes leucocephalus, Kittl. Kupfertaf. p. 15, tab. 20. fig. 2, juv. (1833) (Laysan).
Fregata aquila, d'Orb. in La Sagra's Hist. Nat. Cuba, Ois. p. 399 (1840); Gray, List Spec. B. Brit. Mus.,
Anseres, p. 190 (1844); Hartl. J. f. O. 1854, p. 170 (Galapagos); id. Orn. Westafr. p. 260 (1857);
Tayl. P. Z. S. 1858, p. 318 (egg); id. Ann. & Mag. Nat. Hist. iii. p. 150 (1859); id. Ibis, 1859, p. 150;
id. Ibis, 1869, pp. 11, 316; Schlegel, Mus. Pays-Bas, *Pelecani*, p. 2 (1863); Salvin, Ibis, 1864, p. 374
(Brit. Honduras); Scl. P. Z. S. 1871, p. 620 (alive); Bull. B. New Zeal. p. 339 (1873); id. 2nd ed.
p. 182 (1888); Scl. & Salv. Nomencl. Av. Neotrop. p. 124 (1873); Salvad. Cat. Ucc. Borneo, p. 364
(1874); Garrod, P. Z. S. 1876, pp. 335, 341 (anatomy); Lenz, J. f. O. 1877, p. 382 (Saparua); Scl. &
Salv. P. Z. S. 1878, p. 650 (Ascension); Whitmee, P. Z. S. 1878, p. 273 (Ellice I.); Rosenb. Mal. Arch.
pp. 324, 373, 396 (1879) (Ceram, Aru, Mysol, Salvatti); D'Alb. & Salvad. Ann. Mus. Civ. Gen. xiv.
p. 135 (1879) (Katau); Legge, Birds Ceylon, p. 1024 (1880); A. & E. Newton, Handb. Jamaica, p. 112
(1881); Scl. & Salv. 'Challenger,' Birds, pp. 117, 118 (1881); Milne-Edw. Ann. Sc. Natur., Zoologie,
sér. vi. t. xiii. art. 4, p. 37 (1882) (anatomy); Salvad. Ann. Mus. Civ. Gen. xviii. p. 402 (1882); id.
Orn. Papuasia, iii. p. 402 (1882); Salvin, P. Z. S. 1883, p. 427 (Payta, Peru); Pleske, Bull. Acad.
Pétersb. xxix. p. 538 (1884) (Ternate); Salvin, Ibis, 1886, p. 168; Taez. Orn. Pérou, ii. p. 427 (1886);
Goss, Auk, 1886, p. 113 (Kansas, straggler); Saunders, P. Z. S. 1886, p. 335 (Diego Garcia, Chagos
Group); Sharpe, P. Z. S. 1887, p. 516 (Christmas I., 190 miles south of Java); MacFarlane, Ibis, 1887,
p. 210 (Salas y Gomez), & p. 213 (Christmas I., W. Pacific); Scott, Auk, 1887, pp. 137, 139, 281 (S.
Florida); Cory, Auk, 1887, p. 180 (Old Providence), & p. 181 (St. Andrews); id. Auk, 1888, p. 69 (W. I.);
Dutcher, Auk, 1888, p. 173 (Long Island, N.Y., shot Aug. 4, 1886); Scott, Auk, 1888, p. 378 (Gulf
Coast, Florida); Lister, P. Z. S. 1888, p. 529 (Christmas I.); Feilden, Ibis, 1889, p. 501 (Barbados);
Salvin, Ibis, 1889, p. 376 (Cozumel I.); Hartert, J. f. O. 1889, p. 406 (Bay of Bengal); Cory, B. W.
Ind. 1889; Bryant, Proc. Acad. Californ. 1890 (Lower California); Scott, Auk, 1890, p. 307 (Dry
Tortugas); Northrop, Auk, 1891, p. 79 (Andros I.); Cory, Cat. W. Ind. B. p. 85 (1892); Wiglesw.
Mitth. u. Ber. Zool. Mus. Dresd., Aves Polynes. p. 71 (1891).
Atagen aquila, Gray, Gen. B. iii. p. 669 (partim) (1845); Ramsay, Tabul. List, p. 25 (1888).
Attagen aquila, Gould, B. Austr. i. p. c (1848); Brehm's Thierleben, Bd. vi. p. 369 (1892).
Fregata leucocephala, Rehb. Syn. Av., *Natatores*, tab. xxxi. (1848).
Fregatte, Bolle, J. f. O. 1856, p. 26 (Cape Verd Is.).
Tachypetes palmerstoni, Cass. Pr. Ac. Philad. 1856, p. 255; id. U.S. Expl. Exp., Orn. p. 359 (1858); Scl.
P. Z. S. 1864, p. 9 (Huahcine); Finsch, J. f. O. 1872, p. 58; Streets, Amer. Nat. xi. p. 71 (1877)
(Christmas I., Fannings Group).
Attagen aquilus, Jerd. B. Ind. iii. p. 853 (1863); Gray, Hand-l. iii. p. 130 (1871); Holdsw. P. Z. S. 1872,
p. 462 (Ceylon).

Fregetta aquila?, Ramsay, Proc. Linn. Soc. N. S. W. i. p. 375 (1876) (Solomon Is.).
Fregata strumosa, Kittl. MS. teste Hartert, Katal. Mus. Senckenb. p. 235 (1891).

Adult male from Laysan. Above black with metallic gloss, the lanceolate feathers of the back and scapulars strongly glossed with green and purple; median wing-coverts brown with blackish centres; wings and tail black; under surface deep blackish brown without gloss. Total length about 41 inches, wing 23, tail 15·8, bill 4·4.

Female from Laysan. Feathers of the back and scapulars not lanceolate. Above deep blackish brown with very little gloss, much paler on the lower hind neck; median wing-coverts pale brown with dark brown centres; throat and sides of neck dull whitish brown; breast and sides down to the flanks white. Wings 23·8 to 24 inches, tail 16, bill 4·7.

Immature birds from Laysan. The head white; beneath the throat rufous cinnamon; breast brown; abdomen white; under tail-coverts black. Another specimen has the neck and head cinnamon-rufous, and the whole under surface white.

In appearance and measurements my specimens agree with Frigate-birds from other localities (Pacific Ocean and Madagascar, whence only the quite distinct *Fregata minor* (Gm.) has hitherto been recorded), but two adult males from the West-Indian seas (one south of Florida and one from Aruba) have the lower surface deeper black, and no sign of the pale upper wing-coverts.

The cinnamon-rufous tinge of the usually white head and neck of immature birds is very strongly developed in my specimens from Laysan, but a specimen of *Fregata minor* (Gm.) from Borneo shows it equally well, and the same has been recorded of specimens from other countries.

The *nestling* is covered with very soft and fluffy white down.

Kittlitz (see Mus. Senckenb. i. p. 121, and Kupfertafeln, pl. 20 and text p. 15) believed the young bird with the white head to belong to a different species, which, however, is not the case.

The eggs from Laysan measure 2·8 by 1·86 and 2·65 by 1·9 inches, and are dull white in colour. The shell is covered with a chalky crust like those of the Gannets, Pelicans, Cormorants (but thinner than in most of the latter), clearly indicating their belonging to the Steganopodes.

THERE are numerous rookeries of the Frigate-birds on Laysan and most of the other islands visited, almost wherever there is scrub. Their breeding-season is from the middle of May to the middle of July. Their nests are made of twigs and used year after year, till they become often mere masses of guano and dead birds. These nests are built on the top of the bushes, and from half a dozen to fifty are found in one rookery. The male is as much on the nest as the female. They lay only one egg. They live principally on fish, but often eat the disgorged food and young of other birds.

It is very interesting to watch them when a squall is passing over; they then sail up in the air in large flocks, keeping just out of the reach of the squall, whichever way it may be passing. It is very interesting, too, to see them chasing other birds until they disgorge their prey.

(The photograph of this bird fighting with the Gannet and that of the young Frigate-bird are unique as illustrating the life-history of this interesting bird; also the illustrations of the old bird on the nest and the female creeping in the scrub depict subjects not generally known. The engraving of Palmer among the Frigate-birds (done from a photograph) illustrates the great tameness of the bird in the breeding-season.)

12. SULA CYANOPS (Sundev.).

MASKED GANNET.

Sula dactylatra, Less. Voy. 'Coquille,' ii. p. 494 (1826) ; id. Tr. d'Orn. p. 601 (1831) (Ascension) ; G. R. Gray,
Gen. B. iii. p. 666 (1845) ; Pucher. Rev. et Mag. Zool. 1850, p. 626 (crit. of Lesson's type) ; Hartl.
J. f. O. 1853, p. 420 (crit. ; spec. bon.) ; Bp. Consp. ii. p. 165 (1855) ; Scl. Ibis, 1859, p. 352 (note) ;
Bryant, J. f. O. 1861, pp. 57, 59 (Bahamas) ; Hume, Stray Feath. v. pp. 311, 312 (crit. remarks) ; Milne-
Edw. Ann. Sc. Nat., Zool. sér. vi. t. xiii. art. n. 35, pl. xiii. (1882) (Antarctic Regions).
Sula nigrodactyla, " Less.," in the synonymy of *S. dactylatra*, Bp. Consp. ii. p. 165 (1855).
Dysporus cyanops, Sundev. Physiogr. Sällsk. Tidskr. 1837, p. 218, tab. 5 (Atlantic) ; id. Isis, 1842, p. 858 ; id.
Ann. & Mag. Nat. Hist. xix. p. 236 (1847) ; Hartl. & Finsch, Vög. Ost-Afr. p. 843 (1870) ; Sund. P. Z. S.
1871, p. 125 (Galapagos) ; Gigl. Viaggio Magenta, pp. 843, 844, 884 (1875) (Peru) ; Schrk. J. f. O.
1879, p. 410 (egg) ; Finsch, Ibis, 1880, pp. 431, 434 (Gilbert Is.).
Sula cyanops, G. R. Gray, Gen. B. iii. p. 666 (1845) ; Bp. Consp. ii. p. 166 (1855) ; Scl. P. Z. S. 1856, p. 145 ;
id. Ibis, 1859, p. 352 ; Schleg. Mus. P.-B., *Pelecani*, p. 39 (1863) (Torres Str.) ; Shelley, B. Egypt, p. 294
(1872) ; Heugl. Orn. N.O.-Afr. p. 1180 (1873) ; Salvad. Ucc. Borneo, p. 367 (1871) ; Lawr. Proc.
Boston Soc. N. H. xiv. p. 302 (Socorro Is.) ; Hume, Stray Feath. v. p. 307 (criticism) ; Scl. & Salv.
P. Z. S. 1878, p. 652 (Raine I.) ; Scl. P. Z. S. 1879, p. 310 (egg) ; Penrose, Ibis, 1879, pp. 275, 283
(Ascension) ; Legge, B. Ceylon, p. 1110 (1880) ; Salvad. Ann. Mus. Civ. Gen. xviii. p. 404 (1882)
(distribution) ; id. Orn. Papuas. iii. p. 416 (1882) ; Ridgw. Man. N.-Am. B. p. 75 (1887) ; Cory, Auk,
1888, p. 71 (W. Indies) ; id. Auk, 1889, pp. 31, 32 (Cayman Brack and Little Cayman Is.) ; Murray,
Stray Feath. 1887, p. 165 (Kurrachi) ; Cory, Cat. W. Ind. B. p. 84 (1892) ; Wiglesw. Abh. u. Ber.
Mus. Dresd. 1890-91, p. 72 (1892, in " Aves Polynesiæ ").
Sula personata, Gould, P. Z. S. 1846, p. 21 (N. Australia) ; id. B. Austr. vii. pl. 77 (1848) ; Macgill. Voy.
Rattlesn. ii. p. 359 (1852) ; Cass. U.S. Expl. Exp. p. 368 (1858) ; Jerdon, B. Ind. iii. p. 852 (1864) ;
Cab. & Reichen. J. f. O. 1876, p. 329 (Timor) ; Crowfoot, Ibis, 1885, p. 269 (Norfolk I.).
Sula bassana, Thomps. (nec Linn.), Allen's Niger Exped. ii. p. 175 (1848).
Sula melanops, Hartl. Ibis, 1859, p. 351, pl. x. figs. 2, 3 (Bur da Rebschi, Somali coast) ; Heugl. Peterm.
Geogr. Mitth. 1869, p. 418 (in Orn. N.O.-Afr. Heuglin admitted the identity with *S. cyanops*).
Dysporus dactylatra, Gundl. J. f. O. 1875, p. 403 (Cuba).

Although there can be hardly any doubt that *Sula dactylatra* of Lesson is referable to the
present species, and that, according to the strict modern rules of nomenclature, his name
should take priority over *Sula cyanops*, Sundev., I refrain from accepting it, because
Lesson's description is so bad (and even wrong) that it does not well diagnose the species ;
besides, it is one of the worst hybrid names ever given to a bird.

Adult. Upper mandible yellowish horn-colour in life and in dried skin, the lower mandible
shaded with dusky. Face and naked skin on throat blackish in skin, blue-black in life.

E

Iris bright yellow. Plumage pure white, remiges and greater wing-coverts dark blackish brown. Tail dark sooty brown, the central tail-feathers whitish at their basal part. Legs and feet deep brown in skin, a yellowish brown in life. Total length about 27 inches, wing 17 to 18, tail 8, culmen 4 to 4·4, tarsus 2·3, middle toe with claw 3·9 to 4.

One specimen (not quite adult) has the back streaked with brown and the wing-coverts spotted with brown.

The *first plumage* is quite different from that of the adult bird. Head and neck above, and upper parts generally, dark greyish brown; lower neck and entire lower surface white, flank streaked with brown; back and rump streaked with white. The *iris* is dark brown.

The *nestling* is covered with white down.

The eggs are two in number, ovate or elongate ovate, covered with a dull white chalky crust, and beneath this chalky covering very pale bluish; they measure 2·84 by 2 inches, 2·6 by 1·8, and 2·55 by 1·88.

This Gannet frequented all the islands visited, but not in great numbers. There is a large rookery on French Frigate Island, but only a few were observed on Laysan, where they invariably frequented the shore and never went inland. Although Palmer always found two eggs, he never saw more than one young in a nest, if such a term may be used, for they build no nest whatever, but simply deposit their eggs on the sand. The old birds have a harsh note, similar to the quacking of a Duck. When sitting on its eggs or with the young, the old bird is easily killed with a short stick, for they only bite at the stick or hand and never attempt to fly off.

The birds and eggs described above are from French Frigate and Brooks or Midway Islands.

(On the Plate representing the breeding-place of this species on Laysan Island it has been named *Sula australis* by mistake.)

13. SULA PISCATRIX (*L.*).

RED-FOOTED BOOBY.

Pelecanus piscator, Linn. Syst. Nat. ed. x. p. 134 (1758); id. Syst. Nat. ed. xii. i. p. 217 (1766); Gm. Syst.
  Nat. ii. p. 578 (1788); Vieill. Enc. Méth. i. p. 49 (1823); Kittlitz, Reise, i. p. 351, ii. p. 79 (1858)
  (Brown's I. and Ualan); id. Mus. Senck. i. p. 123 (1834) (Laysan and Lisiansky).
Le Fou brun, Briss. Orn. vi. p. 499 (1760) (young).
Le Fou blanc, Briss. Orn. vi. p. 501 (1760) (Les côtes d'Afrique et de l'Amérique); Buff. Hist. Nat. Ois. viii.
  p. 371.
Pelecanus fiber, Linn. Syst. Nat. ed. xii. i. p. 218 (1766); Gm. Syst. Nat. ii. p. 579 (1788); Vieill. Enc.
  Méth. i. p. 48 (1823).
Morus piscator, Vieill. Nouv. Dict. xxi. p. 40 (1817).
Dysporus piscator, Licht. Verz. Doubl. p. 87 (1823); Sundev. Phys. Sällsk. Tidskr. 1837, p. 220; id. Isis,
  1842, p. 857; id. Ann. & Mag. Nat. Hist. xix. p. 235 (1847) (Indian Ocean); Hartl. & Finsch, Orn.
  Centralpolynes. p. 255 (1867); Hartl. P. Z. S. 1867, p. 831 (Pelew Is.); Buller, B. N. Zeal., Introd.
  p. xv (1873); Layard, P. Z. S. 1876, p. 504 (Fiji); Nehrk. J. f. O. 1873, p. 410 (egg); Finsch, Journ.
  Mus. Godeffr. viii. p. 47 (1875).
Sula piscator, Gray, Gen. B. iii. p. 666 (1845); Gould, B. Austral. vii. pl. 79 (1848); Blyth, Cat. B. Mus.
  As. Soc. Beng. p. 297 (1849) (Bay of Bengal, Maldives); Hartl. Orn. W.-Afr. p. 258 (1857); Cass. U.S.
  Expl. Exp., Ornith. p. 365 (1858); Hartl. J. f. O. 1860, p. 176 (Madagascar); Albrecht, J. f. O. 1862,
  p. 207 (Jamaica); Salvin, Ibis, 1864, pp. 378, 379 (Brit. Honduras); Jerd. B. Ind. iii. p. 852 (1864);
  Newt. Ibis, 1866, p. 200 (Guatemala); Ramsay, Proc. Linn. Soc. N. S. Wales, ii. p. 203 (1877) (S. coast
  of New Guinea); Scl. & Salvin, P. Z. S 1878, p. 651 (off Cape York); Scl. P. Z. S. 1879, p. 310 (egg);
  Hume, Stray Feath. viii. p. 116 (1879); Saund. P. Z. S. 1880, p. 163 (Trinidad); Layard, Ibis, 1880,
  p. 233 (Loyalty Is.), 1881, p. 134 (New Caledonia); Salvad. Orn. Papuasia, iii. p. 419 (1882); Ridgw.
  Man. N.-Am. B. p. 86 (1887); Cory, Auk, 1887, p. 180 (St. Andrews); id. Auk, 1888, p. 72 (W. Indies);
  Feilden, Ibis, 1889, p. 501 (Barbados); Stejn. Proc. U.S. Nat. Mus. xii. p. 382 (1890) (Kauai); Milne-
  Edw. & Grandidier, Hist. Nat. Madagascar, xii. p. 695 (1885); Cory, Cat. W. Ind. B. p. 84 (1892).
Sula leucophæa, Steph. Gen. Zool. xiii. p. 106 (1826) (Cayenne); G. R. Gray, Gen. B. iii. p. 666 (1845).
Sula fiber, Steph. Gen. Zool. xiii. p. 105 (1826); Gray, Gen. B. iii. p. 666 (1845).
Sula erythrorhyncha, Less. Tr. d'Orn. p. 601 (1831).
Sula rubripes, Gould, P. Z. S. 1837, p. 156 (Nova Cambria, Austr.); Hartl. J. f. O. 1855, p. 360 (Gold
  Coast).
Sula rubripeda, Peale, U.S. Expl. Exp. p. 274 (1848); Hartl. Wiegm. Arch. 1852, p. 125; id. J. f. O. 1854,
  p. 170 (Pacific Ocean).
Piscatrix candida, Reichenb. Av. Syst. Nat. p. vi (1852); Bp. Consp. ii. p. 166 (1855).
Sula piscatrix, Scl. P. Z. S. 1856, p. 145 (Ascension); Schleg. Mus. P.-B., *Pelecani,* p. 40 (1863) (Ind. Arch.,
  Torres Strait); Schleg. & Poll. Rech. Faun. Madagascar, Ois. p. 140 (1868); Swinh. Ibis, 1873, p. 231
  (Hainan); Salvad. Cat. Ucc. Borneo, p. 368 (1874); Rosenb. Reist. Geelvinkb. p. 9 (1875) (Ternate);
  Hume, Stray Feath. iv. p. 483 (1876); Hartl. Vög. Madagascar, p. 397 (1877); Sharpe, P. Z. S. 1879,

E 2

p. 353 (Borneo) ; Legge, B. Ceylon, p. 1180 (1880); Nichols. Ibis, 1882, p. 70 (Cocos Keeling Is.) ;
Milne-Edw. Ann. Sc. Nat. Zool. sér. vi. tab. xiii. art. no. 4, p. 36 (1882); Seeb. B. Japan, p. 213 (1891) ;
Wiglesw. Abh. u. Ber. Mus. Dresd. 1890–91, p. 72 (1892, in "Aves Polynesiæ").
Piscatrix piscator, G. R. Gray, Hand-l. iii. p. 126 (1871).
Sula plumigula, Pelz. Ibis, 1873, p. 52, ex Natt. MS. (Australia?) (type examined by Salvadori).
Disporus hernandezi, Gundl. J. f. O. 1878, p. 298 (Cuba) ; id. J. f. O. 1881, p. 401.

Adult. Bill horn-colour, overspread with dull blood-red in dried skin, light blue with the
bare space on the forehead and a mark near the base of lower mandible purplish red in
life. Bare space on chin and throat black in dried skin, blue with blackish markings
in life. Plumage pure white, with a more or less developed buff tinge on the top of head
and hind neck. Greater upper wing-coverts and quills dark greyish brown. Shafts of
white tail-feathers straw-yellow. Feet purplish red in dried skins, red with purple tinge
in life. Claws horny white. Iris dark brown, with a circle of greyish and brownish lines
on the outer edge. Total length about 28 inches, wing 14·5 to 16·5, tail 8·3 to 9·3,
culmen 3 to 3·5.

Immature birds are spotted with brown above, and the young in the first plumage are
dark brown above and lighter smoky grey beneath, with quills and tail-feathers much
paler. The young birds vary extremely in coloration, and it is difficult to find several
specimens of exactly the same colour.

The eggs are like those of other Gannets, covered with a chalky crust of a dull white, and the
shell under it is light greenish blue. They measure 2·56 by 1·57 inches, 2·4 by 1·6,
2·25 by 1·47, and 2·76 by 1·77.

This bird resembles the Common Booby (*Sula sula,* Linn.) in building its nest on the top of
a bush and not on the ground like *Sula cyanops,* Sundev. The nest is made of twigs, and
only one egg is laid. Its cry is loud and harsh, and it vigorously attacks anyone trying to
drive it off the nest.

It is very plentiful on Laysan, and was caught and skinned at sea near Niihau, on
Lisiansky and Midway Islands, also often seen at sea.

The species of *Sula* do not intermix, and the breeding-places of the different species
are always separate.

These islands are about the northernmost places where the widely-spread *Sula piscatrix* (L.)
occurs, but in America it goes as far north as Southern Florida and Lower California.
This species has (*l. c.*) been noticed from Kauai by Stejneger.

(NOTE.—The Plate given with this species is a photograph of the palm-tree mentioned
by Kittlitz, and which has thus been continuously used by the birds for more than 70 years.)

14. SULA SULA (L.).

BOOBY.

Le Fou, Briss. Orn. vi. p. 495 (1760) (coasts of Africa and America).

Pelecanus sula, Linn. Syst. Nat. i. p. 218 (1766) ; Raffl. Trans. Linn. Soc. xiii. p. 330 (Sumatra) (1822) ; Vieill. Enc. Méth. iii. p. 47, pl. 15 (1823).

Petit Fou, Buff. Hist. Nat. Ois. viii. p. 374 (1781).

Le Fou commun, Buff. Hist. Nat. Ois. viii. p. 368 (1781).

Fou de Cayenne, Daubent. Pl. Enl. no. 973.

Pelecanus leucogaster, Bodd. Tabl. Pl. Enl. p. 57 (1783) (ex Daubent. Pl. Enl.).

Pelecanus parvus, Gm. Syst. Nat. ii. p. 576 (1788) (ex Buffon) ; Vieill. Enc. Méth. i. p. 48 (1823).

Dysporus sula, Ill. Prodr. p. 280 (1811) ; Wied, Beitr. iv. p. 890 (1832) ; Bp. Consp. ii. p. 164 (1855) ; Heugl. Ibis, 1859, p. 351 (Red Sea) ; A. & E. Newt. Ibis, 1859, p. 369 (St. Croix) ; Verr. & Des Murs, Rev. et Mag. Zool. 1860, p. 442 (New Caledonia) ; Hartl. & Finsch, Orn. Centralpolynes. p. 260 (1867) ; Hartl. P. Z. S. 1867, p. 831 (Pelew Is.) ; Dohrn, J. f. O. 1871, p. 8 (Cape Verde Is.) ; Finsch, J. f. O. 1872, pp. 33, 58 (Samoa) ; Walden, Trans. Zool. Soc. viii. p. 106 (1872) (Celebes) ; id. op. cit. ix. pp. 216, 252 (Philippines) ; Finsch, Mus. Godeffr. viii. p. 47 (Pelew Is.) ; id. P. Z. S. 1877, p. 777 (Eua) ; Layard, P. Z. S. 1875, p. 441 (Fiji) ; id. P. Z. S. 1876, p. 498 (Friendly Is.) ; id. Ibis, 1876, p. 393 ; David & Oust. Ois. Chine, p. 350 (1877) ; Finsch, P. Z. S. 1879, p. 16 (Duke of York Is.) ; A. B. Meyer, Ibis, 1879, p. 115 (Celebes, Minahassa) ; Nehrk. J. f. O. 1879, p. 410 (egg) ; Finsch, P. Z. S. 1880, p. 577 (Carolines) ; id. Ibis, 1880, pp. 431, 434 (Gilbert Is.).

Morus sula, Vieill. Nouv. Dict. xii. p. 38 (1817) (New Guinea).

Morus parvus, op. cit. p. 41 (1817).

Sula brasiliensis, Spix, Av. Bras. ii. p. 83, pl. 107 (young) (1825) ; Burm. Syst. Ueb. Th. Bras. iii. p. 458 (1856).

Sula fusca, Vieill. (nec _Sula fusca_ (Fou brun) of Briss. Orn. vi. p. 499, pl. xliii. (1760), which is the young of _Sula piscator_), Gal. Ois. p. 194, pl. 277 (1825) ; Aud. Orn. Biogr. iii. p. 63 (1835) ; Sundev. Phys. Sällsk. Tidskrift, 1837, p. 220 ; Gosse, B. Jamaica, p. 417 (1847) ; Gould, B. Austral. vii. pl. 78 (1848) ; Temm. & Schleg. Fauna Japon., Av. p. 131 (1850) ; Lembeye, Aves Cuba, p. 135 (1850) ; Hartl. J. f. O. 1854, pp. 168, 170 (Pacific Is.) ; Scl. P. Z. S. 1856, p. 144 (Ascension) ; id. op. cit. 1857, p. 237 (St. Domingo) ; Gray, P. Z. S. 1859, p. 167 (New Caledonia) ; Swinh. P. Z. S. 1863, p. 325 (China) ; id. Ibis, 1868, p. 317, and 1870, p. 367 (Hainan) ; id. P. Z. S. 1870, p. 603 (Formosa) ; Garrod, P. Z. S. 1876, pp. 335, 341, 1877, p. 451 (anatomical remarks) ; Ramsay, Proc. Linn. Soc. N. S. Wales, iv. p. 84 (1879) (Savo, Solomon Is.) ; Finsch, Ibis, 1881, p. 549 (New Britain) ; MacFarlane, Ibis, 1887, p. 213 (Christmas I.) ; Oust. Miss. scient. Cap Horn, p. B 301 (1891).

Sula fuliea, Less. Tr. d'Orn. p. 601 (1831).

Dysporus parvus, Sund. Physiogr. Sällsk. Tidskr. 1837, p. 220; id. Ann. & Mag. Nat. Hist. xix. p. 237 (1847); Bp. Consp. ii. p. 164 (1855).

Pelecanus piscator, Forst. Descr. An. p. 278 (ed. Licht. 1844).

Sula fiber Blyth (nec Linn.), Cat. B. Mus. As. Soc. Beng. p. 296 (1849); Hartl. J. f. O. 1854, p. 137 (Bonin Is.); Gray, P. Z. S. 1858, p. 188 (Key Is.); Scl. & Salvin, Ibis, 1859, p. 233 (San Salvador); Gray, P. Z. S. 1860, p. 366 (Gilolo); Taylor, Ibis, 1864, pp. 96, 173 (Venezuela); Hartl. Ibis, 1861, p. 232 (Fiji); Salvin, tom. cit. pp. 381, 385; Rosenb. Reis Zuidoostereil. pp. 54, 82, 83 (Aru and Key Is.); Salvad. Ucc. Borneo, p. 369 (1874); Beavan, Ibis, 1868, p. 455 (Bay of Bengal); Rosenb. Reist. naar Geelvinkb. p. 9 (1875) (Ternate); Salvad. Ann. Mus. Civ. Gen. x. p. 167 (New Guinea); id. op. cit. xiv. p. 669 (1879) (Ternate); Bocage, Orn. Angola, p. 521 (1881).

Sula parva, G. R. Gray, Gen. B., App. p. 30 (1849); Albrecht, J. f. O. 1862, p. 207; Jerd. B. India, iii. p. 852 (1864); Jouan, Ibis, 1865, p. 338 (New Caledonia).

Sula sinicadvena (?), Swinh. Ibis, 1865, p. 109.

Diomedea, sp. ?, Gräffe, J. f. O. 1870, p. 411 (Tongatabu).

Dysporus leucogaster, Sundev. P. Z. S. 1871, p. 125 (Galapagos).

Sula leucogastra, Gray, Hand-l. iii. p. 126 (1871); Salvin, Trans. Zool. Soc. ix. p. 496 (1875); Blak. & Pryer, Ibis, 1878, p. 216 (Japan); Scl. P. Z. S. 1879, p. 310 (egg); Penrose, Ibis, 1879, pp. 275, 281 (Ascension); Legge, B. Ceylon, p. 1177 (1880); Salvad. Orn. Papuasia, iii. p. 422 (1882); Coues, Key N.-A. B. p. 720 (1884); Sceb. Ibis, 1890, p. 107 (Bonin Is.); id. B. Japan, p. 212 (1890); id. Ibis, 1891, p. 192 (Volcano Is.); Wiglesw. Abh. u. Ber. Mus. Dresd. 1890-91, p. 72 (1892, in "Aves Polynesiæ").

Dysporus fiber, Gundl. J. f. O. 1874, p. 134 (Pto. Rico); id. op. cit. 1875, p. 402 (Cuba); id. op. cit. 1878, pp. 163, 191 (Pto. Rico).

Sula sula, Ridgw. Man. N.-Am. B. p. 75 (1887); Cory, Auk, 1888, p. 72 (W. Indies); Feilden, Ibis, 1889, p. 504 (Barbados); Hartert, Kat. Vogels. Mus. Senckenb. p. 235 (1891) (Red Sea); Cory, Cat. W. Ind. B. p. 84 (1892); id. Auk, 1892, p. 229 (Mona I., near Pto. Rico).

Adult. Head, entire neck, and the whole upper surface and breast dark sooty brown, lower parts from the breast backwards pure white. Iris greyish brown; bill horn-grey with a bluish tinge in the male, dull grey with a yellow tinge in the female; naked base and gular sac bluish in male, yellow with a bluish spot before the eye in female; feet and legs pale yellow. The young have the bill greyish brown, the iris brown; feet and legs flesh-coloured, assuming a yellow tinge the older they become. (*From Palmer's notes.*) (Probably the differences in colour in the soft parts, noted by Palmer, are not peculiar to the sexes but due to age.)

Total length about 30 inches, wing 15·6 to 16·5, tail 8·5 to 9, culmen 3·75 to 4·2, tarsus 1·6 to 1·8, middle toe 3·3 to 3·5.

Young birds are similar to the adult above, but have the lower parts also brown, the feathers of the abdomen, however, having white bases and whitish edges, sometimes broader, sometimes narrower.

The *nestling* is covered with pure white down.

I have received some eggs from Midway Island. They are similar to those of the other *Sulæ*, and measure 2·76 by 1·75, 2·56 by 1·68, 2·35 by 1·68, and 2·6 by 1·76 inches.

THE Brown Gannet, or Booby, was noticed when passing Niihau, or Bird, Island and was very plentiful on Lisiansky and Midway Islands, but it was altogether the rarest of the three species in these waters and was absent from Laysan.

This species builds a nest of twigs on scrub and lays two eggs, but on Midway Island, where only grass is found, it builds on the ground. They are very shy and when sitting let their necks hang down over the edge of the nest as if they were dead.

15. PHAËTON RUBRICAUDA, *Bodd.*

RED-TAILED TROPIC-BIRD.

Le Paille-en-Queue à brins rouges, Buff. Hist. Nat. Ois. viii. p. 357 (1781) (Isle de France); Vieill. Tr. d'Orn., Atl. pl. 114 (1831).

Le Paille-en-Queue de l'isle de France, Daubent. Pl. Enl. 979 (1783).

Phaëton rubricauda, Bodd. Tabl. Pl. Enl. p. 57 (1783) (ex Buff. & Daubent. Pl. Enl.); Gray, Gen. B. iii. p. 663; Peale, U.S. Expl. Exp. (1848); Gould, B. Austral. vii. pl. 73 (1848); Hartl. Wiegm. Arch. f. Naturg. 1852, pp. 127, 137; Cass. U.S. Expl. Exp. p. 395 (1858); E. Newton, Ibis, 1861, p. 180 (breeding, Round I. north-east of Mauritius); G. R. Gray, Ibis, 1862, p. 250 (List B. New Zealand); Jerd. B. India, iii. p. 819 (1864); Finsch & Hartl. Faun. Centralpolynes. p. 248 (1867); Schleg. & Poll. Faune Madagascar, p. 140 (1868) (observed once, Cardagos I.); Sperling, Ibis, 1868, p. 295 (Mozambique Channel); Beavan, Ibis, 1868, p. 405 (obs. on way to the Andamans); Dole, Proc. Bost. Soc. Nat. Hist. 1869, p. 308 (Niihau and Kauai); Holdsw. P. Z. S. 1872, p. 482 (Ceylon); Hengl. Orn. N.O.-Afr. ii. p. 1172 (1873); Hume Stray Feath. 1876, p. 481 (in the text is stated that it has occurred in the Bay of Bengal, but that the only species Mr. Hume found in the Laccadives was *P. indicus*, Hume, an entirely different species); Salvad. Ibis, 1876, p. 266 (New Hebrides); Layard, Ibis, 1876, p. 393 (Fiji Is.); Marie, Ibis, 1877, p. 363 (Fly River, New Caledonia); Streets, American Naturalist, 1877, p. 71 (Fannings Group); Layard, Ibis, 1878, p. 265 (New Caledonia, breeding); id. Ibis, 1880, p. 233 (Loyalty Is.); id. Ibis, 1882, pp. 542, 544 (New Caledonia); Sharpe, ed. Layard's B. S. Africa, p. 775 (1875-1884) ("It is with some hesitation that we include this species among the birds of S. Africa. A single red tail-feather was picked up on the beach at Port Elizabeth by our corr. Mr. Rickard, who believes that it was freshly cast there"); Crowfoot, Ibis, 1885, p. 268 (breeding, Norfolk Islands, eggs); MacFarlane, Ibis, 1887, pp. 210, 213 (W. Pacific, breeding); Finsch, Mitth. orn. Ver. Wien, 1884, p. 125; Buller, B. N. Zealand, 2nd ed. ii. p. 186 (1888); Seeb. Ibis, 1890, p. 107 (Bonin Is.); id. B. Japan, p. 213 (1890); id. Ibis, 1891, p. 192 (San Alessandro, Volcano Is.); Lister, P. Z. S. 1891, pp. 293, 298 (Phœnix Is.); Sibree, Ibis, 1892, p. 271 (Madagascar); Wiglesw. Abh. u. Ber. Mus. Dresd. 1890-91, p. 73 ("Av. Polynesiæ") (1892).

Phaëton phœnicurus, Gm. Syst. Nat. i. p. 583 (1788); Shaw, Nat. Misc. v. pl. 177 (1794); Sonnini, Sonnerat's Voy. Indes, 2nd ed. v. p. 380 (1806); Rüpp. Vöy. N.O.-Afr. p. 110 (1845) (Southern Red Sea); Brandt, Monogr. Phaëtons in Mém. Ac. Pétersb. Sc. Nat. iii. p. 253, pls. i., v. (anatomy); Gould, B. Austral. vii. pl. 73 (1848); Jard. Contr. Orn. v. p. 35, pl. 84 (egg) (1853); Hartl. J. f. O. 1854, p. 170 (Pacific Is.); Tschudi, J. f. O. 1856, p. 149 (Pacific Is.); E. Newton, Ibis, 1860, p. 201 (Mauritius); Pelz. Ibis, 1860, p. 422 (Norfolk Island); Hartl. J. f. O. 1860, p. 175 (Mauritius); Krefft, Ibis, 1862, p. 192 (Australia, breeding on shoals); Walker, P. Z. S. 1863, p. 379; Layard, Ibis, 1863, p. 247 (Rodriguez, north of Mauritius); Jouan, Ibis, 1865, p. 338 (New Caledonia); Gould, Handb. B. Austr. ii. p. 501 (1865) (not personally observed, but gives excellent account by Macgillivray); Bennett, P. Z. S. 1869, p. 472 (Lord Howe's I.); Hartl. Vög. Madagascar, p. 303 (1877); Lister, P. Z. S. 1888, p. 520 (Christmas I.).

Red-tailed Tropic Bird, Lath. Gen. Syn. iii. p. 618; id. Gen. Hist. B. x. p. 447, pl. 183 (1824).

F

Phaëton aethereus, Bloxham, Voy. Blonde, App. p. 251 (nec Linn.!) (1826). (As Bloxham tells us that the
　　red tail-feathers are much valued by the Sandwich Islanders there can be no doubt that he chiefly or
　　only referred to *Ph. rubricauda*, Bodd.)
Phaenicurus rubricauda, Bp. Consp. ii. p. 183 (1856).
Phaëton rubricaudata, Finsch & Hartl. Vög. Ost-Afr. p. 389 (1870).
Phaëton rubricaudatus, Finsch & Hartl. J. f. O. 1870, p. 122 (Tonga Is.); Finsch, J. f. O. 1872, pp. 33, 57
　　(Samoa Is.).
Phaëton rubricaudus, Cab. & Reichen. J. f. O. 1876, p. 329 (S. Gazelle); Finsch, J. f. O. 1880, p. 296
　　(Ponapé); id. Ibis, 1881, p. 115 (Ponapé); Ridgw. Man. N.-Amer. B. p. 74 (1887).

Adult. Whole plumage silky white, with an indication of a rosy tinge above in some specimens
and a distinct rosy tinge on the quills and axillaries. A large crescent-shaped black spot
in front of the eye and a smaller one behind the eye. The feathers of the back and
breast have a delicate scaly appearance, the borders being of a somewhat different shade
of white and not so silky. Most of the feathers are white throughout, but those of the
head and hind neck have blackish bases, becoming more longitudinal and arrow-
shaped towards the back and disappearing in the interscapular region. Quills, rectrices,
and axillaries have the shafts black from above, white at the tips and from beneath.
Innermost secondaries with very large black spots in the middle of the feathers. The
very much elongated feathers on the flanks white, with broad longitudinal blackish spots
along their central portion. The two elongated central tail-feathers bright crimson with
black shafts. Bill scarlet, with a black stripe running through the nostrils; tarsus and
base of feet bluish flesh-colour, greater part of toes and webs black. Iris dark brown.

The *nestling* is covered with uniform grey down. Its bill is of a brownish flesh-colour; legs
flesh-colour with purplish tinge; webs bluish brown; iris brown.

Total length of the adult birds about 30 inches with the long tail-feathers, about 18 without,
wing 12·4 to 12·7, culmen 2·35 to 2·55 (2·45 on an average).

Eggs ovate, ground-colour a dirty white, speckled, sprinkled, spotted, and blotched, and
sometimes almost entirely covered with a dirty brownish claret-colour and some darker
and browner spots; by transparent light yellow. They measure 2·45 by 1·74, 2·4 by 1·77,
2·52 by 1·77, 2·6 by 1·74, 2·6 by 1·77, 2·65 by 1·88 inches.

PALMER found this species breeding all over the island of Laysan. He records:—"These
birds are now (June) breeding. They make a hollow in the ground under the bushes for
their nest, and lay only one egg, and are just losing their long red tail-feathers. If disturbed
they will not leave their nest, but begin a most hideous harsh screaming, and will continue
to scream till they are pulled off their egg or left alone."

16. GYGIS ALBA (Sparrm.).

WHITE TERN.

Sterna alba, Sparrm. Museum Carlsonianum, ii. fasc. i. no. 11 (1786) (Diagn. Sterna toto alba, rostro pedi-
busque nigris) (Hab. India, ad caput bonæ spei, et in ins. maris australis); Gm. Syst Nat. i. p. 607 (ex
Sparrm.) ; Lath. Ind. Orn. ii. p. 808; Less. Tr. d'Orn. p. 623 (1831); Forst. Descr. An. p. 179 (ed. Licht.
1844); Peale, U.S. Expl. Exp. p. 389 (1848); Hartl. Wiegm. Arch. f. Naturg. 1852, p. 126; Schleg.
Mus. P.-B., Sternæ, p. 35 (1863) ; Schleg. & Poll. Rech. Faune Madagascar, ii. p. 150 (1868).
Gygis alba, Cass. U.S. Expl. Exp. p. 389 (1858); Blasius, J. f. O. 1866, p. 73; Hartl. P. Z. S. 1867, p. 382;
Finsch & Hartl. Orn. Centralpolynes. p. 232 (1867); iid. P. Z. S. 1868, pp. 9 & 118 (Pelew Islands);
Dole, Proc. Bost. Soc. Nat. Hist. 1869, p. 306 (Sandwich Is.); Finsch & Hartl. J. f. O. 1870, pp. 122 &
140 (Tonga Islands); iid. P. Z. S. 1871, p. 32 (Rarotonga); iid. op. cit. 1872, p. 114 (Pelew Is. and Uap);
Finsch, J. f. O. 1872, p. 56 (Upolu, Samoa); id. Journ. Mus. Godeffr. 1875 (viii.), p. 43 (Pelew Is.); id.
op. cit. 1876 (xii.), p. 40; id. P. Z. S. 1877, pp. 776 & 786; Layard, P. Z. S. 1875, p. 30 (Viti Levu); id.
P. Z. S. 1876, pp. 497 & 504; id. Ibis, 1876, p. 393 (distribution of Fijian B.); Hartl. Vög. Madagascar,
p. 389 (1877); Streets, American Naturalist, 1877, p. 70 (Palmyra, Fannings Group); id. Bull. U.S. Nat.
Mus. 1877, p. 28; Hume, Stray Feath. vii. p. 417 (1878) (Bay of Bengal, Leyden Mus., from Dussumier);
Finsch, J. f. O. 1880, pp. 295, 309 (Ponapé, Kushai); id. Ibis, 1881, pp. 105, 109, 115, 216 (Kushai,
Ponapé, and Nawodo); id. Mitth. orn. Ver. Wien, 1884, p. 125; Wiglesw. Abh. u. Ber. Mus. Dresd. (Av.
Polynes.) 1890-91, p. 78 (1892).
White Tern, Lath. Gen. Syn. iii. p. 363, no. 17.
Sterna candida, Gm. Syst. Nat. i. p. 607 (ex Latham) (Hab. ins. nativit. Christi, aliisque maris australis; visa
quoque in ins. S. Helenæ; an vere distincta ab alba spec.?).
Gygis candida, Wagl. Isis, 1832, p. 1232 (genus Gygis established) ; Gray, List Laridæ, p. 180 (1844) ; id.
Gen. B. iii. p. 660; Gould, B. Austr. vii. pl. 30 (1848) ; Hartl. J. f. O. 1854, pp. 169, 170; Kittl. Reise, i.
p. 382 (1858), ii. pp. 39, 60 (Ualan) 452 (St. Helena; Kittl. believes those from St. Helena are different
from those from the Carolines); Gray, B. Tropical Is. p. 59; Pelz. Reise Novara, Vög. p. 155 (1865) ;
Saund. P. Z. S. 1876, p. 667, 1877, p. 797; Milne-Edw. & Grand. Madagascar, p. 660 (1879) ; Finsch, Ibis,
1880, pp. 220 (Taluit), 330, 431, 434 (Marshall and Gilbert Is.) ; Tristr. Ibis, 1881, p. 251 (Marquesas) ;
Saund. Voy. 'Challeng.' ii. p. 136 (1881) (Ascension and Tahiti) ; Salvad. Orn. Papuasia, iii. p. 152 (1882)
(doubtful New Guinea) ; Tristr. Ibis, 1883, p. 48; MacFarlane, ibid. 1887, pp. 211 & 213; Buller, B. New
Zealand, 2nd ed. ii. p. 338 (1888) (Kermadec Is.) ; Tristr. Cat. Coll. B. p. 10 (1889) ; Lister, P. Z. S.
1891, p. 297.
Sterna nirea, F. Bennett, Whaling Voy. i. p. 370 (1840) (Caroline Islands south of Equator, west of
Marquesas).
Gygis alba, caudida (et ? napoleonis), Bp. Compt. Rend. 1856, p. 773.
"Eki-aki," schneeweisse Seeschwalbe, Gräffe, J. f. O. 1870, p. 410 (Tongatabu, interesting account of habits),
p. 403 (is stated that Eki-aki is Gygis alba (Sparrm.)).
Gygis alba kittlitzi, Hartert, Kat. Vogels. Mus. Senckenb. p. 237. (The form from the Carolines (Ulea) has
been distinguished subspecifically from the Australian one.)

White, with a narrow black circle round the eyes. Shafts of remiges and rectrices dark brown, white at the tips. Bill, tarsi, toes, and claws black, webbing between the toes pale flesh-colour. Total length about 13 inches, wing 9·5, tail 5, culmen 0·4 to 0·5, height of bill at base 0·4, tarsus 0·5.

The first plumage of the young is white, like that of the adult bird, only on the scapulars some pale brown margins are visible. The nestling is covered with brownish down.

Mr. HARTERT (*l. c.*) has separated the *Gygis* from the Caroline Islands (collected by Kittlitz) subspecifically under the name of *Gygis alba kittlitzi*, chiefly on account of its smaller bill. My specimens from Laysan and Lisiansky are of exactly the same measurements as Mr. Hartert's type in the Frankfort Museum, and I must admit that they are very much smaller than a series of specimens from the Kermadec Islands in my collection. Mr. Saunders, however, will not admit the smaller subspecies, and as he has a greater material from very many localities in the British Museum and in his own collection than I have at present, I refrain from positively giving an opinion about it at present, though I feel sure, as all the small specimens known have been obtained north of the Equator, and the large ones south of it, that Mr. Hartert's opinion will prove correct. (I have given two photographs of this bird illustrating its positions at rest.)

Palmer found the White Tern in great abundance on Laysan and Lisiansky Islands. They generally sat about on the ground in pairs, and deposited their eggs in a very careless manner anywhere on the rocks or among the scrub; but a peculiarity of this bird is that it often lays its single egg in the forked branches of bushes. Its note is a low deep croak.

The eggs are longish oval, mostly equal at both ends, dull white or pale buff, light green by transparent light. They are spotted and blotched with dark brown, often nearly blackish brown, and with underlying spots of a pale purplish grey; some specimens are marked with hair-like lines, and most of them have a more or less developed band around one end. They measure 1·68 by 1·25, 1·7 by 1·25, 1·7 by 1·22, 1·75 by 1·22 inches, thus varying very little in size and shape.

17. HALIPLANA LUNATA (Peale).

GREY WIDEAWAKE; GREY-BACKED TERN.

Sterna lunata, Peale, U.S. Expl. Exp., B. p. 277, pl. 74. fig. 1 (1848) (Vincennes I.) ; Hartl. Wiegm. Arch. f. Naturg. 1852, p. 125 ; id. J. f. O. 1854, p. 170 ; Cass. U.S. Expl. Exp., Orn. p. 382 (1858) ; Gray, Cat. B. Trop. Is. p. 59 (1859) ; Schleg. Mus. P.-B., *Sternæ,* p. 27 (1863) (Halmahera !) ; Finsch & Hartl. Orn. Centralpolynes. p. 231 (excellent descr.), pl. xiii. fig. 3 (poor figure) ; Hartl. P. Z. S. 1867, p. 831 (Pelew Is.) ; Hartl. & Finsch, P. Z. S. 1868, pp. 4, 9, 118 (Pelew Is.) ; Scl. P. Z. S. 1869, p. 124 (Salomon Is.) ; Hartl. & Finsch, P. Z. S. 1872, pp. 90, 113 (Pelew Is.) ; Finsch, Journ. Mus. Godeffr. viii. p. 41 (Vög. Palau Gruppe, 1875) ; Rosenb. Reist. naar Geelvinkb. p. 9 (1875) ; Saunders, P. Z. S. 1876, p. 665 (distrib.) ; Layard, Ibis, 1876, p. 393 (Fiji) ; Nehrk. J. f. O. 1879, p. 409 (egg ?) ; Tristr. Ibis, 1880, p. 144 (Salomon Is.) ; Stejn. Proc. U.S. Nat. Mus. xii. p. 379 (1889) (Kauai) ; Lister, P. Z. S. 1891, p. 296 (Phœnix Is.).

Haliplana lunata, Coues, Ibis, 1864, p. 392 ; Blas. J. f. O. 1866, p. 80 ; Gray, Hand-list, iii. p. 122 (1871) ; Coues, B. N.W. Amer. pp. 698, 703 (note) (1874).

Onychoprion lunatus, Salvad. Ann. Mus. Civ. Gen. xviii. p. 409 (1882) ; id. Orn. Papuasia, iii. p. 451 (1882) ; Wiglesw. Aves Polynes. in Abh. u. Ber. Mus. Dresd. 1890–91, p. 76 (1892).

The dark-coloured Terns have been kept generically distinct by many of the most critical ornithologists, but Mr. Saunders (in 1876) united them with *Sterna,* and until lately, with a few exceptions (Salvadori for example), most of the leading ornithologists have followed him. I am glad to learn that Mr. Saunders has altered his opinion, and is going to keep them distinct in his forthcoming Monograph on the Laridæ in the 'Catalogue of Birds,' which, I think, is very wise, especially in such a huge genus as *Sterna* will be. On the other hand, I regret to learn that Mr. Saunders will not accept for the genus the name *Onychoprion* of Wagler, which was published 945 pages before *Haliplana,* and which should stand, according to all modern rules of nomenclature, notwithstanding its meaning. I accept the name of *Haliplana* only because, for the sake of convenience, I like to follow in my writings that great systematic work of our time, the 'Catalogue of Birds,' as far as I can.

Adult. Forehead and a broad stripe over the eye, entire underparts, and under wing-coverts white. A broad loral stripe running from the base of the upper mandible through the eye, top of head and nape black. Above dark ashy, with a slight brownish tinge, distinctly paler on the hind neck. Outer webs of primaries blackish, inner webs white with a blackish stripe along the shaft, occupying the whole web near the tip, and with a narrow grey fringe on the inner web, except on the basal half; shafts of primaries brown above and white below. Outer webs of tail-feathers grey, inner webs white, greyish brown at the tip; outermost pair of rectrices entirely white, brownish at the tips. Iris dark brown, bill and feet black. Total length about 16½ inches, wing 10·6, lateral tail-. feather 6·8, central 3, culmen 1·6, tarsus 1·8.

The plumage of the *young*, which has never been described before, is spotted like that of other Terns. The head looks white spotted with blackish, all the feathers being blackish with very broad whitish borders. All the feathers of the upper parts are grey, with buffy white borders and deep brown subterminal bands. Tail-feathers with buffy white terminal spots, lower parts entirely white.

THE Grey-backed Tern was met with on all the islands, except Midway Island, and was breeding in great numbers on Laysan Island. They lay one single egg only. The laying of one egg is a peculiarity of all the members of the genus *Haliplana* (or rather *Onychoprion*), for also *H. anæstheta* (Scop.) and *H. fuliginosa* (Gm.) lay only one egg [1], while the members of the genus *Sterna* lay (so far as is known) two or three eggs. The eggs of *Haliplana lunata* (Peale) are deposited on the sand, like those of many other Terns. These birds are, compared with the other birds of these islands, very shy, and leave their egg or young before one can approach them. The eggs are in colour quite similar to the well-known eggs of *H. fuliginosa* (Gm.), but smaller as a rule: those from Laysan measure 1·8 by 1·3, 1·8 by 1·23, and 1·85 to 1·3 inches. The Plate gives exact representations of them.

[1] See Young, Ibis, 1891, p. 145; Hartert, Ibis, 1893, p. 310, &c.

18. HALIPLANA FULIGINOSA (Gm.).

WIDEAWAKE; SOOTY TERN.

Hirondelle de mer à grande envergure, Buffon, Hist. Nat. Ois. viii. p. 345 (1781) (Ascension I.).

Sooty Tern, Lath. Gen. Syn. vi. p. 352; id. Gen. Hist. x. p. 102 (Ascension I.).

Sterna fuliginosa, Gm. Syst. Nat. i. p. 605 (1788); Lath. Ind. Orn. ii. p. 804 (1790); Less. Tr. d'Orn. p. 622 (1831) (Iles Malouines); Licht. in Forst. Descr. An. p. 276 (1844) (note); Cass. U.S. Expl. Exp., Orn. p. 386 (1858); Finsch & Hartl. Orn. Centralpolynes. p. 225 (1867); id. Vög. Ost-Afr. p. 831 (1870); Saund. P. Z. S. 1876, p. 666; Dresser, B. Europe, viii. p. 307, pl. 587 (1877); Legge, B. Ceylon, p. 1036 (1880); Tristr. Ibis, 1881, p. 252 (Marquesas); Milne-Edw. & Grand. Hist. Nat. Madagascar, Ois. p. 660 note (1885); Ridgw. Man. N.-Amer. B. p. 45 (1887); Stejn. Proc. U.S. Nat. Mus. xii. p. 379 (1889) (Kauai); Cory, B. Bahamas, p. 214 (ed. ii., 1890); id. Cat. B. W. Ind. p. 83 (1892); Oates, ed. Hume's Nests & Eggs Ind. B. iii. p. 303 (1890); Young, Ibis, 1891, p. 145 (number of eggs).

Sterna infuscata, Licht. Verz. Doubl. p. 81 (1823). (The types in the Berlin Museum have been examined by H. Saunders. Rüppell's *Thalassipora infuscata* (Syst. Ueb. no. 519, p. 140, 1845) belongs to *Haliplana anœstheta*: *cf.* Hartert, Kat. Vogels. Senckenb. Museum, p. 238, note 462.)

Sterna oahuensis, Bloxh. Voy. Blonde, p. 251 (1826); Stejn. Proc. U.S. Nat. Mus. xii. p. 379 (1889) (critic.).

Sterna serrata, Reinh. Forst. apud Wagl. Natürl. Syst. d. Amphib. p. 89 (note) (1830).

Onychoprion serratus, Wagl. Isis, 1832, p. 277.

Planetis guttatus, Wagl. Isis, 1832, p. 1222.

Haliplana fuliginosa, Wagl. Isis, 1832, p. 1224; Bp. Compt. Rend. xli. p. 1112 (1855) (Marquesas); Coues, Ibis, 1861, p. 392; id. Proc. Ac. Philad. 1862, p. 556; Blas. J. f. O. 1866, p. 81; Gray, Hand-list, iii. p. 122 (1871); David & Oust. Ois. Chine, p. 528 (1877); Hartl. Vög. Madagascar, p. 386 (1877).

Hydrochelidon fuliginosum, Bp. Comp. List, p. 61 (1838); Gosse, B. Jamaica, p. 433 (1847).

Sterna guttata, Forst. Descr. An. ed. Licht. p. 211 (1844) (juv.) (I. Paschali).

Anous l'herminieri, Less. Descr. Ois. p. 255 (1847).

Onychoprion fuliginosus, Gould, B. Austral. vii. pl. 32 (1848); Macgill. Voy. ' Rattlesnake,' ii. p. 359 (1852); Scl. P. Z. S. 1856, p. 144 (Ascension); E. Newt. Ibis, 1865, p. 153 (Rodriguez); Scl. & Salvin, P. Z. S. 1871, p. 572 (distrib. S. America); Salvad. Ucc. Borneo, p. 373 (1874); id. Orn. Papuasia, iii. p. 418 (1882); Wiglesw. Av. Polynes. in Abh. u. Ber. Mus. Dresd. 1890-91, p. 75 (1892).

Sterna gouldii, Reichb. Syn. Av. Natat. pl. xxii. fig. 829 (1848).

Anous fuliginosus, Finsch, Neu-Guinea, p. 184 (1865).

Sterna luctuosa, Phil. & Landb. Wiegm. Arch. f. Naturg. 1866, p. 126 (Chili).

Adult. Forehead, sides of head, and entire lower parts, including lower wing-coverts, white, with a very delicate bluish tinge on the abdomen, under wing-coverts, and under tail-coverts when the birds are alive or quite fresh. Lores and upper parts, including the hind neck (which is whitish in *H. anœstheta* (Scop.)), uniform sooty brown. Primaries black, but the shaft and outer web of the first primary white below, except on the outermost tip. Tail-feathers sooty black; all the shafts white below, and the shafts of the outer pair, as well as their outer web and basal part of inner web, white. Total length about 17 to 17½ inches, wing 11·6 to 12, outer rectrices (if not abraded) 7·5 to 8, central pair 3, culmen 1·7 to 1·8, tarsus 0·85. (Specimens from America and the Kermadec Islands are exactly similar.) Iris dark brown; bill and feet black.

Young. Dark sooty brown above and below, somewhat paler below; under wing-coverts pale greyish; anal region greyish white; feathers of the back and rump narrowly, upper wing-coverts, scapulars, and rectrices broadly tipped with white.

WITH the exception of the breeding-season being a month later, the habits of the Sooty Tern are the same on Laysan as on French Frigate Shoals. They lay *one egg*, and are quite fearless, pecking at anyone, but never flying off the nest. They are extremely noisy, and seem never to sleep. The Americans here call them " Wideawake Terns."

The very well-known eggs are extremely variable and white, or creamy white, spotted with chestnut and brown, with some underlying spots of purplish grey. They measure 2 by 1·4 inches. One beautiful variety is cream-colour, covered with rich chestnut patches, and measures 2·2 by 1·31.

(The lithograph of this bird breeding was done from a photograph taken by Mr. Walker, the negative being too dense to print in collotype.)

19. ANOUS STOLIDUS (*Linn.*).

NODDY.

Sterna stolida, Linn. Syst. Nat. ed. x. i. p. 137 (1758); id. ed. xii. p. 227 (1766); Gm. Syst. Nat. i. p. 605
(1788); Wied, Beitr. iv. p. 874 (1832); Kittl. Kupfertaf. iii. p. 27 (Caroline Is.), pl. 36. fig. 2 (not fig. 1,
as erroneously given in the text); id. Reise, i. p. 364, ii. pp. 77, 86 (Carolines, Senjawin Is., Ualan); Burm.
Syst. Ueb. Th. Brasil. iii. p. 53 (1856); Schleg. Mus. P.-B., *Sternæ*, p. 36 (1863); id. P. Z. S. 1866,
p. 426; Schleg. & Poll. Rech. Faun. Madagascar, p. 119 (1868).
La Mouette brune (Gavia fusca), Briss. Orn. vi. p. 192, pl. xviii. fig. 2 (1760) (N. America) (adult).
Hirondelle de mer brune (Sterna fusca), Briss. Orn. vi. p. 220, pl. xxi. fig. 1 (1760) ("Dominicensi Insula")
(young).
Sterna fuscata, Linn. Syst. Nat. ed. xii. i. p. 228 (1766) (founded on Brisson's *Sterna fusca*); Gm. Syst. Nat.
i. p. 605 (1788) (young).
Sterna senex, Leach, in Tuckey's Exped. Congo, App. p. 408 (1818) (obtained by Cranch).
Anous niger, Steph. in Shaw's Gen. Zool. xiii. p. 140, pl. 17 (1825), and *Anous fuscatus*, ibid. (young of the
same species).
Anous spadicea, Steph. Shaw's Gen. Zool. xiii. p. 143 (1825) (young).
Megalopterus stolidus, Boie, Isis, 1826, p. 980; ? Gould, Voy. 'Beagle,' p. 145 (1841) (Galapagos).
Sterna unicolor, Nordm. in Erm. Verz. d. Thiere u. Pflanzen, p. 17 (1835).
Anous stolidus, Gray, Gen. B. p. 100 (1841); Gould, B. Austral. vii. pl. 33 (1848); Hartl. J. f. O. 1854,
p. 186 (Caroline Is.); Cass. U.S. Expl. Exp. p. 391 (1858); Scl. & Salv. Ibis, 1859, p. 233 (Centr.
America); Swinh. Ibis, 1863, p. 430 (Formosa); Scl. P. Z. S. 1861, p. 9 (Hualeinc); Newt. Ibis, 1865,
p. 153 (Rodriguez); Finsch & Hartl. Fauna Centralpolynes. p. 234 (1867); Dole, Proc. Bost. Nat. Hist. xii.
p. 307 (1869) (Sandwich Is.) (? partim); Finsch & Hartl. Vög. Ost-Afrika's, p. 835 (1870); Melliss, Ibis,
1870, p. 107 (St. Helena); Hartl. & Finsch, J. f. O. 1870, pp. 122, 402 (Tonga); iid. P. Z. S. 1872, p. 113
(Pelew); Heugl. Orn. N.O.-Afr. ii. p. 1159 (1873); Salvad. Ucc. Borneo, p. 379 (1874); Finsch, Mus.
Godeffr. viii. p. 42 (Vögel der Palau Gruppe) (1875); Saund. P. Z. S. 1876, p. 669; Tristr. Ibis, 1876,
p. 266 (New Hebrides); Hartl. Vög. Madagascars, p. 391 (1877); David & Oust. Ois. Chine, p. 529
(1877); Legge, B. Ceylon, p. 1043 (1880); Salvad. Orn. Papuasia, iii. p. 453 (1882); Tristr. Cat. Coll. B.
p. 10 (1889); Lister, P. Z. S. 1891, pp. 296, 300 (Phœnix Is.); Oates, ed. Hume's Nests and Eggs Ind.
B. iii. p. 315 (1890); Young, Ibis, 1891, p. 146 (number of eggs one only); Wiglesw. Av. Polynesiæ, in
Abh. u. Ber. Mus. Dresden, 1890-91, p. 76 (1892); Cory, Cat. B. W. Ind. p. 82 (1892).
Megalopterus tenuirostris, Rüpp. (nec Temm. !), Syst. Uebers. p. 140, no. 520 (1845).
Anous pileatus, Gray, Gen. B. iii. p. 561 (1846).
Anous tenuirostris, Heugl. (nec Temm.), Ibis, 1859, p. 351; Hartl. J. f. O. 1861, p. 273.

Adult. Top of the head light grey, getting darker and merging gradually dark into the sooty
brown of the hind neck. All the rest of the plumage dark sooty brown, like the hind

G

neck. Quills and tail-feathers brownish black. Bill black; feet blackish brown; claws blackish brown; iris dark brown. Sexes similar.

THE young bird has the top of the head the same colour as the back; and this is proved to be true by a specimen in my collection. According to Palmer's observations the young, when hatched, is of a light grey colour, which soon changes into a dark brown down.

Specimens from Laysan and French Frigate Island have the following measurements:— wing 10·75 to 11·15 inches, tail 6·4 to 6·6, culmen 1·65 to 1·75, tarsus 1.

The eggs are well known. They differ from the eggs of the true Terns in being less sprinkled and blotched, but rather sparsely covered with roundish spots of a dark brown, and underlying spots of a pale purplish grey; they are not much pointed. Specimens from French Frigate Island measure 2·13 by 1·34, 2·13 by 1·46, and 2·2 by 1·5 inches, and vary but little. Their ground-colour is white, with a faint buffy tinge. From the eggs of *Haliplana fuliginosa* (Gm.) they are also distinguishable by being darker and more green inside by transparent light.

The Noddy makes its nest on the ground, and mostly under the tussocks of grass and scrub; but Palmer has found a few on some of the low scrub on Laysan Island. It lays one egg, and is very fearless, boldly attacking the Albatrosses which disturb them.

It was met with plentifully on French Frigate and Laysan Islands.

(The Plate representing this species and *A. melanogenys* was only added to help readers to appreciate the distinguishing characters between these two species and *A. hawaiiensis*.)

20. ANOUS HAWAIIENSIS, *Rothsch.*

NOIO (Sandwich Islands).

Anous hawaiiensis, Rothsch. Bull. B. O. C. no. x. p. lvii (July 1893).

Adult. Forehead and entire top of head ashy white, this gradually merging into the sooty blackish back and rump, so that the hind neck and the upper part of the interscapular region are light ashy green. Back and rump sooty black; wings with their upper and under coverts uniform sooty black. Tail pale grey; the outer rectrices a shade darker, the middle ones palest. Beneath sooty black; the under surface of the neck very slightly washed with grey. Iris very dark brown; bill black; legs dark yellowish brown. Total length about 13½ inches, wings 8·4 to 8·75 (average 8·65), tail 4·8 to 5 (average 4·8), tarsus 0·7, culmen 1·5 to 1·7 (average 1·6).

The *immature* bird is more brownish and quite uniform above and below; only the forehead and crown whitish ash-colour.

The *nestling* is covered with whitish down.

THIS interesting new species is a northern representative of the *Anous melanogenys,* Gray, to which it is closely allied. It is, however, without difficulty distinguished by the whitish colour of the crown being spread over the nape, by the hind neck being light ashy grey, instead of sooty black as in *A. melanogenys,* and by the light grey tail, which is sooty black in *A. melanogenys;* the tail in this latter species is almost, or quite, of the same colour as the wing, while it distinctly contrasts with that of the wing in *A. hawaiiensis.* The wing seems to be a little shorter on an average, but, as it varies somewhat, this character does not hold good; also the bill is somewhat shorter and less pointed on an average, but varies a little like that of *A. melanogenys.*

The "Noio," as the Hawaiian Noddy is called on the Sandwich Islands, was first found rather common on the shores of Kauai, and breeding-places were discovered on Laysan, Lisiansky, and Midway Islands. On Laysan they were observed in some numbers on the north side of the island, sitting about in clusters. They lay *one* egg. Two eggs sent from Laysan are cream-buff, with some dark brown and pale purplish-grey underlying more or less roundish spots; they measure 1·78 by 1·2 inch, and are quite of the character of the eggs of other species of *Anous*[1].

[1] The photograph of the nesting-colony of this fine new Tern is one of the best of Mr. Williams's unique series.

21. PUFFINUS NATIVITATIS, *Streets.*

BLACK SHEARWATER.

Puffinus, n. sp., Streets, Amer. Nat. xi. p. 71 (1877) (Christmas I., south of Fanning Is.).
Puffinus nativitatis, Streets, Bull. U.S. Nat. Mus. 1877, no. 7, p. 29 (Christmas I.) ; Lister, P. Z. S. 1891,
   pp. 295, 300 (Phœnix and Canton Is., Krusenstern I., west of the Hawaiian Is.) ; Ridgw. Man. N.-Am.
   B. p. 62 (1887) (Christmas I.).

Sooty black above, deepest on head, wings, and tail; sooty brown below. Iris brown ; bill
   black; legs and feet brown. Total length about 14½ inches, wing 9·5 to 9·85, tail 3·5,
   culmen 1·25, tarsus 1·65 to 1·7, middle toe with claw 1·95 to 2.

THE Black Shearwater was met with on French Frigate and Laysan Islands. It lays its
egg on the surface of the ground under grass and scrub, and does not burrow like its congeners. The bare apology for a nest consists of a little grass, and they lay *only one* egg.
Palmer often saw the parents sitting together on the nest side by side. The eggs before
me resemble those of *Puffinus cuneatus*, Salvin, but are much smaller and more elongate,
measuring 2·24 by 1·45 inches and 2·37 by 1·35.

22. PUFFINUS CUNEATUS, *Salvin.*

WEDGE-TAILED SHEARWATER.

Puffinus cuneatus, Salvin, Ibis, 1888, p. 353 (Krusenstern I.) ; Stejn. Proc. U.S. Nat. Mus. xii. p. 377 (1890)
(Kauai) ; Seeb. Ibis, 1891, p. 191 (Sulphur I., Volcano Is.) ; Wilson, Aves Hawaiienses, pt. iv. pl. &
text (1893) (very good figure, but coloured too light above).
Puffinus knudseni, Stejn. Proc. U.S. Nat. Mus. 1888, p. 93 (Kauai).

Adult. Above deep sooty brown, darkest on the head, rump, and wing-coverts ; feathers of
the upper back with paler borders ; quills and tail-feathers sooty black. Beneath white,
sides of the head and body grey ; lower vent and belly shaded with brownish grey ;
under tail-coverts dark brown ; under wing-coverts white, brown near the bend of the
wing. Bill horn-grey with dark hook ; iris brown ; feet pale pink. Total length about
17 to 18 inches, wing 11·5 to 11·8 ; tail cuneate, central rectrices 5·5 to 5·8, lateral 3·7 ;
bill from gape 2·2 to 2·3, from the nostrils to tip 1·2, exposed culmen 1·6 ; tarsus 1·8 to 1·9,
middle toe with claw 2·3 to 2·4.

The above refers to the regular plumage of the adult bird. In one of my specimens the
entire head is white with a few brown feathers ; this is either an albinism or a very
aged specimen. In *young individuals* the underparts, including the under wing-coverts,
are brownish grey.

THIS very rare Petrel lives in pairs and lays only one egg in a rude nest made of grass in
a burrow in the sand ; it has a low moaning cry, and was observed on all islands except
Midway, but was universally very scanty in numbers. Breeding-season, May and June.

The eggs elongate-oval, the shell very thin and smooth but without gloss, white in colour.
They measure 2·53 by 1·62 inches, 2·53 by 1·7, 2·48 by 1·58, and 2·45 by 1·74.

23. ŒSTRELATA HYPOLEUCA, *Salvin*.

SALVIN'S WHITE-BREASTED PETREL.

Œstrelata hypoleuca, Salvin, Ibis, 1888, p. 359 (Krusenstern I., N. Pacific) ; Seeb. B. Japan, p. 269 (Bonin Is.); id. Ibis, 1890, p. 105 (Bonin Is.).

Adult. Feathers of the forehead up to the middle of the head deep slate-colour, broadly margined with white, and white at base ; feathers just above the bill, lores, and entire under surface white ; occiput and hind neck deep slate-colour ; feathers of the inter-scapular region, back, and upper part of rump slate-colour, with pale cinereous margins ; lower rump deep slaty ; upper wing-coverts slaty black ; quills and tail-feathers slaty black ; primaries with narrow greyish margins on the *inner* webs ; secondaries lighter, and with light grey margins on the *outer* webs ; under wing-coverts white in the central portion of the wing, blackish all along the bend of the wing. Iris dark brown ; bill black ; tarsus and basal portion of toes and webs dark flesh-colour, lower portion brownish black. Total length about 13 inches, wing 9, tail 4·6, bill from gape 1·4, tarsus 1·1, middle toe with claw 1·4. (Measurements taken from Salvin, as my specimens are more or less in moult.)

PALMER only met with this rare Petrel on Laysan, where he found four moulting specimens in the daytime in deep burrows. They were completely dazed when taken out of their hiding-places, and behaved as if they were quite blind. According to Mr. Freeth, their breeding-season was now over, but they came ashore in large numbers during their breeding-time. They are quite nocturnal like other species of this genus.

(The photograph showing these birds nesting is extremely interesting, as it shows their burrow close to the nests of the White Albatross, of which they seem to have no fear.)

H

24. BULWERIA BULWERI (*Jard. & Selby*).

BULWER'S PETREL.

Procellaria bulwerii, Jard. & Selby, Ill. Orn. pl. 65 (1830) (Madeira and adjacent islands).
Thalassidroma bulweri, Bp. Comp. List, p. 64 (1838) ; Gould, B. Europe, pl. 448; Gray, Gen. B. iii. p. 648 ;
  Newton, Man. N. H. Greenland, 1875, p. 108; Godman, Ibis, 1872, p. 223 (Desertas).
Œstrelata bulweri, Coues, Proc. Ac. Phil. 1866, p. 158 ; Ridgw. Proc. U.S. Nat. Mus. 1880, p. 200.
Bulweria bulweri, Bp. Cat. Met. Ucc. Eur. p. 81 (1842) ; id. Consp. ii. p. 194 (1856) ; Ridgw. Water-B.
  N. Amer. p. 398 (1884) ; id. Man. N.-Amer. B. p. 69 (1887) ; Coues, Check-l. N.-Amer. B. p. 103
  (1886) ; Stejn. Proc. U.S. Nat. Mus. xii. p. 378 (1890) (Kauai).
Puffinus columbinus, Moq.-Tandon in Webb & Berth. Orn. Canar. p. 44, pl. iv. (1841) ; Bolle, J. f. O. 1855,
  p. 178, & 1857, p. 345 (Canary Is. and Madeira).
Bulweria columbina, Brehm, Vogelfang, p. 354 (1855) ; Dresser, B. Eur. viii. p. 551, pl. 614 (1881) ; M.-Waldo,
  Ibis, 1889, p. 517 (Canaries) ; Koenig, J. f. O. 1890, p. 463 (Canaries).

Entire plumage sooty brownish black, the greater wing-coverts much paler and the underparts browner; chin and throat with a greyish tinge. Bill black; iris dark brown; feet
brown ; webs between the toes dusky. Total length about 11¼ inches, wing 7·8 to 8,
tarsus 1·05; tail cuneate and much graduated, central feathers 4·4 to 4·55, lateral pair 2·9;
bill from gape 1.

BULWER'S PETREL is very common on the Canary and Madeira Islands and in their
neighbourhood, and is stated to be an accidental visitor to the Bermudas and near the coast
of Greenland, and has occurred once in Great Britain.

The occurrence of this species beyond the Atlantic has been noted by Stejneger (*l. c.*),
who received specimens from Kauai.

Palmer found this bird very common on French Frigate Islands, whence he sent me
eleven skins and half a dozen eggs. They were breeding under dead turtle-shells that
had been heaped up by a shipwrecked crew. They were also met with on Laysan; but there
the breeding-season appeared to be over, for not many were seen ashore during the day,
while many came at night.

The eggs are white, and measure from 1·63 by 1·27 and 1·075 by 1·25 inch to 1·7 by 1·28
and 1·74 by 1·3.

25. OCEANODROMA CRYPTOLEUCURA (Ridgw.

HAWAIIAN STORM-PETREL.

? *Thalassidroma, sp. innominata*, Dole, Proc. Boston Soc. Nat. Hist. xii. p. 308 (1869) ; id. Hawaiian Almanac, 1879, p. 55 (Hawaiian Is.).
Cymochorea cryptoleucura, Ridgw. Proc. U.S. Nat. Mus. iv. p. 337 (1882) (Kauai) ; Ridgw. Water-B. N. Amer. ii. p. 406 (1884) (Sandwich Is.).
Oceanodroma cryptoleucura, Ridgw. Manual N.-Am. B. p. 71 (1887) ; Stejn. Proc. U.S. Nat. Mus. 1887, p. 78 ; Wilson, Av. Hawaii. pt. iv. pl. & text (Feb. 1893).

Fuliginous above and below; the greater wing-coverts and outer webs of tertials paler, inclining to ashy; quills and tail-feathers sooty black, the latter (except the middle pair) white at base; upper tail-coverts white, the longest broadly tipped with blackish; anal region mixed with white. Total length about 8 inches, wing 5·75 to 6·3; tail very slightly forked, the outer feathers 2·9 to 3·2 in length, the middle ones 0·2 to 0·3 shorter; tarsus 0·85 to 0·9; length of bill from base of culmen 0·6.

(Of my specimens the females have the largest measurements.)

As the above synonymy shows, this rare species has until now only been known from the Sandwich Islands.

This little Petrel was only observed on French Frigate Island. According to Mr. Gay's observations on Niihau, their habits are very similar to those of the European Storm-Petrel.

26. DIOMEDEA CHINENSIS, *Temm.*

BROWN GOONEY.

— — ⸺ ⸺

Albatros de la Chine, Daubent. Pl. Enl. 963 (1770).
Diomedea chinensis, Temm. Man. d'Orn. 2nd ed. i. p. 110 (1820) (based on Pl. Enl. 963).
Diomedea brachyura, Temm. Pl. Col. v. livr. 94, pl. 554 (1835) (based on Pl. Enl. 963, but figure and description of *D. albatrus*).
Diomedea derogata, Swinh. P. Z. S. 1873, p. 786 (China).

The *quite adult* bird is dark sooty brown above; the forehead much paler, dirty white in most of them, the hind part of the crown is darker, with a blackish line behind the eye. Below much paler and more greyish; the feathers more or less distinctly bordered with sandy buff; anal region, and in some specimens a few of the upper tail-coverts, dirty white, or at least much paler. Quills and tail-feathers blackish brown, the shafts of the quills straw-yellow. Bill dark brown, base and tip blackish; legs and feet black; iris dark brown. Total length about 33 inches, wing 19 to 19·6, tail 5·7, tarsus 3·3, bill from gape 4·4.

THERE is no doubt whatever that Temminck's name *D. chinensis* has the precedence over the same author's *D. brachiura* (corr. *brachyura*), bestowed also on the Pl. Enl. 963 in 1835, that is fifteen years later. In the Pl. Col., moreover, Temminck figures and describes another species, *D. albatrus*, Pall.

I am much obliged to Mr. Salvin, our great authority on Tubinares, for pointing out to me the differences between *D. albatrus* and *D. brachyura* (or rather *D. chinensis* as it should be called), for these two species have hitherto been considered to be the old and young of one species by most ornithologists, although Swinhoe clearly said that his *D. derogata* were fully adult birds. There is, indeed, no doubt that the dark form is not an immature bird.

On French Frigate Island, Laysan, and all the other islands visited by my collector, the dark Albatross was fairly numerous, but far less abundant than the white *D. immutabilis.* The two species always kept apart from each other. On Laysan the breeding-place of the dark Albatross was on the south side of the island, where they sat on the beach with their young. The young feed by putting their beaks crossways into the old birds' mow and catching the cast-up fish.

(The photograph shows the breeding-colony on the south of Laysan Island.
Unfortunately the Plate of this species was lettered before the synonymy was quite finished; and so on the Plate the name of *Diomedea brachyura*, instead of *D. chinensis,* appears, *D. chinensis* being undoubtedly the older and therefore correct name.)

27. DIOMEDEA IMMUTABILIS, *Rothsch.*

GOONEY.

Diomedea (an *exulans?*), Kittl. Mus. Senckenb. i. p. 120 (1834) (Gardner, Moller, and Lisiansky).
Diomedea immutabilis, Rothsch. Bull. B. O. C. no. ix. p. xlviii (June 1893).

Adult. Head, neck, lower rump, and entire under surface pure white; space in front of the eye sooty black; wings and wing-coverts blackish brown; interscapular region, back, and upper part of rump paler and more smoky brown; tail black, fading into white at base; under wing-coverts mixed blackish brown and white.

Sexes entirely similar. "Bill grey, darker at base, tip blackish brown, base of under mandible pale yellow; iris brown; tarsi and feet fleshy pink." Total length about 32 inches, wing 18·6 to 19, bill 4, tarsus 3·2, middle toe with claw 4·3.

The first plumage of the *young* (which is dark in most Albatrosses) is similar to that of the adult bird; the breast and entire underparts pure white.

The *nestling* is covered with brown down; its bill is blackish brown, and its iris brown also. Palmer did not send me eggs, as the time to get them was over.

On Lisiansky two young albino Albatrosses of this species were found. They are white throughout, wings and tail of a delicate pearl-grey (the tail in one more brownish); the back pale grey; feet pale flesh-colour. The down, which is still visible in some parts of the birds, is pure white.

THIS Albatross, of which a few only were seen on French Frigate Shoals, literally covers the island of Laysan, the young in some places being as thick as they could stand. It is also fairly numerous on Lisiansky. It is very curious to watch the love-making antics of these birds: first they stand face to face, then they begin nodding and bowing vigorously, then rub their bills together with a whistling cry; after this they begin shaking their heads and snapping their bills with marvellous rapidity, occasionally lifting one wing, straightening themselves out and blowing out their breasts; then they put their bill under the wing or toss it in the air with a groaning scream, and walk round each other, often for fifteen minutes at the time.

They are quite fearless, and do not move out of the way. When Mr. Freeth was going to the guano-fields on his tramway-line he had to send a boy ahead to clear the track of the

I

young Albatrosses. Mr. Freeth protected most vigorously the birds on his island; but the photograph representing the train laden with Albatross-eggs shows how they were treated after he had left Laysan.

(The four photographs and the lithographs illustrate much better the immense numbers of this bird, and its utter fearlessness and tameness, than any description possibly could.)

ERRATA.

1. On plate 2 of the coloured plates read:
   freethi instead of "*fraithii.*"
2. On the black plate of Gannets read:
   Sula cyanops instead of "*Sula australis.*"
3. On the first plate with nests and eggs read:
   Telespiza cantans, Wils., instead of "*Telespiza flavissima.*"
4. On the second plate with nest and eggs read:
   Acrocephalus instead of "*Acrulocephalus.*"
   Bulweria bulweri instead of "*Bulweria columbina.*"
   rubricauda instead of "*rubicauda.*"

THE Author's collectors have been actively engaged for some years past in exploring the Islands of the Hawaiian Archipelago, and many species of birds, new to science, have been discovered by them; these, with others, will be figured in a series of about 46 Hand-coloured Plates,—most of which will be delineated by that master-hand Mr. J. G. KEULEMANS, and others, including tinted Plates, are entrusted to Mr. F. W. FROHAWK's careful treatment.

In addition to the above, a most interesting series of Collotype Photographs, showing various phases of bird-life and landscape, will be included in the Volume.

The size of the book will be imperial 4to, and will be issued in 3 Parts, price **£3 3s.** each, net.

As no separate Parts will be sold, Subscribers are expected to continue their Subscriptions until the work is completed.

The Edition is limited to 250 Copies.

CONTENTS OF PART I.

List and Description of the Species of Birds from Laysan and the Neighbouring Islands.

Two Coloured Plates of Nests and Eggs.

PRINTED BY TAYLOR AND FRANCIS, RED LION COURT, FLEET STREET.

RÉSUMÉ OF PALMER'S DIARY.

HENRY PALMER arrived in Honolulu in December 1890, and on the 24th he left for the island of *Kauai* to begin his collections. Kauai is the most northern of the Sandwich Islands proper, and the channel separating it from Oahu is 64 miles broad. Through its isolated position it is of special interest among the larger islands of the Hawaiian Possessions. Kauai is 28 miles long and about 23 miles in greatest breadth. Like the rest of the archipelago, it is of volcanic formation. Its N.E. and N.W. sides are broken and rugged, but to the south it is more even. The highest mountain of the island, Waialeale, rises above 5000 feet. Kauai is one of the best cultivated islands of the group, and even when discovered by Cook the plantations of the natives were managed with industry and neatness.

The principal harbour is *Waimea*, on the south side. The south point is named Koloa. About 7 miles westward of Koloa and 6 miles to the south-eastward of Waimea is the valley of *Hanapepe*, celebrated for its beauty, with a waterfall at its head. The west point of the island is called Mana, and another harbour, Hanalei, lies on the north side. Palmer landed at Waimea, and soon began to collect what he could get in the neighbourhood of that place.

On December 30th Palmer went into camp some seven or eight miles above Waimea, on the edge of the forest. On January 1st, 1891, he began collecting there. Much help was afforded and great kindness shown to him by Mr. Gay and Mr. Robinson, who had already assisted Mr. Scott Wilson to so great an extent. It was Mr. Gay, too, who gave Palmer the chief, and certainly the most trustworthy, native names which are in use on Kauai.

Collecting in the whole neighbourhood was industriously carried on until February 7th. On this day Palmer departed for Honolulu, leaving his assistant, Mr. Muuro, in the tent on Kauai.

I had ordered Palmer to go to the unexplored islands in a north-westerly direction from the Sandwich Group, and he had to go to Honolulu to get information about this trip, and if possible to hire a vessel and arrange all the necessary details. Unfortunately, the time was very bad, and everybody advised him not to go during this dangerous and stormy period of the year, but to wait some months. As the trip would be a long one, and rough weather could be expected at any time, he was warned not to take too small a vessel, and as many hands as he could get. From all that Palmer heard he could not do better at present than return to Kauai, where he landed again on February 18th. Meanwhile his assistant had collected a number of birds, but nothing particular.

On February 20th Palmer saw a Goose, which he described as follows, but was not able to shoot :—

1

" Bill and forehead dark. Top and sides of head light. Neck dark. Breast scarcely as light as head. Wings dark, but scarcely as dark as neck. Belly as light [as the top of the head. The bill appeared to be rather short. Of course, this was all as seen from a distance beyond the range of gun-shot. The same day I saw also two Ducks which appeared to be smaller and darker than the common native duck."

On the 22nd Palmer went far into the hills with a party for goat-hunting, which gave him a splendid opportunity to become acquainted with another part of the island.

Palmer stayed in his camp till March the 10th. He was very much troubled with rainy weather, and it was very cold in the hills, often only 43° Fahr. and less. Palmer observed that such birds which are really inhabitants of higher elevations come down to the lower parts when rainy, cool, and foggy weather continues for some time.

At the beginning of March a number of very young birds were observed, while *Testiaria coccinea* and *Himatione sanguinea* were just pairing off.

On the 10th of March Palmer stayed at *Waimea*, where Mr. Gay showed him his local collection of birds and some eggs of sea-birds, besides giving him again some hints and information about the island and its birds.

On March 11th Palmer went to *Kekaha* and *Mana*, where he remained eight days, collecting shore- and water-birds in the vicinity.

On March 20th Palmer arrived at a place named *Halemann*, the property of Mr. Knudsen, whose name is so well known to ornithologists through Dr. Stejneger's publications on the birds of Kauai. Palmer speaks of the place as follows :—

" It is quite different up here to any other part of the island I have seen yet. The bush is not very thick, and close by it consists almost entirely of the Kon (*Acacia koa*). Soon after our arrival I tried to shoot wild cattle, having brought no meat up with us, but, although we saw several, were not fortunate enough to kill any.

"My guide seems to know the country well, which will save me much time.

"The next day was so wet and foggy that my guide declared he could not go up to the high mountains, so we tried to collect in the lower elevations, and were fortunate enough to shoot a bird which I believe is a species of the Hawaiian ' Thrushes ' or ' Kamaos '; but much smaller than the common Thrush of Kauai [*Phæornis myiadestina*, Stejn.—W. R.], and also than the species from Hawaii (*Phæornis obscura*, Gm.), which I saw at Messrs. Gay and Robinson's. [Unfortunately this specimen of a very rare bird, which I have described under the name of *Phæornis palmeri*, was partly destroyed by rats.—W. R.]

" Soon after this capture we shot a wild bull, which provided us with meat for some time.

" On March 23rd we made a trip to the hills above Kalalau. It rained a little when we started, but we thought it would clear up. The higher we came, contrary to our hopes, the thicker fell the rain, and therefore we had no success. The bush above Kalalau is not so thick as below, but the dense fern-vegetation makes up for it, and it is extremely difficult to find birds when shot down.

" On the following day, in the morning, I missed the new Thrush, and you may imagine my feelings! I at once suspected rats, so I crawled all under the cottage on my stomach, and finally succeeded in finding it in a rat's hole, although much damaged, yet partly preserved. They had indeed not touched one of the common birds! I am glad nobody heard the prayers uttered for the benefit of the rats."

During the last week of March Palmer remarks that several birds must be breeding or shortly about to do so, for he saw *Chlorodrepanis stejnegeri* (Wils.) pairing, and noticed the ovaries enlarged in *Loxops cæruleirostris* (Wils.) and in *Psittirostra*.

Palmer deplores his ill-luck in losing a horse and the damage done by rats to the legs of the goose. After his experience with the new *Phæornis*, this latter ought to have been avoided!

At the beginning of April Palmer collected at Makaweli and made several smaller excursions, and on April 7th he went to *Hanalei*. This place is situated right at the foot of the most rugged mountains, but the season was an unfortunate one, it being rainy and foggy. The forest here Palmer believes to be maiden forest, the undergrowth is very thick, and the trees high and lofty. No pass runs up to any distance, so that one had to be cut.

On April 20th Palmer went to *Hanakapie*, but for some time was much embarrassed by a sore heel. After getting better, a fair series of the Hawaiian Noddy were collected. This is allied to *Anous melanogenys* but quite distinct, and I have named it *Anous hawaiiensis* (see Part I. p. 43). In vain my collector searched for the small *Phæornis*, which must be very rare.

At the end of April Palmer left Kauai and went to Honolulu, from there he made the trip to Laysan, the Diary of which is given in the first Part of this work.

After his return from the Laysan trip Palmer fell ill with influenza, and was laid up for nearly a fortnight. On September 4th he sailed from Honolulu, and reached *Kealakeakua*, on the island of Hawaii, at 3 P.M. on the following day.

HAWAII

is by far the largest island of the Hawaiian Possessions. It is called Owhyhee by Cook, O Wahi by Kotzebue, Owhywi by Freycinet and others. All these words are representations of the same sound. The west side of Hawaii is nearly 100 miles long, the N.E. side about 76 miles, and the S.E. side 60.

The mountains of Hawaii rise gradually and comparatively unbroken, particularly from the southern shore, to the lofty summit of *Mauna Loa*, which is 13,675 feet high. Its appearance is the grandest and most majestic in the Pacific Ocean, although perhaps less romantic and picturesque than that of Tahiti. Great parts of the interior of Hawaii are still uninhabited wilderness. The big mountain in the north of the island, the *Mauna Kea*, is, according to the most recent measurements, even higher than the Mauna Loa, being 13,805 feet high, while the Mount Hualalai, on the west side, is only 8275 feet high.

The *Mauna Kea* (White Mountain) is covered with vegetation up to about 1000 feet from the summit, where frosts prevail.

Mauna Loa has a most extensive crater, about which much has been written of late, it having been visited by many travellers. The eruptions of Mauna Loa are among the most destructive on record. Terrible streams of lava have many times devastated the most fertile parts of the island, and their effects on animal life must have been dreadful. The vegetation of Mauna Loa is said to differ greatly from that of Mauna Kea.

The principal places are Hilo (which is next to Honolulu in size and importance), Kailua, Waimea, Kohala, and a few others.

1*

In the lower parts of the district of *Kaawaloa* collections were made until September 21st, when the tent was pitched at about 5000 feet, not very far from the mountain-house in which Mr. Scott Wilson had stayed for some time.

On the following day already Mr. Palmer shot half-a-dozen specimens of *Loxioides bailleui*, Oust., several of *Heterorhynchus wilsoni*, Rothsch., a *Hemignathus obscurus* (Gm.), and a number of *Chlorodrepanis virens* (Gm.), although there was much rain and fog, which is very often the case on these mountains.

On September 28th the first specimen of the big "Finch" which I described under the name of *Rhodacanthis palmeri* was shot.

On October 5th Palmer writes :—" I have to-day been on the slopes of Mauna Loa and succeeded in getting no less than five specimens of the big Finch (*Rhodacanthis palmeri*), besides several other birds. I searched for a place where, according to Mr. Greenwell, there was an old grass-house, which I found all right, but no water close by, so rather useless for camp.

" Generally the weather is quite favourable in the morning, but every afternoon just after dinner fog, and very often also rain, sets in, so making shooting in the afternoon very bad and difficult."

On October 14th Palmer went up *Mount Hualalei*, but did not shoot anything of importance. He was favoured during most of the time he spent on this mountain with very fine weather and went round the country in many directions. He must have gone up to considerable altitudes, as he complained twice of the very thin air, which made walking very trying.

He broke up the camp on October 26th, and went down to a place about three or four miles above Kaawaloa, where he occupied an old house, kindly lent him by Mrs. Greenwell.

On the following days a number of *Moho nobilis* (Merrem) were collected, which were now in much better plumage than before.

On November 6th and 7th collecting was carried on near a place called *Hanamalina*, belonging to a gentleman named Mr. H. Smith: after this the camp was pitched on a place called *Holo-kalili*.

On November 16th Palmer went to "Honaunau," a dairy belonging to a Mr. Johnston, on the slopes of Mauna Loa and, as Palmer thinks, about 6000 feet above the sea and some twenty miles to the south of Pulehua.

" The country here looks different to that above Pulehua. The Koa-trees are very high and in patches, most of their limbs and trunks are covered with moss. Between the Koas are patches of Ohias and large openings.

" Apapane (*Himatione sanguinea*) and Iiwi (*Vestiaria coccinea*) are common, the Ohia being in flower. Amakihi (*Chlorodrepanis virens*, Gm.), as usual, are everywhere.

" Leaving my camp next morning I went up a bullock-path for about a mile or so. All the way nearly was through dense forest of Koa, very thick, with much undergrowth of high fern, intermixed with wild raspberry-vines, which make it almost impenetrable."

On November 20th Palmer writes :—"We had breakfast (Mr. Johnston with his men and I) by sunrise, and at once left the camp and rode up the mountain. We must have been at least 9000 feet high, for we saw Mount Hualalei on a level with us, if not below, when we tied our horses up and proceeded on foot. The first two or three miles the country is very similar to what it is here, that is, forest with a clearing here and there. Gradually the trees get shorter, until all Koas are left behind and Mamane forms the principal portion of the bush, rising from 10 to 30 feet in height. The Mamane (the same tree I used to shoot the *Loxioides bailleui*, Oust., on at Pulehua) was here all in flower, therefore it was a good place for many honey-sucking birds. *Vestiaria coccinea*, *Loxops*, and *Himatione* were on the Mamane in great numbers, but I saw nothing new to me, although I travelled through a great deal of country and looked about all the time. The thinness of the air was felt very much; I did not notice it when hunting about, except that I could not breathe quite freely and my legs seemed very heavy, but as soon as I stopped to rest a peculiar sensation came over me, though from what I had heard I had been afraid it would be much worse. We climbed almost as far as the short trees go, above which there is no higher vegetation.

"Besides the birds already mentioned, I saw some *Loxioides bailleui*, Oust., and *Oreomyza mana* (Wils.). At a height of, I believe, at least 8000 feet we came across a Crow (*Corvus hawaiiensis*) perched on a tree, and we did not notice it until it began to cry. This leads me to think that it lives all over the country.

"Among the Koas I got some *Loxops coccinea* and *Rhodacanthis palmeri*, Rothsch., which seems to keep much to the Koas. This one is the first I saw in this part of the country. When going down we had a rough time. We lost our way and had to ride across a big lava-field, and when we came into the track again it became pitch dark and none of us escaped without bruised and half-skinned legs, while one of Mr. Johnston's men fell and injured his arm very badly. It was indeed the worst riding I ever experienced."

By the end of November Palmer left this part of the island and rode over to *Honakohau*, Mr. Clark's residence, some 15 miles north of Kauwaloa, on the slopes of Mount Hualalei.

A great hindrance to camping out in these parts is the scarcity of water, and before Palmer began to collect he had to procure permission to shoot from some of the landowners, which delayed him several days.

On December 2nd the camp was pitched on the slopes of *Mount Hualalei*. "As there are no cattle about, the undergrowth is very thick. The forest is chiefly composed of Ohia, which are just in flower. Apapane are very numerous, feeding from the flowers of the Ohia, but no other birds were seen to-day."

"During the next few days we went up the mountain. Some distance above the camp we found a broad belt of Koa-trees, with a few Ohias only. An Ou was heard and some other birds seen.

"I did not see any sign of Geese, although they are said to be not uncommon, and although the Ohelo-berries on which they feed are plentiful."

On December 28th the tent was pitched on the *Kohala* Mountains. The first days there heavy fogs and rain prevailed, adding much to the difficulty of collecting. The country was very swampy and but few birds were seen, so that Palmer was greatly disappointed. On December 31st he left this last camp, where he did not collect a single bird, and spent the 1st of January, 1892, under Mr. F. Spencer's hospitable roof.

On the 5th of January Palmer travelled over to Waimea, where he had got permission to occupy an old mountain-house. On the 10th he writes: "Last night the weather was unusually cold and a heavy thunderstorm passed over Mauna Kea and Hualalei. In the morning I saw both mountains for a great part covered with snow. This is nothing unusual on Mauna Loa at this time of the year, but on Hualalei I am told none has been seen since 1858."

Travelling was very bad on leaving this place, and not only the horses fell once, but Palmer himself got a wound on the leg. The natives were rather suspicious about Palmer, as they could not quite see why anyone should take so much trouble and expense to procure some useless birds. They distil a kind of gin from the root of some plant, which they sell claudestinely, as the government does not allow its sale, and they seemed to believe that Palmer was a spy sent by the government. The weather was mostly wet and cool.

"The famous Mamo (*Drepanis pacifica*) is said to have been seen above the place where I stay now the last time; and if this be true I don't see why it might not still exist here, for the forest here is very old and consists of beautiful ohia-trees. No sign of wild cattle is to be seen, and consequently the undergrowth is very dense, although pigs are common enough. The forest on the higher parts you may say is entirely maiden forest. A good deal of the soil here seems to be much more solid and not so much broken up. The Ou (*Psittirostra psittacea*) is very numerous, also *Chlorodrepanis virens*, which one hears singing nearly all the day. Other birds seemed not to be plentiful. I saw a few Iiwi (*Vestiaria coccinea*), some 'Apapane' and *Chasiempis*, and also a couple of *Phaeornis*. We were obliged to cut a trail to go up to the height I desired.

"*January 14th.*—It was raining in the morning, but in the afternoon the rain came down in torrents, accompanied by a strong gale of wind from S.W., known here under the name of 'kona.' If this weather lasts we will run short of provisions, for the road, which was bad enough when coming up in fine weather, must now be terrible after all this rain. I am already living quite like a Kanaka, sleeping on a mat made of the 'Lauhala' or Screwpine, of which the whole house is also built, and having nothing but 'poi' to eat, which will soon be finished.

"*January 15th.*—The weather was better. Just above the house is the finest forest, for higher up it seems to die off, a great number of dead trees being conspicuous, and it gets much lighter. The undergrowth is almost impenetrable, and strongly intermixed with bananas. Birds I saw only very few to-day.

"From what I saw I am sure that the Mamo, if really yet in existence, is not found in the highest parts of the hills.

"*January 19th.*—Our want of food compelled me to go down for provisions, but the trip was terrible. It was not before sunset that I reached the good people in Kohala, who are always very kind and willing to help and assist as much as they can. They were much astonished to see me, and said they should not attempt to travel among the mountains in such weather.

"Heavy rains continued for some days, and it was not before the 25th that it got fine once more, when at once I got into tent again on the Kohala Mountain. This is the only place where, according to what I am told by several people, the Ulaaihawane (*Ciridops anna*) is still living.

"The forest there is not very thick and little or no undergrowth. There are distinct belts of trees distinguishable. First one much like the Koa, then comes the Mamane, with here and there an Ohia, and various others follow, the uppermost parts mostly consisting of dead and dying trees. The introduced *Carpodacus* was common right to the top of the hill and singing beautifully."

Here follows an uninteresting part in Palmer's diary, full of personal matters, troubles with horses, complaints of a severe cold he caught, of uncertainty about where to go, of rainy weather, and of getting nothing rare.

On the 10th of February someone telephoned from Kohala to the next place where Palmer was staying, saying he had got the Ulanihawane (*Ciridops anna* (Dole)), but Palmer did not see the man before the 13th and found he had really got a specimen of this bird, which was believed to be extinct. He bought it and put it in spirits. It was shot near the head-waters of *Awini* on Mt. Kohala.

On the 20th of February Palmer arrived at the place where the rare bird was shot. He noticed only eight of the Loulu palms; and four of them together where the *Ciridops* was shot. After some time spent in search of the Ulaaihawane Palmer travelled along the mountains, pitching the tent on another place nearly every night and intending to stay on the most suitable place he might come to. There was, however, so much rain at the height of 4000 feet and upwards that he soon came down again, because almost living in the water at such heights and at low temperatures was more than he could stand.

By the beginning of March camp was made on Mauna Kea at the sheep-station of Mr. Hanneberg.

On March 12th Palmer saw a bird about the size of *Hemignathus obscurus* but unknown to him, which he could not shoot. It was probably *Viridonia sagittirostris*, Rothsch.

On one of the first days of April Palmer got severely kicked by his horse, an accident which prevented him from collecting for a long time. This happened at Hilo, where he had come round the island. Fortunately Palmer had here met with an old birdcatcher and his brother, who went up for him into the forest with Wolstenholme, Palmer's assistant. While Palmer was still not able to take long walks, on the 18th of April his assistant saw the *Drepanis pacifica*, and the old birdcatcher, Ahulan, caught it alive with a snare. All particulars about this glorious capture are given under the species. Soon after, on April 23rd, Palmer himself made a very interesting discovery, for he shot the first specimen of the then unknown bird which I described as *Viridonia sagittirostris*.

Palmer went up to the place where the Mamo had been caught, and many other places where Mamos had lived in former days, and employed a number of Kanakas to hunt for the Mamo, but in spite of his efforts no more were caught or even seen or heard by all these men.

May 12th.—"I left Mr. Hitchcock's house this morning at 6.30 with my horses and pack, and rode up as far as his mountain-cottage on the edge of the woods, which is called Bougainville. Here I left the horses, the natives I had engaged to go up with me and we ourselves taking the loads on our heads and shoulders. We ascended along an old trail that leads to Laumia, Mauna Kea. As nobody has travelled along this trail for a considerable time, it is completely grown over; so we were obliged to cut our way up till we reached the lava-flow. This made the journey slow and tiresome, as we all carried heavy loads, namely provisions for a week for ourselves and the dogs, besides all necessary camp outfit. I saw nothing remarkable of birds on the way up. The forest is mostly Koa with a few Ohia. To-night we are camped by the side of the great lava-flow of 1880, some seven miles from Hilo. In the evening some Bats were flying about, but I could not shoot them.

"We expect to reach the region where the Loulu palms grow to-morrow, and my old birdcatcher will be there already, I hope."

"*May 13th.*—At 7 A.M. we were packed again and started on the way towards the higher regions. After a very hard journey we reached the place at 3 P.M. All the way was along either an old or the new lava-flow. The weather was very hot, not a drop of water could be had; we therefore felt our heavy loads rather much. After a short rest our first thought was water. The stream of lava had shut up all running streams on this side; we could not find a hole for water. So we commenced to dig holes with our knives, and this proved to be our salvation, for from the three holes we had dug we managed to get enough water for drinking and cooking purposes. We then pitched our tent and enjoyed a hearty supper of wild goat, one of which I had shot a few hundred yards from the place where we camped, while my assistant had killed another and one of the Kanakas had caught a kid.

"The Loulu seems pretty common about here, for I could see no less than eighteen round the camp. I therefore think I must find the Ulaaihawane (*Ciridops anna*) here, if it is true that it feeds chiefly on the Loulu. I am very sorry to say that the old birdcatcher fell down one of the gulches and injured his leg so much that his son had to help him back. The son came up and told me.

"The forest round here is very dense and the undergrowth frightfully thick, but on some parts higher up wild cattle have opened it up somewhat.

"*May 14th.*—We followed up the flow of lava for a couple of miles. Both Loulu and the tree on which the Mamo chiefly feeds were numerous all the way, but I did not see any birds except the common species. On entering the forest, after leaving the lava-flow, we had to cut again. Birds were less numerous among the trees than on the outskirts of the forest. The bush consisted chiefly of Ohias with a sprinkling of Konas and other smaller trees. We returned after a good march and reached the camp at 4 P.M.

"My assistant shot an Io (*Buteo solitarius*). In the stomach I found remains of eggshells, and the bird itself (an old female) had partly developed eggs. So there is no doubt a number of birds are breeding now."

On May 19th Palmer broke the camp up after having no success whatever, and marched down to Bougainville, where he slept. On May 20th he reached Hilo, where he rested a few days. Then Palmer went up to the hills again in search for the Moho-rail, but did not succeed in finding any.

On May 25th Palmer visited the lake of fire, which has so often been described. He says that all the descriptions he had seen do not give an idea of the magnificent aspect. The fire-lake is usually called Madam Pele, and it throws out a pretty thread-like material, called Madam Pele's hair. There is commonly said to be a bird which builds its nest from this hair-like lava, but what kind of a bird it is seems to be unknown. Mr. Hitchcock showed Palmer what he said were the remnants of such a nest, and that it belonged to an Iiwi (*Vestiaria coccinea*).

Palmer stayed on Hawaii until the end of June. Besides several times visiting the fiery lake and the crater, and collecting very interesting specimens of minerals, he shot a few fine specimens of *Nesochen sandvichensis* and some small birds, but nothing new.

At the beginning of June the island of *Maui* was reached by steamer.

Palmer went into camp north of Olinda, Maui, on July 8th, 1892. The country was much cut up by the continued rains, which occur in that district, transforming the narrow valleys and precipitous gulches into roaring cataracts. The forest appears all dead except in the deep gulches, which still look green. Between Wailuku and Haleakala is a vast plain mostly planted with sugar-cane and called Great Sprecklesville Plantation, and in which the only birds to be found are Mynahs, Linnets, and Rice-birds. Most of the time on Maui it rained, and so collecting was extremely difficult and very unpleasant.

July 13th, still continuing to rain. Palmer determined finally to start. Above Mr. Hocking's house the mountain soon became impassable for the pack-horses; so a camp was made, from which centre the country could be explored.

Bird-life was very scarce. The first day in camp the only ones seen were an "Iiwi" (*Vestiaria*) and two "Amakihi" (*Chlorodrepanis wilsoni*).

On July 15th it was very fine all day, and "Apapane" (*Himatione sanguinea*) was seen and some more Amakihis secured. This latter, instead of whistling like the Hawaiian bird (*Ch. virens*), has a low plaintive tweet.

On July 16th the weather was very bad again, but was signalized by the discovery of the new Akikiki (*Oreomyza newtoni*, Rothsch.).

July 17th.—"Was mostly spent in skinning specimens and writing up the log. 6.40 P.M. A stroke of luck has just come in my way. After supper my assistant and I went out to attend to the horses, when we saw a strange-looking beautiful yellow bird sitting on a passion-flower vine. I sent my assistant for the gun, and was fortunate enough to shoot it. It turned out to be an albino Amakihi, being bright canary-yellow instead of green, but, strange to say, the iris, bill, and legs were of the normal colour."

For some days the weather continued fine, but no new birds were procured, though several specimens were added to the collection. Although so much rain had fallen in the mountains, on going down to Sprecklesville for provisions Palmer found everything dried up and the dust almost unbearable. He says several times he could hardly see his horse's head for the clouds of dust.

August 1st.—"I am again in camp, about a hundred yards from the forest, which covers the lower half of the mountains and fills up the numerous gulches. It consists mostly of Ohia-trees (*Metrosideros* sp.). The camp is 5000 feet high on the mountain. After breakfast next day I started exploring, and had hardly entered the woods when I saw a strange bird hanging on the branch of a Koa-tree, and, on shooting it, was delighted to find it a new form. It had a bill not unlike the 'Ou' (*Psittacirostra*), so I shall name it in the log-book the 'Small Ou' to distinguish it. [This was the female of *Pseudonestor.*] When I shot this bird I saw two more, but failed to procure them. Shortly after this I saw a small bird of an old-gold colour, but missed it. Half an hour later, however, I was fortunate enough to shoot the male of the 'Small Ou.' The 'Akikiki' are very numerous here, and now and then I can hear a note like that of an 'Akakane' (*Loxops*) which my assistant is trying to locate.

"My next two birds captured were true Parrots, which, from their numerous colours, are evidently imported birds (*Platycercus palliceps*). There were four of them flying close together into an Ohia-tree, when I shot the first, and the rest were easily killed one after the other. Immediately afterwards we again heard the note of an 'Akakane' (*Loxops*), but it

2

was so low that it took us some minutes to locate it. At last my assistant saw one in an Ohia-tree and instantly shot it, and was lucky enough to secure the female with the second barrel. From the colour of the male I conclude that this is the bird mentioned by Mr. Wilson of an old-gold colour (*L. aurea*, Finsch, nec *Drepanis aurea*, Dole!) in his book of instructions to me.

"*August 2nd.*—I think we secured in all five of these birds to-day. All of them were hopping about among the Ohia-trees. As I was anxious to secure the bird I did not wait to study its habits very closely. The next note we heard, in spite of the uproar made by Akikiki (which are very numerous at this height), was that of an '*Akialoa*' (*Heterorhynchus affinis*, Rothsch.). I immediately went towards the sound, and soon saw the bird on a Koa-tree, the scarcest of the Maui trees, and was, after a lot of trouble, successful in shooting it. This bird most resembles the '*Kauai Nukupu*' (*Heter. hanapepe*, Wils.), but its note is almost identical with that of the Hawaiian species (*Heter. wilsoni*, Rothsch.). In all we shot three Akialoas and two Ou (*Psittacirostra psittacea*), so that I believe to-day every species of bird except the Rice-bird and the Linnet were obtained. Except the Akikiki (*O. newtoni*) they are here all far from numerous, in fact rather rare. The forest is not very dense, and consists chiefly of small Ohias, a few Koa-trees, and a sprinkling of Mamane. Unlike on Hawaii, the Ohias form the upper boundary of the forest, while on the former island the Koas do this.

"*August 3rd* was spent in skinning and preparing specimens, but in the evening I shot three more Akakane. They constantly fly in company with Akikiki, but creep about among the leaves, while the latter hop from twig to twig.

"*August 6th.*—Yesterday morning I left camp and started for the crater of Mount Haleakala, taking 5 days' provisions with me. All went well, and we reached the summit at noon; but not many minutes after rain fell in torrents, with vivid lightning and thunder, which lasted for two hours. On starting again the pack-horse fell down a precipice, and was so knocked about that we could not proceed, and had to take refuge in a cave half a mile below the summit.

"Next morning we explored the crater. As far as we could see no forest is visible, and only a few single patches of low scrub and grass. In the crater, to my astonishment, a flock of Plover flew over us—the first I have seen since they migrated from these islands last May. Except Mynahs there are no other birds in the crater.

"The trail right up to the top is excellent, surpassing most of the so-called government roads. It lies all though open country, covered in places with dwarf scrub and small Ohelo-trees (*Eucalyptus* spec. ?).

"The other portions of the mountain are covered with bunch-grass. Both the sunrise and sunset were so magnificent that they defy description.

"I must herewith return my thanks to Mr. Mossman, of Makawao Store, whose sketch of the crater and trail was most useful to me.

"On the way down, hardly 400 feet below the summit, I shot a small bird, which turned out to be an Amakihi (*Ch. wilsoni*, Rothsch.). This proves that this bird is found at all altitudes, unlike the species of the other islands.

"*August 8th.*—Shot and skinned some more of the previously-mentioned birds. I saw an Akikiki (*Oreomyza*) feeding its young. The latter opened its bill, and the old one put an insect down its throat. I also heard yesterday an Akialoa singing, in quite a different tone from any previously heard (see *Heterorhynchus affinis*).

"*August* 12*th.*—The weather has been delightful all day, quite a change from yesterday. My assistant and I worked different sides of the ridges and at first only shot two Akialoas, but suddenly, after hearing a gunshot, I was summoned by my assistant with wild shouts. He had an entirely new bird![1] I made this note at the time :—Assistant just shot a beautiful specimen which may be an 'Oo.' Its mate is also in the vicinity, but has flown up the mountain. It is undoubtedly the same bird I saw on Wednesday and thought it was a young 'Oo.' Its note was a kind of hoarse 'O-o-o.' Shortly after the other one returned, and following it down hill I was fortunate enough to shoot it.

"I saw a bird fly into an Ohia-tree and begin to suck the flowers. It hopped about so rapidly that I thought it was an 'Apapane' (*Himatione sanguinea*). However, I soon saw it was much larger and darker and showed signs of red on the head. On approaching it, it flew out onto an outside branch, and its colours glittered beautifully in the sun. I immediately shot it, when it turned out to be the young of the former bird. Then my assistant shot another old one, which gave vent to a chuckling note quite unlike the first sound we had heard. We then went further down the mountain, and during the day procured in all eight specimens of this same bird.

"*August* 15*th.*—Yesterday afternoon, for want of anything else to do, my assistant and I took a gun and wandered along the edge of the forest, where I was fortunate enough to shoot two more small Ou (*Pseudonestor*). They were both busily digging in the bark of half-rotten Koa-trees.

"To-day we pushed higher up the mountain, but were very unsuccessful owing to the numerous ridges and deep ravines.

"*August* 16*th.*—We were only out a short time, but got five more Maui Oo (*Palmeria*) and one Okinloa, but as there is nothing new I shall leave here to-morrow.

"*August* 30*th.*—Up to now we have been occupied in camping on new ground and trying the upper forest-region, but owing to bad weather have done very little. To-night sees us in camp no. 2, all day being occupied in reaching the place where we left off cutting the trail. We had the greatest difficulty to find a place level enough to pitch the tent on. When this was found, there came the task to clear it of the network of Te-vines. At last a hole large enough for the tent was made, but we are perfectly hemmed in all round except an opening large enough to reach the water, which is about 20 yards away. Above us is a canopy of Tes, only admitting just enough light for us to know when day breaks. Bird-life is very scarce, only a few Iiwi and Apapane to be seen. All next day was spent in cutting the trail, but with two cane-knives going we only cut one mile the whole day. We then climbed a high tree, and as the undergrowth appeared the same the whole way up the mountain and no birds were to be seen or heard, we determined not to go any further. Where there are tall trees three parts of them are dead. You may really say the forest only consists of undergrowth and small trees about 20 feet high.

"*September* 2*nd.*—Our return journey downwards was much worse than going up. Torrents of rain fell, and what had been the trail was a roaring brook. We were up to our knees in mud and water the whole way down.

[1] *Palmeria dolei* (Wils.).

2*

"*September 9th.*—Went up to the forest above Kipahulu plantation. There is a heavy Koa-forest with dense undergrowth of ferns.   Ordinary birds are plentiful here, so I hope I shall find something good inside the forest.   I saw a bird of a dark red colour and considerably larger than an Iiwi.   I was talking to an old Kanaka, and he described to me a bird somewhat like that of a farmyard-cock.

"In Keanei they told me about a bird with a comb, but as I could get no definite information from them I thought it was a myth, but this native assures me it was often seen in Kipahulu Valley years ago.   He calls it ' Akohekohe.'

"From September 10th to September 16th nothing could be collected and nothing particular was seen.   The days were occupied by cutting the trail, every day it was raining heavily.   We had to build a hut of banana-leaves for the luggage, and it was difficult to light a fire or to do anything.

"On September 16th the lofty Koa-forest was passing over into Ohia-forest, as the country was rising, the place became extremely wet and swampy.   Sometimes we sank so deep into the mud that we could not go on any further.

"When we pitched the tent to-day we had the greatest difficulty to fix the poles, so soft and swampy was the ground, and torrents of water were running down the slopes.   We did not succeed in pitching the tent before we had created an island by digging deep ditches and gutters for the water to go off, and we never had so much difficulty in making up our fire. The forest here has a wild and peculiar aspect, dead fallen trees lying everywhere in one's way and others standing lifeless and leafless between the green ones, so that perhaps only half of all of them are alive.

"*September 18th.*—We moved on to camp no. 5 to-day, and came into a somewhat drier region, where we found better space for our camp than for the last week.

"The Portuguese servant, whom I had sent down yesterday, returned loaded with provisions which we needed very much.

"As the weather was finer more birds were seen.   ' Akikiki' were quite common in certain places, going about in small flocks, and I saw a few Akakane, Apapane, Iiwi, and Amakihi. An Ou I heard at a distance.

"*September 21st.*—The past three days were as usual wet, dreary, and unprofitable, but to-day it is fine, and we reached a spot above the forest-region, I think about 7000 to 8000 feet high, our last camp being about 4000 feet high.   The ascent up to the rim of the crater would have been easy, if I had wished to go there[1].

"Besides the birds already mentioned we found here again the Maui Oo (*Palmeria*), and since observing it more closely I came to the conclusion that it is the bird the natives speak of as having a head like a fowl.   We must have seen eight or ten to-day, but could only procure four.

"*September 22nd.*—Besides six *Palmeria* secured we saw a ' small Ou ' (*Pseudonestor*), so that except ' Akialoa ' (*Heterorhynchus*) we have found every bird we obtained at Olinda, also at Kipahulu and the parts of the mountain just above it.

[1] Palmer speaks of the *Mauna Haleakala* (" house of the sun "). This mountain occupies the greater part of South-eastern Maui and is 10,030 feet high.   It rises from perpendicular cliffs near the sea, in many parts with one unbroken slope to the summit.   Mauna Haleakala has the largest known crater in the world, or it may be that it is formed of several craters.

" *September* 23rd.—We marched down in one long march to *Kipahulu* in the Hana District, where we stayed until September 28th. That day we marched to Kaupo, some eleven miles or so from Kipahulu. I stayed with Mr. Andrews, the schoolmaster of Kaupo, to whom I am much obliged for his kind hospitality. There is a store with very little in it. The natives here are poor, and few of them can afford to buy fish, their chief food being sweet potatoes and salt.

" Mr. Wilson had advised me to go to Kahikinui, but Mr. Andrews warned me not to do so, as every horse or mule was falling ill and many died there, besides there being very little forest. So I started for *Ulupalakua*, which I reached after a long day's travelling. The ride was very monotonous : everywhere one sees lava and stones and bare rocks, only here and there a small tree and ' prickly pears.' It is only within a mile from this ranch that the aspect of the country changes suddenly from that of the dry lava to fresh pasture-green. Lofty Eucalyptus and numerous other trees surround the house and a once beautiful garden. Mr. Buchanan, the manager of this ranch, very kindly gave me accommodation and helped us greatly.

" *October* 1st.—I made a tour to the small island of *Kahoolawe*, 5½ miles off the S.W. point of Maui. [This island is 12½ miles long and 5 miles wide, and almost destitute of every kind of verdure or shrub excepting a kind of coarse grass. It is used as a sheep-pasture.—W. R.] I stayed under the hospitable roof of Mr. Gay, the governor and manager of Kahoolawe.

" On this and the following days I rode over the island. There is no forest on the island, and I think not more than two dozens of trees to speak of. The upper portions of the island are quite barren, while all around it and in all the lower parts two kinds of grass are growing, called ' pili ' and ' manauea.' Several hundreds of acres consist of a good, deep, rich volcanic soil, intermixed with a deep reddish-coloured sand. This sand is often carried off by the wind when dry, and can often be seen drifting over the sea in large clouds for miles. Mr. C. Kinsley runs cattle and sheep besides a number of horses and mules, all being under the management of Mr. Gay, my kind host. I have not found any land-birds whatever except the introduced *Carpodacus*, and all I saw were Plovers and Akekeke (*Strepsilas interpres*, L.).

" *October* 6th.—I returned to Maui again and reached Ulupalakua at noon. The afternoon I searched for birds in the vicinity, but I found nothing except introduced birds, as ' Linnets' (*Carpodacus*), Mynahs, and Peafowls.

" *October* 7th.—Started from Ulupalakua to *Waikupu*. Much of the country is covered with the famous *Lantana*, which became a pest to the island, spreading everywhere and over-powering all other vegetation. On a lagoon I shot some Plovers of a kind not yet met with by me on these islands, numbers 1779 to 1781 of my collection [*Heteropygia acuminata—*W. R.].

" The hills here are very steep and the country cut up with innumerable deep gulches, close to each other and with extremely narrow ridges. This makes travelling and hunting difficult, but a far greater trouble is that there is no fresh water to be found to make a camp. I saw very few birds and nothing else but ' Apapane ' and ' Amakihi,' which astonished me very much, as there were a great many Ohia-trees in full flower.

" *October* 11th.—My assistant had a bad cold, and therefore I did not send him into that rough part I had intended to let him make collections in, but sent him to the tracts above Olinda to get some more Akialoas and small Ous.

"I called on the Catholic Brothers again, who have a local collection of birds. They gave me in exchange a Tropic-bird that had been captured not far from Honolulu. I think it is different from the one I collected before.

"I had another look at the Iiwi (*Vestiaria coccinea*) of Brother Matthias, with the short lower mandible, and now it seems to me a freak of nature, although the bill seems more slender. [This is a specimen with the half of the lower mandible wanting, so that the bill looks like that of a *Heterorhynchus*.—W. R.]

"To-night I depart for Hilo.

"*Thursday, October 13th.*—Tuesday night I left Maalaea Bay, Maui, in the steamer 'Kinau.' and after calling at Makena we made a start to cross the channel, reaching Maukona 8.30 on Wednesday morning. Here we stayed till noon, when another start was made for Kawaihae, and we finally reached *Hilo, Hawaii*, at 3.30 this morning, after a fair passage. About 11 o'clock I came across Ahulan, my birdcatcher, who tells me he could not find another Mamo (*Drepanis pacifica*), although they have been up several times. He also says that 'Aku' is not in flower yet; so after all I have come over too early, but I intend to go up for a few days, to see what changes have taken place in the vegetation and try to secure some more Great Amakihi (*Viridonia sagittirostris*).

"*Wednesday, October 19th.*—After reading the last report this date will surprise you, but unfortunately through an accident I have not been able to write my diary up. No sooner had I mounted my horse to start up the mountain, when the horse reared up and threw himself over backwards on the top of me. Yesterday only I managed to commence my journey on a mule kindly lent to me by a Mr. Pladen.

"I had engaged two natives as carriers, and we reached the banana-houses on the slopes of Mauna Kea at 2 o'clock. I was very glad of this, as my spine seems strained and troubled me very much. However, to-day I have been out hunting, though I have not walked far. I killed an 'Akialoa' and an 'Akakane,' thinking they were something else.

"It is valuable to be here at the time when the high trees are flowering, for so many birds come to the flowers to suck the nectar from them. I was afraid I would come too late for the best flower-season, but see now that I am too early. The flower 'aku' only a few are in bloom yet. Ahulan, the old birdcatcher, thinks the Mamo might be seen again on these flowers in about three weeks. He was up last week and did not hear or see one; none the less he is full of hopes, but I am not.

"*October 20th.*—There is much rain now, and I must cut out trenches all round my camp to keep the water off.

"Mongooses are unfortunately very numerous here and cannot be kept out of the house. One of them ran away with two bird-skins I made yesterday."

All the time until October 28th Palmer tried to get new information about *Drepanis pacifica* and to interest a number of natives in this bird, but did not collect anything worth speaking of.

From October 21st to November 2nd Palmer tried to collect at several places on Hawaii, but did not have any success at all.

On the 3rd of November he landed at Maunalei on the island of Lanai.

LANAI or RANAI lies 7 miles to the west of West-Maui. It is a dome-shaped island, 16 miles in length and 4 to 10 miles in width, much higher than Kahoolawe, but neither so high nor broken as some of the other islands, its greatest altitude being about 3000 feet.

The greater portion of Lanai is barren, and the island in general suffers from the long droughts which frequently prevail; the ravines and glens, notwithstanding, are filled with thickets of trees. According to Palmer the trees are nowhere very lofty, and the forest seems to be dying out. The cause is attributed to the great number of goats and cattle that eat up the undergrowth. There are, however, gulches and ridges, where the undergrowth is thick enough to prevent climbing without constant use of a cane-knife, and there is especially a kind of climbing staghorn fern, as Palmer calls it, that forms a great obstacle; besides, the loose soil and stones are very troublesome, chiefly when going down hill.

Lanai, too, is volcanic; the soil is hollow and by no means fertile.

From a little peak near Lanaihale a beautiful view was obtained. The whole of Lanai could be viewed : conspicuous right below to the south the deep Palawai Valley, with here and there a cottage standing and an abundance of green. To the S.E. lies the island of Kahoolawe, and beyond it the two giants Mauna Kea and Mauna Loa, just showing their heads above the clouds. East of these stands Haleakala in bold relief, and little farther to the north Mt. Eke. Between N.E. and N.W. in a long line lies Molokai, her mountain peak hidden by clouds and fog.

On the same day (November 15th, 1892) Palmer shot the first specimen of the very rare Lanai Akialoa (*Hemignathus lanaiensis*, Rothsch.).

Bird-life is very scarce, according to Palmer, and he thinks the native birds are dying out fast, partly on account of the supposed disappearance of the forest, partly from the great number of introduced cats. Unfortunately some introduced birds, as the *Carpodacus* and the Rice-birds, were found in numbers, certainly not to the advantage of the native birds.

Palmer was very much assisted during his stay on Lanai by Mr. Hayselden and Mr. Henry Gibson, to whom he is much indebted, also Messrs. Macfarlane and Grennel.

On December 8th, a few minutes before midnight, Palmer landed on Molokai, at a place called Pukoo.

MOLOKAI or MOROTOI lies 7 miles north of Lanai. It is a long, irregular island, apparently formed by a chain of volcanic mountains, 35 miles in length, E. by N. and W. by S., and only 2½ to 9 miles broad. The mountains are nearly equal in elevation to Mount Eke on Maui, and are broken by numerous deep ravines and watercourses, the sides of which are frequently clothed with verdure, and ornamented with shrubs and trees. There is but little level land on Molokai, but several spots are rather fertile.

One third of the island to the west is a barren waste. The remainder, to the east, is almost one entire mountain, rising gradually from the south to the height of 2500 feet, while to the north it is almost perpendicular.

The people are mostly very poor and comparatively ill-provided with necessaries.

Palmer speaks of his first impressions as follows :—

" My first glances of and experience on Molokai did not very favourably impress me. As far as I can see, there appears to be a narrow strip of land, running all along this side of the island, only a few feet above the level of the sea, and then at once the mountain rises abruptly at places almost perpendicular, cut up with innumerable and very deep ravines and razor-back ridges."

Here at first the name "*Kakawaheia*" for the *Oreomyza flammea* was heard and generally known. From below no forest can be seen. Lantana, like near Kona on Hawaii, is overgrowing the lower portions of the land, oppressing the native vegetation. Along the coast were scattered here and there a few houses, partly empty and almost one-half of the others were occupied by Chinamen.

On December 9th Palmer made his first trip to the hills, but did not procure any birds. The forest consists, he says, all of dwarfish Ohias with much undergrowth of climbing ferns.

The 10th and 11th of December rain fell so constantly and heavily, that Palmer did not go out for collecting; on the 12th of December a trip was made, but with no success. "Apapane" was seen in numbers and a few "Iiwi" and "Amakihi" (*Himatione wilsoni*), but nothing collected. The very dense undergrowth made collecting very difficult.

On December 16th Palmer was guided by Dr. Mouritz, who showed much kindness to him in many ways, to one of the highest peaks of the island, and there it was where he shot the first specimen of that beautiful new "O-o" which I named *Moho bishopi*.

Owing to the extremely dense undergrowth and the low trees, together with the deep and sometimes inaccessible ravines, it is very difficult to collect this fine species, and it was not before the 26th of December that a second specimen of it was procured. Stormy and rainy days with some bright and fine ones between prevailed to the end of 1892, and the collection increased but little. Nearly all the collecting was done on a trail cut for the purpose.

The highest hills on Molokai were, in Palmer's opinion, not high enough to produce a different fauna; from which surmise he concluded that all the birds are inhabitants of the forests from the lowest valleys up to the highest peaks. This assumption was probably the reason for Palmer's failure to discover *Drepanorhamphus funereus*. Of all the birds obtained and observed on Molokai, Palmer says the Apapane (*Himatione sanguinea* (Gm.)) is the most numerous. *Phæornis* is not at all rare, though, on account of the habit of keeping in the low undergrowth, is much oftener heard than seen. The Amakihi is not very rare in the lower parts of the island, the Crimson Kakawahia (*Oreomyza flammea* (Wilson)) more seen in higher elevations. The rarest and finest bird of the island is the O-o (*Moho bishopi*, Rothsch.). It was found in the lower and upper forest-region, but more in the latter. Below on the fish-ponds both *Fulica alai* and *Gallinula sandvicensis* are fairly common.

The first week of January 1893 was almost entirely spent in search of the *Moho bishopi*, and as many as five specimens, besides some other birds, were shot, which Palmer considers a very good success, especially as some days heavy gales and continual rain made collecting almost impossible, and, indeed, Palmer says he was very glad that his tent was not swept away by the water.

From Pukoo Palmer went to Halawa, in the district of *Koolau*, where he collected on the hills above. The forest here was much higher than near Pukoo, and many Ohias were in full bloom. The flower-trees were full of Apapane (*Himatione sanguinea*) and Iiwi (*Vestiaria coccinea*); *Phæornis* too was more numerous here, but of *Oreomyza flammea* (Wilson) and *Psittirostra* only a few were seen. *Moho bishopi* was rare.

Not far from Halawa a projecting and low V-shaped peninsula could be seen, covered with houses and with gigantic perpendicular cliffs behind, the home of the Lepers, the cause of so much of the notoriety which Molokai gained through the deplorable death of Father Damien. (See Schauinsland, "Ein Besuch auf Molokai," in Abh. nat. Ver. Bremen, xvi. 3, 1900.)

Palmer was told by two natives that "*Mamo*" was plentiful on Molokai a few years back, but they were extremely rare now, though still existing on the island. Palmer was informed by Mr. R. W. Meyer that it is *Palmeria dolei* which bears this name on Molokai, and not *Drepanis pacifica*, which is called "*Mamo*" on Hawaii. [I, however, believe Mr. Meyer was wrong and that the bird called Mamo on Molokai is the *Drepanorhamphus funereus* (Newt.).]

The Mauna Loa, a mountain on the west side of the island (not to be confused with the great Mauna Loa on Hawaii) has no forest, consisting of red volcanic soil with big boulders, and with patches of grass here and there.

Along the shores *Himantopus knudseni* and *Nycticorax* are not rare, and the commonest shore-bird is *Charadrius fulvus*, Gm.

Several days were lost for collecting through rain and thunderstorms, and especially on the 7th of February an enormous quantity of rain fell. It was said that for the last eight years such a quantity of rain had not fallen on one day.

On February 15th and the following days collections were made near Pelekunu and Makakupaia, and a number of specimens of *Palmeria dolei* (Wils.) were procured, which were not seen before.

On February 25th, Palmer left Molokai and reached Honolulu on the island of Oahu after a rough passage.

Oahu or *Woahoo* is the principal island of the Hawaiian Possessions as regards trade and maritime affairs, inasmuch as it contains the principal town and residence, Honolulu, and the port chiefly frequented by the shipping of the North Pacific. It is divided from Molokai, the nearest island to it, by the Kaiwi Channel, 22 miles in width. It is 40 miles long by 21 miles in its greatest breadth. It is, like the rest of the islands, of volcanic formation. Civilization is more advanced and rice-plantations more numerous on this island than on the others, not to the benefit of its native birds, for we know of several species that have become extinct on Oahu. There are nevertheless still extensive forests on the mountains.

Very few native birds, except *Chasiempis*, are seen near Honolulu, but many of the introduced foreigners.

On March 9th, Palmer left Honolulu for a collecting-tour to the district of Koolauloa. He describes his trip as follows :—

"From Honolulu to Pali the road ascends until Pali is reached, from where it descends again. Once below, the country is pretty level. A few miles further Heeia, a sugar-plantation, is reached, situated by the shores of a large bay. Rice-plantations are very numerous in the lower parts. From the hill-range of Pali narrow ridges run up in straight lines to the high mountains. After another 10 miles through level country, Laie is in sight, the landing-place for the Kahuka sugar-plantation. All the level parts hereabout are planted with sugar and rice, the lower hills afford good pasture for cattle, while the higher mountains seem to be covered with forest. At 5 P.M. Kahuku Ranch, 38½ miles from Honolulu, was reached, but the manager was away at Wailua and the place was locked up. Having had only a couple of hard biscuits and a bottle of beer all day, my assistant and I were somewhat hungry. Fortunately we found food and lodging at a plantation close by. Next day the manager came, and very kindly gave me a room in the plantation to stay in.

3

" I was at once told that an Alaska Snow-Goose was here, but very wild. It was the first ever seen here. I went out, saw it, and had at last a shot from a long distance, but did not get it. It seems to be a large bird, snow-white, with the wings brown or black at the tips, bill dark. Some time ago three such geese were seen on Maui, of which two were captured, so I suppose this one was the one that escaped the hunters on Maui.

" *March 13th.*—I made my first acquaintance with a forest on Oahu, or at least the outskirts of one. It will hardly be believed if I assure you that we did not see a single native bird to-day, although I think I heard a *Himatione*. I was always told that bird-life was very scarce on Oahu, but this day's experience surpassed all I had expected. Afternoon and evening heavy rains and thunderstorm.

" *March 14th.*—The rain of yesterday had set all the flats under water and was one of the heaviest rainfalls known for many years on Oahu. My assistant and I rode up to the distant hills, where I shot two *Chasiempis* and saw one building a nest. Several *Himatione* were seen, but none shot. Mynahs are extremely common.

" *March 15th.*—I made again a camp on the outskirts of the forest in a deep ravine, surrounded by open forest of Lauhala (Screwpine), Mountain-Apple, Koa, Ohia, and other trees of which I do not know the names. Mosquitoes were plentiful.

" *March 16th.*—We had again to cut a trail on the hills, for there was no way high up as far as we wished it for our purposes. Although we now had a road to walk on and the forest was high and fine, I saw but a few birds and shot nothing except a native duck (*Anas wyvilliana*, Sel.). Never anywhere did I see such scarcity of birds. I don't know whether there have always been so few or whether cats or what else may be the reason of their rarity. [There is no doubt that not very long ago there were many more birds on Oahu, and cats and the foolishly-introduced foreign birds no doubt play a part in the drama of the disappearance of the interesting ornis of Oahu.—W. R.]

" *March 20th.*—On the stream a colony of *Nycticorax* was found. Most of the nests contained two eggs each.

" *March 21st.*—I broke up my camp and rode to Waialua or Wailua, where I was glad to meet Mr. Perkins, the collector. He has collected for a good time and on different places on the island, but nowhere did he find any sign of *Hemignathus* or *Heterorhynchus*. He only found four species of native Passerine birds, i. e. *Vestiaria coccinea*, *Himatione sanguinea*, *Chlorodrepanis chloris* (Cab.), and *Chasiempis gayi*, Wilson. Of the white-backed and rufous-backed specimens of this Flycatcher he says he is sure, from many observations, they are young and old of one species.

" I went up with Mr. Perkins to the mountain-house he was staying in, and found it the most favourable place that could be selected. Therefore I decided to come here with my camp, and so I did.

" After having hunted here for several days, I am again astonished about the scarcity of birds. Why all the birds should have died out I cannot yet sufficiently understand. No doubt the abundance of cats, the disappearance of forest, and the Mynahs, who certainly destroy many other birds' eggs, have to do with it; but there is still good forest enough to keep lots of birds, and cats could not have destroyed all.

" There have been ' Akialoas ' [*Hemignathus* and *Heterorhynchus.*—W. R.] and ' Ous '

[*Psittirostra olivacea.*—W. R.], but neither Mr. Perkins nor I did meet with them anywhere.

"Besides the Mynahs, which are a great nuisance to the islands, and the Doves, Quails, and Pheasants, there are Skylarks here, which were only introduced a few years ago. It was quite an enchantment to hear them singing up in the air, just as in the old country.

"Unfortunately my good dog, which was always so useful to me, especially in finding specimens even in the thickest jungle, fell very ill."

Only on April 10th the first specimen of Akikiki (*Oreomyza maculata* (Cab.)) was seen and shot, and of which a fine series was brought together during the rest of the stay on Oahu,

About the middle of April Palmer states that on several days he saw a certain number of nests, but all on inaccessible trees.

Even in considerable heights in the mountains Palmer saw a great number of rats, and on one day killed three up in the trees. There can be no doubt, he thinks, that these, too, are very destructive to bird-life.

On the 20th April the only specimen of *Loxops* was shot by Palmer's assistant Wolstenholme, and therefore I named it in honour of its discoverer *Loxops wolstenholmei*, not having found out at the time that it really was *Loxops rufa* (Bloxam).

On May 3rd Palmer rode over to *Waianae* on the other side of the island, that is the west side. On this excursion Palmer saw a specimen of *Circus hudsonius* (Linn.).

After having collected on several more places in the hills, Palmer left Oahu on June 13th, 1893. He has made a most interesting collection of the native birds he found on that island, all in sufficiently large series for complete descriptions of the plumage, except the small *Loxops*, of which he could not find a second specimen, although he evidently tried his best to do so.

Over and over again Palmer mentions the rarity of native birds. He expressly says that there are still great forests full of lofty trees which might be the home of many more birds. He learned that the birds were much more numerous about 20 years ago, and in this he is undoubtedly right. The reasons why the birds became so rare are many. Much of course is due to the destruction of forest, which covered most of the pasture-grounds of the present day, but there is still so much forest that this alone cannot be the cause of the supposed disappearance of more than one species. There are the enormous number of introduced birds, especially the troublesome Mynahs, which do much harm to all sorts of small birds; there are the numerous rats, the many cats that run wild everywhere; the more numerous population of Oahu; but all this does, in Palmer's opinion, not sufficiently account for the disappearance of so many birds. There is, as Palmer justly believes, one more reason, and this one we do not quite know. It is not impossible that some climatic or other change took place, of which we know nothing definite at present, though I suspect that it is due to inherent weakness in the stock.

Palmer also says that he is told collecting was much easier before, because the undergrowth was much less thick than now, but this lacks proof. Further on, Palmer concludes that it is impossible for anyone to say whether those species, that neither Wilson, nor Perkins, nor Palmer himself have found, are really extinct or not; for, as he says, "*there are birds which are so rare that one might hunt the mountains for twelve months without discovering*

3*

them, and in the thirteenth month might find one, and this applies especially to Akialoas" [Hemignathus and Heterorhynchus.—*W. R.*].

I believe this remark of Palmer to be very just, especially as it is conspicuously illustrated by the *Loxops rufa*, of which one single specimen only was obtained, while neither Wilson nor Perkins have found it.

As to the *Psittirostra*, Palmer says he firmly believes it no longer exists on Oahu at present, for this is a very conspicuous bird, generally easily seen in the guava-trees in the gardens, and it is not so much an inhabitant of the higher elevations and lonely forest as the "Akialoas" are. Moreover, Palmer says this bird is slow of movement and less shy, so it falls a much easier prey to all sorts of enemies. On Lanai, however, where really there appears to be less food and much fewer of suitable localities for it, the "Ou" is still common enough.

On the 15th of June, 1893, Palmer landed again on

KAUAI.

It was at my special request that he went to this island once more, to clear up a few points.

On the coasts Palmer noticed that the "Foreign Ducks" and "Plovers" had already left the island, for they are merely winter visitors.

On June 16th Palmer went to the forest again, and in his diary the following remark occurs :—"Once more amongst the forest and its feathered tribe ! What a striking contrast from Oahu ! Everywhere here the trees appear to be alive with birds. The noise of their various songs was quite deafening this afternoon. The 'boom' of the 'O-o' (*Moho braccata*) and the powerful song of the 'Kamao' (*Phæornis myiadestina*, Stejn.) were very noticeable above the others. A decided difference I can hear between this 'Kamao's' song and its near ally's on Hawaii (*Phæornis obscura*, Gm.). It strikes me Mynahs are much more numerous than during my first stay on Kauai, and on places where I am sure I did not see any before."

Palmer stayed on Kauai until July 12th, and shot two young specimens of the rare *Phæornis palmeri*, but nothing else of special interest. He lost some days through a severe cold, and was not able to collect in the district of Hanalei on account of a company of Lepers having retreated to those hills. These were, according to the law of separation, ordered to go to the Lepers' home on Molokai, but had refused, and shot several men who tried to force them away. Martial law was then proclaimed in the Districts of Waimea and Hanalei, and soldiers ordered from Honolulu to fight against the murderers. No one was allowed to leave the low lands, and of course collecting could not be carried on there.

On the 15th of July Palmer landed on the island of

NIIHAU.

Niihau (Oneeow or Oneeheow) lies 17½ miles W.S.W. of Kauai. It is about 15 miles long and 2 to 7 miles broad. The eastern side is rocky and unfit for cultivation, nor is there any anchorage on it. It is comparatively low and destitute of wood, with the exception of carefully cultivated fruit-trees.

In olden times Niihau was famous for its yams, fruits, and mats, and was the property of the king, but was purchased by a Mr. Sinclair, who settled here and used the island exclusively as a sheep-walk.   In 1877 there were about 75,000 sheep on the island, and the inhabitants, 233 in number, were mostly in the employ of the owner, Mr. Sinclair.   (See Findlay, North Pacific Ocean Dir. pp. 1103 & 1104.)

At present the island belongs to Messrs. Gay and Robinson.

On crossing over to Niihau Palmer saw several Petrels, a Brown-breasted Gannet, Grey and Black Terns, and a Black Storm-Petrel with a white bar across the tail.   He was very cordially received by the manager, Mr. E. K. Bull, and by the natives.   On the flat lagoons a number of Coots and Stilts were found breeding, and a collection made of these and other shore- and water-birds, but no small land-birds of the Passerine Order were found, except the introduced Mynahs.

On July 28th Palmer left Niihau again and sailed to Kauai once more to try and get a good specimen of an adult *Phæornis palmeri*, but he was unsuccessful.

Here, in August 1893, ended Palmer's collecting-trip, after an unsuccessful attempt to land on the small island of *Lehua* or Egg Island, off the north end of Niihau.   It is a rugged, naked, barren rock, but is the breeding-place of many sea-birds.

PART II.]   [NOVEMBER 1893.

THE

AVIFAUNA OF LAYSAN

AND THE

NEIGHBOURING ISLANDS;

WITH A COMPLETE HISTORY TO DATE OF THE

BIRDS OF THE HAWAIIAN POSSESSIONS.

BY

THE HON. WALTER ROTHSCHILD.

*Illustrated with Coloured and Black Plates by Messrs. Keulemans and
Frohawk, and Collotype Photographs.*

LONDON:
R. H. PORTER, 18 PRINCES STREET, CAVENDISH SQUARE, W.
1893.

(To be completed in Three Parts, price £3 3s. each, Net.)

HISTORY

OF

THE BIRDS

NOTICE.

＋━・

Owing to unexpected delay and to the return of my collector, Henry
Palmer, from the Sandwich Islands, I have been compelled to omit the
Plates of two species. as well as the extracts from Henry Palmer's
diary, which were intended for this Part; but the Plates and the
Text will appear in the Third Part, which I trust will make amends
for this Part being smaller than the first.

WALTER ROTHSCHILD.

food, which consists of fruits and insects, as the observations of Perkins and Palmer show
(though not of fruits alone as Dr. Gadow says), which seem to prove its *Turdine* relations.
Moreover, many anatomical details. carefully pointed out by Dr. Gadow in 'Aves Hawaiienses,'
pt. ii., are in favour of its *Turdine* relationship. The relation to the *Prionopidæ*, suggested
by Sharpe in the 'Catalogue of Birds,' has very little to commend it, especially since that
author no longer upholds the family *Prionopidæ*, as he shows in a footnote on p. 86 of his
excellent 'Review of recent Attempts to Classify Birds' (1891). His opinion that some of
the so-called *Prionopidæ* might be united with the Flycatchers, while others might belong
to the *Laniidæ*, but that more and closer investigations of the subject are needed, seems
to be very just.

Unfortunately none of the recent explorers have found the nest and eggs of *Phæornis*,
by means of which might be shown its true place in the system.

K

HISTORY

OF

THE BIRDS

OF

THE HAWAIIAN ISLANDS.

THE GENUS PHÆORNIS.

THE type of this *exclusive Hawaiian genus* (so far as at present known) is the *Muscicapa obscura* of Gmelin. Peale and Cassin placed this bird in an American genus which belonged to the *Tyrannidæ*, with which it is in no way connected. Sclater rightly established for it a new genus, *Phæornis*, placing it among the *Muscicapidæ*. There is indeed much in the external appearance of *Phæornis* to justify this position, and the recent biological notes of Perkins ('Ibis,' 1893, p. 110) are much in favour of it. There is, on the other hand, the truly *Turdine* foot, and the tarsus with its unbroken laminæ, the powerful song, and the food, which consists of fruits and insects, as the observations of Perkins and Palmer show (though not of fruits alone as Dr. Gadow says), which seem to prove its *Turdine* relations. Moreover, many anatomical details, carefully pointed out by Dr. Gadow in 'Aves Hawaiienses,' pt. ii., are in favour of its *Turdine* relationship. The relation to the *Prionopidæ*, suggested by Sharpe in the 'Catalogue of Birds,' has very little to commend it, especially since that author no longer upholds the family *Prionopidæ*, as he shows in a footnote on p. 86 of his excellent 'Review of recent Attempts to Classify Birds' (1891). His opinion that some of the so-called *Prionopidæ* might be united with the Flycatchers, while others might belong to the *Laniidæ*, but that more and closer investigations of the subject are needed, seems to be very just.

Unfortunately none of the recent explorers have found the nest and eggs of *Phæornis*, by means of which might be shown its true place in the system.

K

Bill more or less broad and flattened, with a distinct notch; rictal bristles few and not very strong. Nostrils in front of a big groove.

The plumage is soft and full. The primaries are ten in number, the secondaries nine. The fifth, sixth, and seventh primaries are about equal and longest; the first is well developed and half as long as the second. The tail has 12 rectrices and is about equal. The tarsus is covered in front by a long and unbroken lamina. The alimentary organs are essentially those of a frugivorous bird. For further anatomical details see Gadow, *l. c.*

Key to the Genus Phæornis.

A. Spotted above: *Young of all species.*
B. Uniform above: *Adults of all species.*
  a. Very conspicuous white marks on outer rectrices.
    a′. Wing not less than 4 inches; feet dark . .   *P. myiadestina.*
    b′. Wing much less than 4 inches; feet light .   . *P. palmeri.*
  b. No conspicuous white marks on outer rectrices.
    c′. Wing longer, below darker . .   . . *P. obscura.*
    d′. Wing shorter, below lighter . .   . *P. lanaiensis.*

1. PHÆORNIS OBSCURA (Gm.).

OMAO or OMAU.

Dusky Flycatcher, Lath. Gen. Syn. ii. p. 344 (1783) (Sandwich Islands) ; id. Hist. B. vi. p. 211 (1823).
Muscicapa obscura, Gm. Syst. Nat. i. p. 945 (1788) ; Lath. Ind. Orn. i. p. 479 (1790) ; Steph. Shaw's Gen.
    Zool. x. p. 405 (1817) ; Vieill. Nouv. Dict. xxi. p. 465 (1818) ; id. Enc. Méth. p. 809.
Tyrannula obscura, Peale, U.S. Expl. Exp., B. p. 310 (1848).
Chasiempis obscura, Hartl. Arch. f. Naturg. 1852, p. 133; Finsch & Hartl. Orn. Centralpolynes. p. xxxvi
    (Hawaii).
Tænioptera obscura, Cass. U.S. Expl. Exp. p. 155, pl. ix. fig. 3 (1858) ; Dole, Proc. Bost. Soc. N. H. 1869,
    p. 300 ; id. Hawaiian Almanac, 1879, p. 48.
Phæornis obscura, Scl. Ibis, 1859, p. 327 (the genus _Phæornis_ established) ; id. Ibis, 1871, p. 360 ; id. P. Z. S.
    1878, p. 347 (♂ and ♀ from Hilo received; colours of eyes and feet); Pelz. Ibis, 1874, p. 462 (on a
    specimen at Vienna) ; Sharpe, Cat. B. iv. p. 5 (1879) ; Wilson, Ibis, 1890, p. 195 (habits) ; id. Aves
    Hawaiienses, pt. i. (Dec. 1890) (fig. ad. bird) (Excellent notes on habits and history. In the notes on the
    latter the author expresses his belief that since Peale's Expedition no collector had obtained specimens,
    while he had before him the notes and descriptions by Sclater and Sharpe on the specimens collected on
    Hawaii by the collector of the 'Challenger') ; Gadow, Aves Hawaiienses, pt. ii. p. 2, and pl. i. figs. 1-5
    (anatomical details and notes on systematical position) ; Perkins, Ibis, 1893, p. 110 (habits).
Eopsaltria (Chasiempis) obscura, Gray, Cat. B. Trop. Is. p. 22 (1859) (Hawaii); id. Hand-l. i. p. 390 (1869).
    (I am of opinion that the "Sandwich Thrush" of Latham, Gen. Syn. iii. p. 39, and Hist. B. v.
    p. 117, 1822 ; _Turdus sandvicensis_, Gm. Syst. Nat. i. p. 873, and Lath. Ind. Orn. i. p. 338 ; _Col-
    lyriocincla sandwichensis_, Gray, Hand-l. i. p. 385, 1869, is a _Phæornis_, and most likely _Phæornis
    obscura_.)

Adult male. Above dusky olive-brown, more greyish on the head and especially the forehead.
Quills dark brown, outer webs strongly washed with the colour of the back, basal part
of inner webs pale buff. Tail-feathers brown, strongly washed with the colour of the
back on the outer webs and on both webs of the central pair. Lower surface ashy grey,
rusty olive-brown on the flanks, and shading into white on the vent. Under wing-coverts
light rusty brown, under tail-coverts buff. Bill black ; legs and feet greyish horn-brown ;
iris dark brown ; soles pale yellow. The total length was measured by Palmer from
7 to 8 inches, in a series of skins it is about 7 inches or little more up to 7·4, while
Wilson gives only 6·75 and Sharpe 7·2 inches. The wing of the male measures in eight
adult and unmoulting males 4 to 4·15 inches, in nine good females 3·85 to 3·95, average
3·9. Tail in the males 2·85 to 2·9, in the females 2·8 to 2·85. Tarsus 1·3 to 1·35,
culmen 0·7 to 0·65, breadth of bill just above nostrils 0·3, middle toe without claw
0·75 to 0·77 inch.

к 2

The *adult female* does *not* differ in colour from the male, as already stated by Sharpe, Cat. B. iv. p. 5. I have before me nearly twenty fully adult specimens, and they do not show the difference in the grey of the underparts described by Scott Wilson.

The *young bird* does not seem to have been fully described till now. I therefore give a figure of it as well as of the young of the other species. The young of *Phæornis obscura* (Gm.) are spotted like young Thrushes; each feather above is bordered with blackish and before the blackish border is a more or less triangular buff spot; the feathers below are buffy white and broadly bordered with blackish brown.

Hab. Island of Hawaii.

THE Hawaiian Thrush is not uncommon on Hawaii and is especially numerous in the district of Kona. Its habits are very familiar. Palmer, Perkins, and Wilson describe the song as somewhat resembling that of the Common Thrush of England, and they all admire it much, although it appears to be inferior to that of the Thrush. On the Sandwich Islands it is said to be by far the finest songster. It also often sings on the wing, as noticed by Perkins and Wilson. Both these observers also mention a peculiar habit this bird has of shivering or quivering with its wings when perched on a branch. The call-note is a clear "tweet," and Wilson mentions a remarkable hissing sound which these birds uttered when approached.

The food consists chiefly of different kinds of berries, and, according to Wilson, particularly those of the Kopiko (*Straussia hawaiiensis*), but also of insects, which Perkins saw them catching on the wing.

No observer seems to have found the nest, but Palmer says he shot a female on September 17th with an egg "almost ready for expulsion," of which, however, he gives no description. This fact does not contradict Palmer's statement that " the breeding-season on the Sandwich Islands is no doubt about the same as in Europe, or a little earlier, namely from February to May; and the above-stated fact is either an anomaly, or perhaps these birds, as an exception to the rule, may breed all the year round."

2. PHÆORNIS MYIADESTINA, Stejn.

KAMAO.

Phæornis myadestina, Stejneger, Proc. U.S. Nat. Mus. 1887, p. 90 (excellent minute description) ; id. op. cit. 1889, p. 383 ; Wilson, Ibis, 1890, p. 195.

Phæornis myiadestina, Scl. Ibis, 1888, p. 143 ; Wilson, Aves Hawaiienses, pt. ii. text & pl.

Adult male. Above dull hair-brown with an olive tinge, different from the colour of the back of *Phæornis obscura,* and somewhat more olive ; sides of the head dull tawny, bordered with dusky. Quills deep brown, edged externally with rusty olive, but the second to the fifth greyish brown near the tips. The inner primaries and most of the secondaries have a very peculiar pattern ; the shaft is brown above, whitish below ; the inner webs are dark brown, except on the basal third, which is very pale buff. The outer web is bright russet, except near the tip, where it is only margined with russet, and crossed behind the middle by a blackish band. Lower parts of a light smoky grey, lighter than in *Ph. obscura,* and somewhat motley in appearance, of which character nothing is seen in *Ph. obscura,* shading into white on the vent. Middle tail-feathers like the back ; three outer tail-feathers with a white mark at the tip, very long and gradually shading into the brown on the outer pair, very small on the third pair. Under tail-coverts pure white. Bill black ; iris brown ; tarsi and feet dark brown, soles pale yellow. Total length nearly 8 inches, wing 4·05 to 4·25, tail 3·3 to 3·6, culmen 0·57 to 0·6, breadth of bill just behind nostril 0·35, middle toe without claw 0·7, tarsus 1·2 to 1·25.

The *adult female* is entirely similar to the male in colour, and my series does not bear out any constant difference in size.

Young birds are similar in colour to those of *Phæornis obscura* (Gm.), but are distinguished from them by the whitish marks on the outer tail-feathers and the much broader bill, the latter characterizing the Kauai bird.

The young bird is figured on the Plate. It is a little older than that of *Ph. obscura,* I believe, but the youngest I possess. That of *Ph. lanaiensis,* Wilson, figured on the same Plate, is the youngest of the three.

Hab. Island of Kauai.

THE Kauai Thrush is not rare on Kauai.   According to Mr. Knudsen it sometimes sings on the wing, as before stated of *Phæornis obscura* (Gm.).

Palmer describes it as being a quiet bird and not shy.   Its rather melancholy and often repeated call-note attracts one's attention.   Its song reminded him of that of an English Thrush, but it was less powerful, although it could be heard at a great distance, and, in Palmer's opinion, was sweeter.

Its food consists of berries, but caterpillars and insect-remains were also found in its stomach.

3. PHÆORNIS LANAIENSIS, *Wilson.*

OLOMAO or OLOMAU.

Phæornis lanaiensis, Wilson, Ann. & Mag. Nat. Hist. ser. 6, vol. vii. p. 460 (1891) (Lanai); id. Aves Hawaiienses, pt. ii. text & pl. (1891).

Adult male. Above dusky olive-brown, the head more blackish brown. The wings like those of *Ph. myiadestina,* Stejn., but the outer edges of the primaries more brown and less rusty, the bases of the same less rufous. Chin, throat, and breast pale grey, with a faint isabelline tint, passing into the nearly pure white abdomen. Under tail-coverts pale buffy white. Outer pair of rectrices brown; outer webs brownish buff with a very narrow outer marginal line and a small tip of pale cinereous, but without white. Middle rectrices the same colour as the back, the following ones edged with the same colour on the outer webs. Iris hazel; bill black; tarsi and feet dark brown, soles yellowish. Total length about 7½ inches and more as measured by Palmer, but only about 7 in skins. In fourteen specimens from Lanai the wings measure 3·49 to 3·75 inches (average 3·65), while in thirteen from Molokai they measure 3·65 to 3·9 (average 3·8), tail 2·9 to 3·1, tarsus 1·3 to 1·35, culmen 0·6, breadth of bill just behind nostrils 0·27 to 0·3, middle claw without toe 0·75.

This species is easily distinguished from *Ph. obscura* (Gm) by its much smaller size and whiter underparts, and from *Ph. myiadestina,* Stejn., by the absence of the conspicuous white markings in the tail and the smaller size. The bill is somewhat between those two species in width, but varies a little, and it is not so much pointed as in *Ph. obscura.*

The *adult female* is not to be distinguished from the male if my birds are properly sexed.

The *young bird* is like those of the allied species, and I trust Mr. Keuleman's well-executed figure will give a better idea of it then a long description.

Hab. Lanai and Molokai.

THE specimens from Lanai, the island from which Wilson's type came, are as a rule much whiter below, and the majority of them have the brown of the back somewhat less bright. As the measurements of their wings show, there is also a decided tendency to longer wings in the Molokai birds, but the longest of those from Lanai surpass several of those from Molokai

There is nothing extraordinary in it if we assume that the *Phæornis*, inhabiting also low-lying regions, crosses from Lanai to Molokai, and therefore is the same species on both islands, while some other birds of the higher regions have developed into different species by isolation; but at the same time the forms from both islands could be different, since several birds of one island are not to be found on the other. The birds from each island may still be in the process of developing into different forms, and might be separated into subspecies of little value; but the differences being so slight and not quite constant at all, I prefer to use Wilson's name for both.

The Olomao, as it is called, both on Lanai and Molokai, is not rare on both these islands, and Palmer saw it in the lowland as well as at the highest elevations. In the stomachs he found seeds and berries of different plants. When seen on a tree they were generally shaking their wings or "trembling," as Palmer calls it. They have that clear call-note peculiar to this group, and also another deep hoarse cry. Their song is " of a jerky nature," and consists of several clear notes. Nest and eggs unknown.

4. PHÆORNIS PALMERI, spec. nov.

PUAIOHI.

Differt ab omnibus speciebus generis *Phæornis* alis multo brevioribus, rostro graciliore, pedibus pallidis.

Adult female. Above dusky olive-brown, like *Ph. obscura*, a light loral spot before the eye. Chin, throat, and sides of the body ashy brown, somewhat hair-brown; middle of abdomen whitish. Wing-pattern similar to that of *Ph. myiadestina*, Stejn.; outer edges of primaries yellowish brown. Outer pair of tail-feathers with the inner web whitish except the edge near the base, outer web pale drab; second pair drab, with a long white mark on the inner web, becoming quite narrow on approaching the base; third pair with a small white tip; central pairs of the colour of the back. Iris dark brown; bill black; legs and feet light flesh-colour. The total length is given by Palmer as, measured in the flesh, 7,¹₆ inches, but in skins it is apparently not more than 6½ inches. Wing 3·45 inches, tail 2·4, tarsus 1·15, middle toe without claw 0·65, culmen 0·7, breadth of bill just above nostrils 0·25.

Young. Feathers above brown, with a broad blackish border at the tip, and before this a creamy buff more or less triangular spot. Feathers below ashy brown at base, then light cream-colour and broadly bordered with blackish brown, these borders broadest and blackest on the breast, while they become indistinct towards the throat and chin and again near the anal region. Under tail-coverts creamy buff. Iris dark hazel; tarsi grey; soles pale yellow; bill black. Wing 3·45 inches, tarsus 1·26, culmen 0·7, breadth of bill just above nostrils 0·25.

Hab. Halemanu, Kauai.

THE type was shot on March 21st, 1891, at Halemanu, Kauai, on the property of Mr. Knudsen. I was in hopes of getting some more adult birds in better condition, as the one I had was badly damaged by rats; in 1893, therefore, I sent Palmer again to Kauai, where he procured two more specimens, but both young spotted ones. These birds must be very rare, as they were only found at Halemanu.

Palmer says in his diary in connection with this bird :—" This Thrush is much smaller than the common Kamao. When I first saw it it was sitting on a Koa tree not more than five feet from the ground, and keeping very quiet; I for a moment thought it was a

L

big 'Elepaio' (*Chasiempis*), but soon saw it was something else. When I had shot it I was aware it was something new to me. Between the bill and the eye is a white mark." Of this white mark only an indication is left, as the head is partly destroyed. The faulty condition of the skin accounts for my having left some parts of it undescribed; the bill, wings, tail, feet, the abdomen, breast, back, and rump, however, are complete.

In the stomach of the adult bird Palmer found seeds and caterpillars, in those of the young birds insects and green pieces of some seed.

The adult bird is the same species as that mentioned by Wilson, in 'Aves Hawaiienses,' Part II., under *Phæornis myiadestina*. He says:—"In a letter received lately from my friend Mr. F. Gay, he raises the question of there being another species of *Phæornis* found on Kauai, and his remarks on a skin recently obtained by a collector are as follows :— 'It appeared to me to be a species of Kamao, the only difference being a narrower bill, lighter coloured feet, and a smaller body, and, according to the collector, lighter coloured feathers about the head. Our natives always said there was a different variety called the 'Puaiohi,' which they said had a different note from the common Kamao. I never believed much in what the natives said about it, as the Kamao varies so much in colour and spots. [Here Mr. Gay is obviously right in referring to the differences between old and young in 'colour and spots,' but, as my description shows, my *Phæornis palmeri* is quite distinct.—W. R.] This bird may be more common on the windward side of the island, as the name of Puaiohi is more commonly used there than here.' Mr. Gay adds : 'The single skin I saw was a poor one, having been partly eaten by rats.'"

The skins of the two young birds, I may add, are in very good condition.

THE GENUS CHASIEMPIS.[1]

THIS genus was established by Cabanis in the 'Archiv für Naturgeschichte,' 1847, p. 207.

The members of this genus are true *Flycatchers*, having broad and soft bills, and the gape beset with long and strong bristles. The tarsus is long and slender. The first primary is about half as long as the second, the second about one quarter of an inch shorter than the third; the fourth, fifth, and sixth are equal and longest. The tail is about as long as the wing, and the rectrices are pointed.

The sexes are similar, but the plumage of the young differs remarkably from that of the adult bird.

The differences between the old and young birds have led many authors to describe them under different names, so that as many as five or six species were believed to exist, while some other distinguished ornithologists referred all to one species. I have already pointed out that three species are found on three different islands (Bull. B. O. C. 1893, vol. i. p. lvi). No species of the genus is known to occur out of the Hawaiian Islands, so we may fairly regard it as entirely peculiar to this Group.

Sharpe (Cat. B. iv. p. 232, 1879) includes in the genus *Chasiempis* the *Monarcha dimidiata*, Hartl. & Finsch, from Rarotonga, Samoa (P. Z. S. 1871, p. 28), but with our present knowledge of the genus *Chasiempis* he would certainly alter his opinion. The *Monarcha dimidiata* of Hartl. & Finsch has the peculiar short scaly-looking feathers on the forehead which occur in several genera of Flycatchers distinctly developed (though not so much as in other genera), while nothing of the sort is seen in *Chasiempis*. The sexes in *Monarcha dimidiata*, H. & F., are of a totally different colour, while they are alike in *Chasiempis*. The bill is stronger and higher in *Monarcha dimidiata*, H. & F., than in *Chasiempis*; the tail-feathers in the former seem to be rounded, while they are distinctly pointed in the latter. Through the kindness of Professor A. Newton I was enabled to compare the two specimens in the Cambridge Museum, which also served Dr. Sharpe for his descriptions in the 'Catalogue of Birds.'

The nest is open and the eggs are spotted.

[1] Not *Chasiempsis*, as erroneously spelt by Hartlaub, Gray, and Dole.

Key to the Genus Chasiempis.

A. Wing-coverts spotted with tawny ochraceous; base of lower mandible
   light. (Young.)
      a. Lighter tawny ochraceous below. . . . . *C. sandwichensis*, juv.
      b. Deeper tawny ochraceous below. . . . . *C. gayi*, juv., & *C. sclateri*, juv.
B. Wing-coverts spotted with white; base of lower mandible dark and con-
   colorous with tip. (Old.)
      c. Above greyish . . . . *C. sclateri*, ad.
      d. Above brown.
         a'. White tips on outer tail-feathers more than half an inch long . (*C. sandwichensis*.)
           a². Forehead white, and breast almost entirely white . . . *C. sandwichensis*, ad.
           b². Forehead and breast tawny ochraceous . . . . *C. sandwichensis*, med.
         b'. White tips on outer tail-feathers less than half an inch long . *C. gayi*, ad.

5. CHASIEMPIS SANDWICHENSIS (Gm.).

ELEPAIO OF HAWAII.

Sandwich Flycatcher, Lath. Gen. Syn. iii. p. 344 (1783) (Sandwich Islands, in coll. Sir Joseph Banks); id. Hist. B. vi. p. 211 (1823).

Spotted-winged Flycatcher, Lath. Gen. Syn. iii. p. 345 (1783) (Sandwich Islands, Mus. Lev.); id. Hist. B. vi. p. 212 (1823).

Muscicapa sandwichensis, Gm. Syst. Nat. i. p. 945 (1788) (ex Latham); Gray, Gen. B. i. p. 263 (1817).

Muscicapa sanduicensis, Lath. Ind. Orn. ii. p. 479 (1790).

Muscicapa maculata, Gm. Syst. Nat. i. p. 945 (1788) (ex Latham) (nec Muscicapa maculata, P. S. Müller, 1776); Lath. Ind. Orn. ii. p. 480 (1790); Gray, Gen. B. i. p. 263 (1847); Dole, Proc. Bost. Soc. Nat. Hist. xii. p. 299 (copy of Gmelin's diagn. and remark "This is perhaps an uncertain species"); Hawaiian Almanac, 1879, p. 48.

Cnipolegus sp. 1238, Scl. Cat. Amer. B. p. 203 (1862).

Eopsaltria (Chasiempis) sandwichensis, Gray, Cat. B. Trop. Is. p. 21 (1859); id. Hand-l. i. p. 390 (1860); Dole, Proc. Bost. Soc. N. H. 1869, p. 300; id. Hawaiian Almanac, 1879, p. 48.

Eopsaltria (Chasiempis) maculata, Gray, Cat. B. Trop. Is. Pac. Or. p. 22 (1859); id. Hand-l. i. p. 390 (1860).

Chasiempis sandvicensis, Scl. Ibis, 1871, p. 360; id. P. Z. S. 1876, p. 316 (recorded from Hilo, Hawaii); id. Rep. Voy. 'Challenger,' Zool. ii. pt. viii. p. 94 (1887) (Hilo, Hawaii); Sharpe, Cat. B. iv. p. 232 (1879).

Chasiempis sandwichensis, Finsch & Hartl. Orn. Centralpolynes. p. xxxiv; Scl. Ibis, 1885, p. 18, pl. i.; Stejn. Proc. U.S. Nat. Mus. 1887, p. 89; Scl. & Saund. Ibis, 1888, p. 143; Wils. P. Z. S. 1891, p. 166; Perkins, Ibis, 1893, p. 109.

Chasiempis ridgwayi, Stejn. Proc. U.S. Nat. Mus. 1887, p. 88 (based on fig. 1, Ibis, 1885, pl. i., which is drawn from a specimen from Hawaii); Berl. & Leverk. Ornis, vi. p. 23 (1890) (united with the species from Oahu); Wils. P. Z. S. 1891, pp. 164, 165, 166; Rothsch. Bull. B. O. C. i. p. lv (1893); id. Ibis, 1893, p. 570.

Chasiempis ibidis, Stejn. Proc. U.S. Nat. Mus. 1887, p. 88 (based on fig. 2 in Ibis, 1885, pl. i., which is that of a young bird); Scl. & Saund. Ibis, 1888, p. 143 (criticism); Berl. & Leverk. Ornis, vi. pp. 2–4 (1890) (united with the form from Oahu); Wils. P. Z. S. 1891, p. 166 (stated to be from Oahu, but the young bird in the British Museum, from which the figure in Ibis, 1885, pl. i. fig. 2, is taken, had no exact locality, and after having compared it with my series from the different islands, I find it agrees best with the young Hawaiian bird, so there is no doubt it really came from Hawaii).

Final plumage of adult birds. Forehead, lores, and superciliary stripe white, more or less spotted, the bases of the feathers being black. Above from the head to the back dark olive-brown ("bistre" in Ridgway, Nomencl. Colors, pl. iii. fig. 6), spotted with white on the hind neck and lower back. Rump and upper tail-coverts pure white, base of the feathers black. Wing-coverts (except primary-coverts) and inner secondaries black, broadly tipped with white. Primary-coverts black. Quills blackish brown, narrowly edged on the outer webs with olive-brown, distinctly edged with white on the inner webs, the first ones only at the basal parts. Rectrices black; outermost pair with half of the outer web to the tip white, and with the tip of the inner web for one-fourth to at least one-third white. The following feathers with a large portion of the inner web and a much smaller portion of the outer web white, these spots decreasing in size until the central pair is reached, where only quite narrow white tips are visible. Feathers of the underparts black at their bases, white at the tips; the chin remains quite black, then the white tips appear, so that the throat is varied white and black, then the white tips become so broad that the entire lower throat, breast, abdomen, and under tail-coverts are pure white. Sides of breast and body more or less washed with tawny olive. Under wing-coverts spotted brown and white, the bases being deep brown, the tips broadly white. Feathers of the thighs black with white tips. Iris dark brown; upper mandible slaty black, under mandible slaty blue; legs and feet slaty blue.

Intermediate plumage. (The birds in this plumage are adult, inasmuch as they are generally known to breed already in this stage, but it is not the final plumage described above.) Above dark olive-brown with a rufous shade, thus looking a faint bit brighter than the birds described above, which may be slightly faded, from not being so fresh moulted; lores, forehead, and a more or less distinct line above and behind the eyes tawny ochraceous, sometimes intermixed with whitish or white; rump and upper tail-coverts white, the bases of the feathers black. Quills dark brown, narrowly margined with pale tawny on the outer webs, with creamy buff on the inner webs, more so towards the bases; secondaries tipped with white. All upper wing-coverts, except the primary-coverts, broadly tipped with white. Rectrices broadly tipped with white, in the same manner as described above in the final plumage, but the white colour does not extend so far. Chin and throat spotted black and white, in younger specimens appearing almost quite white; in most specimens the chin almost black; this variation is caused by the feathers being black at base and more or less broadly tipped with white. Underparts below the throat dark tawny ochraceous, with a broad more or less irregularly-defined white patch from the upper breast to the under tail-coverts, which are also white. Feathers of the thighs black, tipped with white. Under wing-coverts deep brown and white. Iris dark brown; upper mandible slaty black, under mandible slaty blue; feet slaty blue.

Young. Above tawny ochraceous brown, and pale tawny on rump, browner on the head and upper tail-coverts. Quills dark brown, with pale borders on outer webs and bordered with buff on inner webs. Wing-coverts deep brown, broadly tipped with bright ochraceous buff. Tail-feathers deep brown, with pale borders to the outer webs; outermost pair with a small white spot on the outer web and a large white spot on inner web, the next pairs with white on inner webs only; all these white spots much less extended than in adult birds and decreasing in size to the middle, so that the central pair has no white at all. Underparts of a tawny buff, passing into the white on the middle of the abdomen; some specimens are much whiter than others. Under wing-coverts buff.

Iris dark brown. Upper mandible deep brown ; lower mandible brown at tip, creamy yellowish at base. Legs and feet slaty blue, but less bright and paler than in the adult birds.

Measurements of adult specimens :—Total length about 5¾ to 6 inches, wing 2·9 to 3·05 (average 2·95), tail 2·5 to 2·7, culmen 0·5, tarsus 0·85 to 0·89, middle toe without claw 0·49.

Hab. Island of Hawaii.

THE "Hawaii Flycatcher" is one of the commonest birds on Hawaii. Perkins says (*l. c.*) :— " The single species of *Chasiempis* (*C. sandwichensis*) found in Kona is one of the commonest birds, extending its range from about 1400 feet to the limits of proper forest on Mauna Loa, and also high up Hualalai. It is a pretty species, with great difference in colour between the young and the old ; in this respect resembling the Oahuan species, in both cases snowy-white feathers in the adult taking the place of rufous feathers in the young. On one occasion I obtained a *rufous* female along with the ordinarily coloured male and their young, just out of the nest ; but this I believe to be a very unusual circumstance, the female probably being a young specimen breeding before it had assumed adult plumage. The male was quite typical. The adult female closely resembles the male, but the latter is black under the gorge, and I believe the feathers become blacker in the breeding-season, just as the hackles of our common Starling lose their white tips at this season, but the change is less marked in *Chasiempis*. I have no doubt all the Kona birds belong to one species, the rufous birds nearly always having an entirely or partially unossified skull. These birds live chiefly on insects and their larvæ. The insects they often take on the wing, their beaks closing with a very audible snap. . . . . They frequently descend to the ground or on to fallen trees, where they get wood-boring larvæ or small myriapods. With reference to this habit I had the following anecdote from a native woman in Kona :—' Of all the birds the most celebrated in ancient times was the *Elepeio*, and for this reason. When the old natives used to go up into the forest to get wood for their canoes, when they had felled their tree the *Elepeio* would come down to it. If it began to peck it was a bad sign, as the wood was no good, being unsound ; if, on the contrary, without pecking, it called out ' *Ona ka ia*,' ' Sweet the fish,' the timber was sound.' The names *Elepeio* and *Ona ka ia* (pronounced *ŏnŏkāiā*) are both creditable word-imitations of the cry of *Chasiempis* under various emotions, here presumably of disgust."

Palmer's observations agree with those of Mr. Perkins, but it seems that he found them even lower than 1400 feet. I do not think there is any remarkable difference between the sexes, at least my series at present does not bear out any. There is no doubt that this species sometimes breeds in the quite immature plumage, as also Palmer saw pairs of which one was white-rumped, the other rufous-rumped. There are several specimens among my birds which distinctly show an intermediate plumage between the first two described above.

Although Sclater and Berlepsch and Leverkühn maintained that the white-rumped and rufous-rumped birds belonged to one species, it was not before the field-observations of Palmer and Perkins on these islands that the truth was brought to light, for even some of the best authorities on Hawaiian birds, namely Stejneger and Wilson, believed them to be distinct species.

I trust my four figures will illustrate the different plumages of this bird.

Palmer did not find any nests or eggs of this species, but Perkins found several nests, of which he says: " It is small, very neat and compact, placed from 10 to 30 feet from the ground, and generally well concealed."

6. CHASIEMPIS GAYI, *Wilson.*

ELEPAIO OF OAHU.

—

Chasiempis sandvicensis, Cab. Arch. f. Naturg. 1847, p. 208 (nec Gmel.) (genus *Chasiempis* established: the specimens here called *Ch. sandvicensis* were collected by Deppe on Oahu) ; Licht. Nomencl. Av. Mus. Berol. p. 19 (1854).

Chasiempis sandwichensis, Pelz. Ibis, 1874, p. 462 (Enero, Oahu, 1837, Deppe coll.) ; Berl. & Leverk. Ornis, vi. p. 2, pl. i. fig. 3 (1890) (The authors had before them a fine series from the Museum at Kiel, collected by Behn on Oahu. As all their specimens came from Oahu, and, as they compared them with the published descriptions and figures of all the other species, they were induced to unite all the different species from the various islands into one. They were right, on the other hand, in decidedly stating that the sexes were alike and that the white-rumped were the adults of the tawny-rumped birds) ; Rothsch. Bull. B. O. C. i. 1893, p. lvi, & Ibis, 1893, p. 570 (the name of *sandwichensis* was by mistake referred to the form from Oahu instead to the one from Hawaii).

Chasiempis gayi, Wilson, P. Z. S. 1891, pp. 165, 166 (Oahu).

Adult male. Above brown, slightly tinged with tawny ochraceous, more distinctly so on the head ; forehead, ear-coverts, and an ill-defined line above the eye ochraceous buff, becoming almost buffy white on the lores and round the eyes. Rump and upper tail-coverts white, formed by the feathers being blackish grey at the bases and largely tipped with white. Quills deep brown, narrowly edged with pale brown on the outer webs and broadly edged with white inwardly. Rectrices blackish brown, largely tipped with white ; this colour occupying both webs at the tips on the outer pair, becoming restricted to the inner webs and less extended to the middle and leaving the central pair without white. Feathers of the chin, throat, and upper breast black, those of the chin and upper breast not or very narrowly, those of the throat broadly edged with white, so that there appears a large white patch on the throat, while the chin and upper breast appear black. Abdomen and under tail-coverts white, the bases of the feathers blackish; breast and sides of body washed with brown. Upper wing-coverts deep brown, all except the primary-coverts tipped with white. Under wing-coverts mixed blackish and white, the bases being blackish, the tips white ; the under wing-coverts often appear almost quite white. Iris hazel ; bill blackish ; tarsi slate-blue. .

Young. Above tawny ochraceous brown, bright tawny ochraceous on the rump and upper tail-coverts. Wing-coverts, except the primary-coverts, tipped with bright tawny ochraceous ; secondaries and inner primaries margined with the same colour on the outer webs, while edged with whitish inwardly. Below white, strongly washed with tawny ochraceous on the throat, breast, and sides of body, as well as on the under tail-coverts; while the middle of the abdomen and vent remains more or less pure white. Tail-feathers as in the

M

adult bird, but the white colour less pure and less in extent.   Iris brown; bill deep blackish brown, base of lower mandible light-coloured.

Nestling. Above similar to the young just described and showing the rufous colour on the rump, but some blackish tips of the feathers produce a somewhat spotted appearance. Feathers below white, more brownish on the breast, and mostly slightly bordered with blackish, producing a somewhat spotted appearance.

The total length was measured by Palmer to be as great as 5¾ in the flesh, while in the skins it is only about 5 to 5½ inches.   Wing 2·6 to 2·8, tail 2·5 to 2·7, tarsus 0·9, culmen 0·45.

I cannot see any remarkable difference between males and females, but perhaps the females have a slightly shorter wing as a rule.

Hab. Oahu.

THE observations of Palmer and his assistant Wolstenholme leave no doubt that the white-rumped and the rufous birds are old and young of the same species, and, moreover, Palmer has sent two specimens which distinctly show a resemblance between these two different plumages. On the other hand, Palmer and his assistant have also seen rufous-rumped birds on the nest and feeding their young, although less frequently than white-rumped birds. From these observations it is obvious that the white rump and white wing-spots (*i. e.* the "adult plumage") are not gained in the first year and that these birds, like the *Ch. sandwichensis* of Hawaii and probably also *Ch. sclateri* of Kauai, sometimes breed before they have gained the adult plumage, as is sometimes the case with other birds elsewhere, especially *Rapaces.* Palmer, in his diary of April 1893, says :—" I saw a pair, one bird rufous-rumped and the other white-rumped, build a nest.   I think they change their rufous plumage into the white in the second year, if not much later."

Chasiempis gayi, Wils., is one of the commonest, if not the commonest, of all the small native birds on Oahu.   It is an inhabitant of forest and is generally seen flitting about from tree to tree, very often taking insects on the wing.   They are frequently seen with drooping wings and the tail elevated almost vertically.   Like the Flycatchers of Hawaii and Kauai, these birds are very tame and often come close up to the observer ; they also eagerly follow the call-note when properly imitated by men.   The call-note is a rather shrill forced note and sounds like "*Pép-pá-ké-o.*"   The food consists of insects, which are mostly captured on the wing, but flightless insects are also found in the stomachs, such as caterpillars, grubs, and spiders.

Our early spring months, March and April, seem to be their breeding-time.   A number of nests were found in March and April.

The nests were found on young, tall, broad-leaved trees, generally at heights of from 10 to 30 feet, placed in a kind of fork among the thin branches.

The nest is deep and cup-shaped, built of very fine roots and moss and lined with roots and fine haulms, outside beautifully ornamented with light-coloured lichens.   The cup is nearly or fully 1½ inch deep, and measures about 2 inches across, the walls being nearly ¾ of an inch thick.   The egg is ovate and measures 1·25 × 1·11 inch: it is white, covered with small spots and blotches of a brownish red (brick-red) ; the white ground-colour is not at all shaded with bluish or greenish but is, if anything, yellowish.   The egg resembles much that of a Common Titmouse.

7. CHASIEMPIS SCLATERI, *Ridgw.*

ELEPAIO OF KAUAI.

Chasiempis sclateri, Ridgw. Proc. U.S. Nat. Mus. 1882, p. 337 (Kauai) ; Scl. Ibis, 1885, pp. 17, 18, 19 (united with *Ch. sandwichensis*) ; Stejn. Proc. U.S. Nat. Mus. 1887, pp. 86, 87 ; id. ibid. 1889, p. 382 ; Berl. & Leverk. Ornis, vi. (1890) pp. 2, 3 (united with *Ch. sandwichensis*) ; Wilson, P. Z. S. 1891, pp. 164, 165, 166 ; Rothsch. Bull. B. O. C. i. 1893, p. lvi ; id. Ibis, 1893, p. 570.
Chasiempis dolei, Stejn. Proc. U.S. Nat. Mus. 1887, p. 90 (Kauai) ; id. ibid. 1889, p. 382 ; Wils. P. Z. S. 1891, pp. 164, 165, 166.

Adult. Above dark smoky grey with a brownish tinge ; forehead and space round the eyes washed with pale buff, more whitish on the lores and round the eyes ; rump and upper tail-coverts white, the feathers being blackish grey, largely tipped with white. Quills blackish brown, narrowly margined with pale brownish ash on the outer webs, and margined with white on the inner webs ; secondaries fringed with white at tips. Tail blackish brown ; outer tail-feathers on both webs, the rest on inner webs only, tipped with white, these white tips decreasing in size towards the middle, so that the central pair has no white tips. Wing-coverts, except the primary-coverts, tipped with white. Below whitish, strongly tinged with brownish buff on the throat and breast and sides of body, darkest on the breast. Under tail-coverts, under wing-coverts, and axillaries white, deep grey at bases like all the feathers. Iris brown ; bill blackish, lower mandible with a bluish tinge ; feet bluish black.

Young. Above deep tawny brown, tawny ochraceous on the forehead and sides of head and round the eyes ; rump and upper tail-coverts bright tawny ochraceous. Quills as in the adult birds, but the outer edges more tawny. Tail-feathers as in the adult birds, but the white at the tips less extended and more or less tinged with brownish. Wing-coverts deep brown, broadly tipped with tawny ochraceous. Chin, throat, upper breast, and flanks tawny ochraceous, shading off into the white of the middle of the abdomen. Under wing-coverts pale ochraceous buff ; under tail-coverts pale buffish white. Iris brown ; upper mandible blackish, lower mandible brown at tip and yellowish cream-colour at base ; feet bluish black or deep brown.

Measurements :—Total length about 5·5 inches ; wing, in adult males (one dozen measured), 2·8 to 2·9 inches, in adult females (five measured) 2·6 to 2·65 ; tail, in males 2·5 to 2·75, in females 2·4 to 2·7 ; tarsus 0·85 to 0·95, culmen 0·46 to 0·49.

Hab. Kauai.

N 2

AFTER the many careful observations of Mr. Palmer and Mr. F. Gay of Kauai, there can be no doubt whatever that the white-rumped and the rufous-rumped birds are old and young of the same species, although called by different names (*Elepaio* and *Apekepeke*) by the natives. They keep together and have the same notes.

This species is regularly distributed over the island of Kauai, and is found in the lower and higher parts of the island. They are tame and fearless.

Palmer found a nest on which he caught the adult *male*. This fact and also that he saw both parents feed their young clearly indicate that both sexes take part in incubating the eggs and feeding the young.

The nest was placed in the fork of some thin branches of a tree, about fifteen feet from the ground, and was rather exposed. It was built of moss and fine ferns, open above, and contained two eggs, which were whitish and spotted with reddish brown. Unfortunately both were smashed on the journey down the hills.

These birds feed on insects, which they often catch on the wing. Palmer saw them swallowing enormous caterpillars, for their size, and pulling to pieces large moths while holding them down with their claws.

ON THE

GENERA HEMIGNATHUS AND HETERORHYNCHUS.

THESE two genera are entirely confined to the Sandwich (Hawaiian) Islands, where they were discovered during Cook's exploration of the islands in 1779.

The first species described was *Hemignathus obscurus* (Gm.), which was diagnosed by Latham, in his 'General Synopsis of Birds' in 1782, as the "Hook-billed Green Creeper," which description was translated by Gmelin in his edition of the 'Systema Naturæ' in 1788, who gave to the bird the name *Certhia obscura*.

In 1839, in the 'Abhandlungen der königlichen Akademie der Wissenschaften zu Berlin, aus dem Jahre 1838,' Lichtenstein described and figured two of these birds, for which he founded the genus *Hemignathus*, and which he called *Hemignathus obscurus* and *Hemignathus lucidus*. Both these species were obtained on the island of Oahu, while Latham's (and Gmelin's) *Certhia obscura* came from Owhyhee (Hawaii).

The long-billed species from Oahu was named *Hemignathus lichtensteini* by Mr. Scott Wilson in 1889; but this name cannot stand, as long before the same species was named *Drepanis (Hemignathus) ellisiana* by Gray in his 'Catalogue of the Birds of the Tropical Islands of the Pacific Ocean.'

In 1839 Lafresnaye, in the 'Magasin de Zoologie,' described Lichtenstein's *Hemignathus lucidus* under the name of *Heterorhynchus olivaceus*, forming for this short-billed species the new genus *Heterorhynchus*.

The type of the genus *Heterorhynchus*, therefore, is undoubtedly *H. olivaceus = lucidus*; but it is perfectly obvious to anyone reading the description of Lichtenstein that the type of *Hemignathus* is his *H. obscurus*, and not *H. lucidus* as stated in the 'Catalogue of Birds,' vol. x., and elsewhere. The whole of the description, translated from the German, will be found under my notice of *H. ellisianus*. See also Stejneger's footnote in the 'Proceedings of the U.S. National Museum,' 1887, p. 93.

Mr. Scott Wilson has unfortunately applied Lafresnaye's name of *Heterorhynchus olivaceus* to the species of this group from Hawaii, but I cannot understand anybody possibly mistaking Lafresnaye's figure and description for a bird having no sign of the heavy yellow superciliary stripe and a perfectly straight under mandible. However, this is fully explained in the description of *Heterorhynchus wilsoni*.

So far as can be ascertained at present, there are four described species of *Hemignathus* and four of *Heterorhynchus*; but I have strong presumptive evidence that at one time there was a fifth species of *Heterorhynchus* on the island of Lanai, while Judge Dole records

specimens of *Hemignathus obscurus* from Maui. Now we know that the *Heterorhynchus* of
Maui is distinct; hence we must believe that both Mr. Scott Wilson and my collector Henry
Palmer failed to procure this bird on Maui, supposing Dole was right in his statement, so
that there would remain a new *Hemignathus* to be described [1].

Dr. Stejneger says in the 'Proceedings of the U.S. National Museum,' 1887, under
" *Hemignathus obscurus*," p. 93 :—" Generally this bird is referred to the same genus as
Hemignathus lucidus, but with doubtful propriety, as I think. The bills in this group of
birds have served as the chief character for the establishment of genera, and if we recognize
more than one genus of Drepanine birds, the two species of *Heterorhynchus*, with their
unique bills, should certainly stand alone. With specimens in hand, Mr. Sharpe would never
have included *H. obscurus* in a genus which he defines as having the ' upper mandible nearly
twice the length of the lower one,' for the species in question has ' both mandibles of nearly
the same length,' the difference being about one-tenth the chord of the exposed culmen."

I quite agree with Dr. Stejneger, and now, as we have eight species, of which four have
both mandibles about equal in length, and four have the upper mandible about twice the
length of the lower one, I think it is quite justifiable to split them into these two well-
defined genera.

These two genera belong to the family *Drepanididæ*, so called from *Drepanis pacifica*,
the Great Hook-billed Creeper of Latham. This family contains the bulk of the very curious
Passerine birds of the Hawaiian Islands, and has been placed in the neighbourhood of the
Cærebidæ and *Dicæidæ*. From observations made on the anatomy of a recently-killed
Drepanis pacifica, I have reason to doubt its close relationship to the other members of the
Drepanididæ; this question I leave others to decide, but I shall again refer to it under
Drepanis pacifica in the third part of this work.

Generic Characters of Hemignathus.

Bill very long and curved, not serrated, the upper mandible slightly longer than the
    lower. The cutting margins of the under mandible curved inwards, and therefore
slightly overlapped by the upper mandible. Tongue as long as the mandible, forming a
tubular brush. Nostrils protected by an operculum and bare of feathers. Rictal bristles
absent. Feet of medium size and strong. Tarsus covered in front with four or five long
scales, which sometimes almost entirely or quite fully coalesce with each other, and also
show much tendency to fuse with the lateral rows.

The tip of the wing is formed by the third, fourth, and fifth primaries (the first being
quite rudimentary); the sixth is slightly shorter than the fifth, and a little longer than
the second [2]. Tail about equal, short and soft.

---

[1] As I find an intimate connection between the avifauna of Maui, Lanai, and Molokai, and as Henry Palmer confirms
my ideas about distribution in respect to altitude, I feel bound to suggest that the species of *Hemignathus* on Maui and
the *Heterorhynchus* on Lanai may be only my *Heterorhynchus affinis* and *Hemignathus lanaiensis*.

[2] I may here say that I count from the outermost primary inwards, as all ornithologists used to do, and that I do not
follow Gadow and other anatomists in counting from the first primary which follows the secondaries. Gadow's method
is quite natural and logical, but wholly unpractical for ornithologists in handling skins and stuffed birds.

Plumage rich and soft, greenish; females and young birds duller.

Food insects; habits arboreal.

Confined to the Sandwich Islands.

Nothing is known concerning the nidification, but a peculiar odour which these birds emit when freshly killed leads to the surmise that they may breed in hollow trees, which seems the more probable as they usually obtain their food from clefts and corrugations in the bark, and spend more of their time about the trunks and branches than amidst the foliage.

8. HEMIGNATHUS PROCERUS, *Cab.*

Hemignathus obscurus, Stejneger, Proc. U.S. Nat. Mus. 1887, p. 93 (nec Gmelin l) (Kauai).
Hemignathus procerus, Cab. J. f. O. 1889, p. 331; Wilson & Evans, Aves Hawaiienses, pt. iii. text & pl. (fig. ♂ and ♀).
Hemignathus stejnegeri, Wilson, Ann. & Mag. N. H. ser. 6, vol. iv. p. 401 (Kauai) (Nov. 1889); id. Ibis, 1890, p. 100, pl. vi. fig. 2 (♂); Stejn. Proc. U.S. Nat. Mus. xii. p. 384 (1889).

Adult male. Above bright yellow-olive (something between Ridgway's "olive-green," Nomencl. of Colors, pl. x. fig. 18, and his "wax-yellow," pl. vi. fig. 7, as justly remarked by Stejneger in Proc. U.S. Nat. Mus. xii. p. 384); the top of the head spotted, caused by the deep brown feathers being edged with the colour of the back. Lores black. A distinct narrow bright yellow superciliary stripe. Quills deep brown, edged with the colour of the back on the outer webs; rectrices similar. Below from the chin to the breast and along the sides of the body olive-yellow, shading into bright canary-yellow on the abdomen. Feathers on the tibia dull white. Under wing-coverts and axillaries white, washed with yellow. Iris brown; bill black, paler at base; legs and toes dark brown; soles of feet pale yellow.

Adult female. Quite different from the male. Above olive-grey, with a shade of green in it; rump and upper tail-coverts yellow olive-green. Top of the head scaly looking. Sides of the head greyish; lores whitish; a yellowish line above the blackish loral spot. Below from the chin to the under tail-coverts yellow, washed with olive-green along the sides of the body. The bill is considerably shorter than in the male, blackish, paler at base.

The *young males* much resemble the females, and the younger the birds are the less of the green and yellow colour is visible in their plumage and they are greyer, the bills also are shorter in younger birds. At an early age the young have yellowish spots to the tips of the wing-coverts.

As neither Dr. Stejneger nor Mr. Wilson measured many specimens, I have thought it worth while to give below the dimensions of eighteen adult males and fourteen adult females carefully taken by Mr. Hartert, after eliminating all specimens doubtfully sexed or not fully adult, or which had the bill shot off, as frequently occurs in all long-billed birds, to the regret of every collector.

N

Adult males.

Wing.	Tail.	Tarsus.	Culmen measured along the curve.	Bill from base to tip in a straight line.
in.	in.	in.	in.	in.
3·6	2·15	1	2·8	2·5
3·6	2·15	1·05	2·85	2·65
3·55	2·1	1	2·72	2·4
3·5	2	1	2·6	2·2
3·55	2·05	1	2·55	2·25
3·5	2	1	2·6	2·4
3·30	2	1	2·65	2·35
3·5	2	1·01	2·45	2·2
3·55	2·05	0·96	2·6	2·3
3·45	2·1	1·1	2·65	2·35
3·6	2·1	0·96	2·7	2·4
3·45	2·1	0·95	2·8	2·35
3·36	2·1	1	2·76	2·4
3·5	2·09	1	2·7	2·5
3·5	2·05	1·1	2·75	2·5
3·49	2·1	1·05	2·75	2·6
3·5	2·1	1·05	2·7	2·4
3·55	2·1	1	2·65	2·30

Adult females.

Wing.	Tail.	Tarsus.	Culmen measured along the curve.	Bill from base to tip in a straight line.
in.	in.	in.	in.	in.
3·5	1·93	0·94	2·4	2·1
3·45	2·1	0·95	2·42	2·1
3·4	2	1	2·1	1·9
3·45	2	0·96	2·06	1·85
3·35	2	1	2·06	1·95
3·31	2·14	0·94	2·1	1·9
3·5	2	1	2·1	1·9
3·37	1·05	1	2	1·85
3·35	2·05	0·94	2·2	1·95
3·3	1·95	1	2·12	1·9
3·4	2	1	2·1	1·9
3·35	2	0·95	2·27	2
3·2	1·95	0·97	1·9	1·7
3·2	2	1	2	1·86

THE long-billed *Hemignathus* was seen by Henry Palmer on the south and south-west sides of Kauai at altitudes of from 900 to probably between 3000 and 4000 feet above the sea. On the west side it was common at 3000 feet, and only a few were seen higher. It was also met with on the north and north-east sides at elevations of about 600 to 2000 and 3000 feet, though not quite so numerous. Palmer believes it inhabits all the upland forests of the island.

Wilson and Evans have already given (in 'Aves Hawaiienses,' part iii.) most of the excellent observations of Henry Palmer and his assistant C. Munro; so the best I can do here is to repeat them, as I have received exactly the same notes from my collector.

"This bird is much more common and enjoys a wider range than the Nukupuu (*Heterorhynchus hanapepe*, Wils.), which bird it much resembles in habits. It seems to inhabit the whole forest-region of Kauai; its food consists of insects, their eggs and larvæ, and we have also seen them sucking honey from the Lehua flowers. Above Makaweli in January and February we found it less common than at other places we visited: there they were mostly on the Koa trees, these being the most suitable hunting-ground for them in this locality. Usually there was a pair in the same vicinity, but not keeping very close together, so that when one was shot we usually got another.

"At Kaholuamano in the latter end of February and beginning of March they were more common, generally, in company with the Akikiki (*Oreomyza bairdi*), feeding on the Lehua trees, the pairs keeping more together. In one instance I shot a female, and the male stopped in the top of the tree calling desperately. I fired at him without effect, and so intent was he in looking for his mate that he immediately returned and was brought down by another shot. At Halemanu, towards the end of March, we found them as plentiful as at the latter place; but the Akikiki not being so common, the Akialoa (*Hemignathus procerus*) were more often found apart from them; here we first heard the Akialoa sing, although it was some time before we knew for certain it was the bird whose sweet note we heard every day; once I heard one sing while flying from one tree to another. Near Hanalei in April we found these birds not uncommon, generally in pairs, chasing each other about or singing in the tops of the trees. Their chirp seemed different here; Mr. Palmer likens both its chirp and song to that of the canary. We watched a pair singing together one day; the smaller and duller bird (probably the female) [no doubt the female—W. R.] seemed to have fewer notes than the other.

"Females that I dissected here (in April) had the ovaries enlarged, which, with before-mentioned notes on the subject, would denote the approach of the breeding-season. I have seen these birds from the branches in the tops to the roots of the trees, probing into holes and under the bark, where they find a harvest of cockroaches' eggs, beetles, and grubs; on one occasion I saw one alight on the ground and insert its long bill amongst mats of dead leaves and bits of wood; have also seen them collecting insects from the bases of the leaves of the halapipi tree; have not often seen them feeding on nectar. In feeding they do not seem to depend much on sight; have never noticed them to *look* into a crevice, as the O-o, before inserting their bill. I saw one send its bill at full length into a hole in a tree; have seen them work about one spot for some minutes, but have not noticed them break off any portions of bark or wood. Like the Nukupuu, it is an active bird, but can be easily approached within gunshot with ordinary caution. They have a strong smell when killed; and some, shot at Makaweli, had sores on their feet like the other birds in that locality at that time."

These swellings on the feet, which were so bad that in some cases the affected birds had lost one or more claws and toes, were also observed in many other birds in Oahu and Hawaii during the rainy season in wet places, and were probably caused by the damp. (See also Perkins, Ibis, 1893, p. 112.)

Palmer says this bird does not fly about in flocks, but keeps singly or in pairs; it is sometimes seen in company with *Himatione stejnegeri*, Wils., and *Oreomyza bairdi*, Stejn.

N 2

The song is described by Palmer as a short and not at all powerful warbling, but at the same time melodious.

In the 'Ibis,' 1890, p. 191, Mr. Scott B. Wilson said this bird was "very scarce and shy of approach." This seems to contradict Palmer's notes; but Palmer expressly wishes me to state that he found it not only rather numerous, but not at all shy. He could always approach them close enough to shoot with a "half-charge cartridge."

9. HEMIGNATHUS ELLISIANUS, *Gray*.

Hemignathus obscurus, Licht. Abh. Akad. Wissensch. Berlin, 1838, p. 440, pl. v. fig. 1 (nec Gmel.) (1839)
(coll. by Deppe on Oahu).
Drepanis (Hemignathus) ellisiana, Gray, Cat. B. Trop. Is. Pac. Oc. p. 9 (1859) (Oahu).
Hemignathus lichtensteini, Wils. Ann. & Mag. N. H. ser. 6, vol. iv. p. 101 (1889) (Oahu, specimen from the
Berlin Museum) ; id. Ibis, 1890, p. 100.

PALMER did not find this species on Oahu, where it was discovered by Deppe ; and having
no specimen of it before me at present, I insert here a translation of Lichtenstein's original
description. It runs as follows :—

"Latham describes in the genus *Certhia* a species from the Sandwich Islands, which
struck him particularly in so far that the lower mandible was a quarter of an inch shorter
than the upper. This species was included in the system under the name of *Certhia obscura* ;
but it is so distinct from all the different subgenera into which this group has rightly been
divided, that one is forced to make a new genus for it, for which I propose the above
name (*Hemignathus*).

"The most characteristic feature of this genus is certainly the unusual length and bow-
shaped curve of the bill and the unequal length of the mandible. But, moreover, we, find in
the great development of the hind limbs (the thickness of the tarsi, and the length of the toes
and claws) differences from the allied groups which cannot be overlooked. The nearest ally
of this new genus is *Drepanis (Melithreptus) vestiaria*. But the beak (irrespective of the
length) is much less thick and robust at the base ; the nostrils also are much narrower and
hardly feathered at the base, while in *Drepanis (Vestiaria)* they are feathered for their whole
length.

"The above mentioned, already described species has very inconspicuous plumage ; the
whole upper surface and wing-feathers on the outside are of a dull olive-green, the inner
webs of the flight-feathers dull brown. The under surface is lighter but still dull ; throat
and the centre of the abdomen, as well as the under tail-coverts, pale buff. A bright
yellow superciliary stripe is very conspicuous, especially as directly underneath it, from the
beak to the eye, there is a dark brown streak. The bow-shaped curved bill, which terminates
in a very fine almost hair-like point, is exactly half as long as the body, and the under
mandible is three lines shorter than the upper. The whole length from the point of the bill
to the end of the tail 7 inches. Bill 1¾ inches, tail 1⅜, tarsus 11 lines, middle toe and claw
9 lines.

"The outer toe is half a line, the inner toe one line shorter than the middle toe.

"Herr Deppe procured several specimens, all coloured alike, from the interior of Oahu."

As neither Palmer, Perkins, nor Wilson found this bird, there is every reason to believe that it is extinct. On the other hand, collecting in the dense forest and rugged mountains of Oahu is so difficult that—as Palmer himself says—one might be months or even years busy in collecting and yet fail to find a very rare bird.

Mr. Wilson places Gray's *Drepanis ellisiana* as a synonym under *Hemignathus obscurus* (Gmel.), and gives as his reason that Gray *partly* referred *H. obscura* to the female of *Vestiaria coccinea* and *partly* to his *Drepanis ellisiana*.

This is entirely erroneous, and proves that Mr. Wilson misquoted Gray, for the latter states, in his 'Catalogue of the Birds of the Tropical Islands of the Pacific Ocean,' that *Certhia obscura*, Gm., is the female of *Vestiaria coccinea*, while *Certhia (Hemignathus) obscura*, Licht. nec Gm., is his *Drepanis ellisiana*; therefore I regret to have to reduce Mr. Wilson's name, *Hemignathus lichtensteini*, to synonymic rank and to reinstate Gray's *ellisiana*.

A figure and accurate description will appear in the Third Part of this work.

10. HEMIGNATHUS LANAIENSIS, *Rothsch.*

Hemignathus lanaiensis, Rothsch. Bull. B. O. C. i. (1893) pp. xxiv & xxxiii; id. Ibis, 1893, pp. 256 & 265 (reprint).

This species is intermediate in size between *Hemignathus procerus* from Kauai and *H. obscurus* from Hawaii, and can easily be distinguished from both. Its nearest ally seems to be *H. ellisianus* from Oahu, but that species appears to be larger, with a shorter and stouter beak, and has a very distinct superciliary stripe.
The results of actual comparison will be given in the next Part of this book.

Description of the Three Skins of Hemignathus lanaiensis, *Rothsch.*

1. ♂ ad. (sexed H. C. Palmer). Above yellowish olive-green, somewhat mixed with greyish brown on the head where the brown bases show through. Quills dark brown, edged with the colour of the back; rectrices same. Below yellowish olive-green, much less bright than on the back and with little yellow in it, shading into olive-buff on the vent, and with a brown pale shade on the chin and throat; under wing-coverts pale buff, with an olive-yellow tint on the edges of the feathers; feathers on the tibia buffy white. Palmer gives the total length as about 7½ inches, but it is only about 6 inches in the skin ; wing 3·3, tail 2·1, tarsus 1·93, middle toe without claw 0·7, culmen along the ridge 2·1 (given as 3·1 by a slip of the pen in the original description, *l. c.*), bill from base to tip in a straight line 1·9 inch.

2. ♀ ad. (sexed H. C. P.; ovaries enlarged). Above more greenish olive and less bright; the superciliary stripe faint and greenish; chin, throat, and middle of abdomen buffish yellow ; yellowish olive-green on the sides of the body and breast. In other respects similar to the male in colour. Wing 3·05 inches, tail 1·9, culmen 1·9, bill from base to tip in a straight line 1·7.

3. ♂ (sexed H. C. P.). Above of a very peculiar dark saffron-olive or ochreous-olive colour, without green in it, less bright on the top of the head, where the deep brown bases of the feathers show through; the superciliary stripe narrow and yellowish. Below the same colour as above, but much lighter, and shading into buff on the middle of the abdomen, and somewhat more brownish on the throat. Wing 3·2 inches, tail 1·85 ; culmen along the ridge 2·1, from base to tip in a straight line 1·8.
Iris dark brown ; upper mandible brownish black, with the base lighter and somewhat greyish; lower mandible deep brown, shading into a paler and greyish colour at base; tarsi and feet bright slaty blue ; soles of a yellowish flesh-colour.

PALMER found this rare bird on the mountains of Lanai, where it inhabits the upper forest-region. It seems to live on insects, which it extracts from the partially decayed bark of the various trees. The stomachs of the three specimens which were procured were full of beetles and other insects. They were also seen sucking the flowers of the "Haha." Palmer found them shy and very rare. They were very quick in their movements from flower to flower. Their flight is quick and straight; no jerky movements were noticed.

Palmer heard the birds utter "a whistle, clear, but yet with something hollow in it, very difficult to express in words." Their note and song resemble those of *Hemignathus obscurus* from Hawaii, but are much clearer.

11. HEMIGNATHUS OBSCURUS (Gm.).

AKIALOA.

Hook-billed Green Creeper, Lath. Gen. Synops. i. 2, p. 703, pl. xxxiii. fig. 1 (1782) ("Sandwich Islands in general") (although the bird is very poorly described, and even wrongly in some respects, the description might be called recognizable, but the figure is abominable) ; id. Suppl. p. 126 (1787).

"*Akuiearooa*," King, Voy. Pacific Ocean, iii. p. 119 (1784).

Certhia obscura, Gmelin, Syst. Nat. i. p. 470 (1788) (ex Latham) ; Lath. Ind. Orn. p. 281 (1790) ; Donndorff, Orn. Beytr. i. p. 621 (1794) ; Shaw, Gen. Zool. viii. p. 227 (1812) ; Tiedemann, Anat. Naturgesch. d. Vög. ii. p. 430 (1814) ; Gray & Griffith, An. Kingd., Aves, vii. p. 358 (1829).

"*L'Akaiearoa*," Vieill. Ois. Dorés, ii. p. 111, pl. liii. (1802) (fig. taken from specimen in the Leverian Museum, now in the Derby Museum at Liverpool) ; Less. Compl. Buffon, ix. p. 155 (1837).

Melithreptus obscurus, Vieillot, Nouv. Dict. xiv. p. 322 (1817) ; id. Enc. Méth. p. 601 (1823) ; Cuv. Règne Anim. 2nd ed. i. p. 433 (1829).

Drepanis obscura, Temm. Man. d'Orn. i. p. lxxxvi (1820).

Hook-billed Green Honey-eater, Latham, Gen. Hist. iv. p. 192, pl. 71. fig. 1 (text and figure the same as in Gen. Synopsis).

"*L'Akaiearoa, l'ieill. Ois. Dor.* pl. 53 *est la femelle d'Héorotaire rouge* (*Mellithreptus vestiarius*)," Lesson, Tr. d'Orn. p. 300 (1831).

Vestiaria akaroa, Less. Rev. Zool. 1840, p. 268 (included amongst the synonyms of *Drepanis coccinea*, Gray, Gen. B. i. p. 96, 1847).

Hemignathus obscurus, Peale, U.S. Expl. Exp., B. p. 153 (1848) ; Hartl. Arch. f. Naturg. 1852, ii. p. 110 ; Reichenb. Handb. speciell. Orn., Tenuirostres, p. 312, pl. 501. fig. 4009 (1853) ; Cass. U.S. Expl. Exp. Orn. p. 178 (1858) ; Dole, Proc. Bost. Soc. N. H. xii. p. 298 (1869) (partim) ; id. Hawaiian Almanac, 1879, p. 15 ; Scl. Ibis, 1871, p. 360 ; id. Ibis, 1879, p. 92 ; Sund. Tentamen, p. 48 (1872) ; Sharpe, Cat. B. x. p. 4 (1885) ("Sandwich Islands") ; Wilson, Ann. & Mag. Nat. Hist. ser. 6, iv. p. 400 (1889) ; id. Ibis, 1890, p. 189 ; id. & Evans, Aves Hawaiienses, pt. iii. text & fig. (first good figure of the species, synonymy, history, and description).

Drepanis (*Vestiaria*) *corcinea* ♀, Gray, Cat. B. Trop. Is. Pac. Oc. p. 8 (in the synonymy of *Drepanis* (*Vestiaria*) *coccinea*) (1859) ; id. Hand-l. i. p. 113 (1869).

Hemignathus olivaceus, Perkins (nec Lafr.), Ibis, 1893, p. 108 (" the long-billed species, *H. olivaceus!*").

Adult male. Above all over of a beautiful bright yellowish olive-green, generally very little darker on the top of the head. Loral spot blackish, above the eye and passing into the colour of the back is an ill-defined more yellow superciliary line. Quills and tail deep brown, bordered with olive-green on the outer webs. Below lighter than above, and becoming more yellow on the lower abdomen. Under wing-coverts and axillaries buff,

o

more or less washed with greenish yellow. Iris brown; bill deep brown, almost black, paler at base; legs and toes slate-colour; soles grey.

Adult female. Above olive-green, with a faint greyish tint, more greenish on the rump and upper tail-coverts. A blackish spot on the lores, above which begins an ill-defined yellowish pale superciliary line. Below from the chin to the under tail-coverts pale yellow, tinged with olive along the sides of the body. Iris dark hazel; bill blackish, brown at base; legs and feet slaty blue; soles pale greyish orange.

Younger birds in both sexes resemble the female. Nestlings, eggs, and nest are unknown.

Measurements of adult and sexed specimens.

Wing.	Tail.	Tarsus.	Culmen.	Length of bill from base to tip in a straight line.
		A. *Males.*		
in.	in.	in.	in.	in.
3	1·65	0·85	1·86	1·66
3·15	1·72	0·87	1·62	1·33
3	1·6	0·9	1·6	1·4
3·2	1·73	0·86	1·83	1·5
3·06	1·7	0·9	1·7	1·5
3·1	1·9	0·9	1·7	1·5
3·1	1·8	0·9	1·66	1·5
3	1·9	0·86	1·72	1·5
3·05	1·95	0·86	1·6	1·4
3·05	1·75	0·83	1·5	1·3
3	1·8	0·73	1·62	1·4
		B. *Females.*		
2·95	1·8	0·83	1·52	1·35
2·85	1·7	0·74	1·1	1·2
3	1·8	0·81	1·30	1·25
2·9	1·75	0·82	1·45	1·3

WILSON writes concerning this bird:—"It occupies the lower forest-zone from about 1100 to 2500 feet, and is most plentiful among the tall Ohia trees. . . . . . It prefers decayed timber in which to search for its food, and invariably chooses a rotten or half-dead tree for its hunting-ground, no doubt on account of its slender bill, which requires soft material to work upon. It is also very partial to the great tree-ferns which in the forests of Hawaii reach a height of more than 30 feet, and, as the sombre colour of its plumage is very nearly that of their foliage, it is most difficult to observe, and is at the same time more quiet and unobtrusive in its habits than any other member of the genus; in fact, had it not been for its clear and characteristic call-note, I doubt whether I should have noticed it at all. It must—at least in the several localities I visited and at the time of year I saw them—be considered a scarce bird; and whilst I was at Olaa, in the district of Puna, an old native, Hawelu, an excellent observer and well skilled in the almost forgotten art of bird-catching, told me that it was extremely rare. During a long stay in the higher forest-region in Kona I did not notice it, and believe, as I remarked above, that it is confined to the lower forest-zone.'

Palmer believes that these birds inhabit pretty well all the upland forests on Hawaii. He found them common in the district of Kona, on the south-west side of the island, at altitudes of about 2000 to 6000 feet. He saw none in Kohala district (north side) or Kohala mountain. They were comparatively numerous in Hamakua district, north-east, between 2000 and 5000 feet. Numbers were seen in Hilo district above Hilo, at heights of 1400 to 3000 feet, and a few in the upper parts of Puna district on the east side.

They inhabit dense forest, where they were generally not at all rare, although other birds were more numerous.

Their principal food consists of insects, which they capture in the bark and decaying stems and branches. In collecting their food they were seen to put their bills into the crevices of the bark up to the base.

They were also (above Hilo near Wailuku) seen sucking nectar from the flowers of the Ilaha, of which they seemed to be extremely fond.

THE GENUS HETERORHYNCHUS.

THE type of this genus is *H. olivaceus*, which is identical with *H. lucidus*. In all its chief characters it agrees with *Hemignathus*, except in the formation of the bill. The upper mandible is strong at base, but narrow and thin at tip, not so long as in *Hemignathus*, and quite different from the lower mandible; the latter is only about half as long as the upper, and rather strong. "The tongue is consequently shorter than in *Hemignathus*, and less tubular, being intermediate in structure between those of *Himatione* and *Vestiaria*. The stomach of *H. wilsoni* is, as in *Hemignathus procerus*, quadrangular and strong. The convolutions of the intestinal canal are like those of *H. procerus*; but the central spiral has one twist more, owing to the greater length of the gut, the total length being (in *H. wilsoni*) as much as 25·5 centim., giving the relative length of 6." (See Gadow, in 'Aves Hawaiienses,' part ii. p. 15.)

The species of the genus *Heterorhynchus* can be divided into two groups—one, containing only *H. wilsoni*, with the under mandible quite straight and very strong, much like a Woodpecker's under mandible; and the other, containing three species, with the under mandible curved and following the curve of the upper.

One species is known from each of the islands of Hawaii, Maui, Oahu, and Kauai. From its geographical position one would expect that Kauai, being most isolated and separated from the other islands by a deep channel, much more than twice as broad as the Alenuihaha Strait between Hawaii and Maui, would have the most aberrant form; but instead of this we find that the species from Kauai and Maui are the most nearly allied, between them a much different form occurs on Oahu, and the most aberrant on Hawaii.

There is no reason why Lanai and Molokai should not have been, or still be, the habitat of a species of *Heterorhynchus*, which might be the same as my *H. affinis*, or an undescribed species.

Key to the Genus Heterorhynchus.

A. Under mandible straight . . . . *H. wilsoni.*
B. Under mandible curved.
    a. A very distinct superciliary stripe; head green . . . . *H. lucidus.*
    b. No distinct superciliary stripe; head yellow.
        a'. Colour of head gradually merging into that of the back . . . . *H. hanapepe.*
        b'. Colour of head separated somewhat abruptly from that of the back, which is
           quite different from that of the head . . . . *H. affinis.*

12. HETERORHYNCHUS WILSONI, *Rothsch.*

? *Hemignathus olivaceus*, Cassin (nec Lafr. ?), U.S. Expl. Exp. p. 179 (1858) (as no sufficient description is given nor the exact habitat, it is absolutely impossible to say whether this is the Hawaii form or not; only an examination of the specimens could solve the question).

Hemignathus olivaceus, Wilson (nec Lafresn. !), Ann. & Mag. Nat. Hist. ser. 6, vol. iv. p. 400 (1889) (Hawaii); id. Ibis, 1890, p. 191 (Hawaii); id. & Evans, Aves Hawaiienses, pt. iii. text & pl. (first plate of the Hawaiian species); Gadow, op. cit. pt. ii. p. 15 (1891).

Hemignathus obscurus, Perkins (nec Gm.!), Ibis, 1893, p. 106 (Kona, Hawaii).

Mr. Wilson refers the *Heterorhynchus* of the island of Hawaii without hesitation to the *H. olivaceus* of Lafresnaye, although he confesses to not having seen the type. On looking at Lafresnaye's plate we are at once struck by the fact that his bird has a distinct yellow superciliary line, which is wanting in the Hawaii bird. Moreover, the latter has a perfectly straight lower mandible, while the plate of *H. olivaceus* distinctly shows the under mandible curved downwards to fit the outline of the upper mandible[1].

I have carefully examined the type of Lafresnaye in the Paris Museum; it is a fine adult male and no doubt identical with *Heterorhynchus lucidus* of Lichtenstein, with a remarkably bright yellow breast, so I have renamed the Hawaii bird in honour of Mr. Wilson.

Adult male. Above bright olive-green, lighter and brighter on the rump and upper tail-coverts, distinctly tinged with yellow on the top of the head. Black loral spot. Below bright yellow, deeper golden yellow on the chin and upper throat, washed with olive on the sides of the breast and flanks. Under wing-coverts and under tail-coverts whitish yellow. Iris dark hazel; bill black, base brown; tarsi slaty grey; soles pale orange.

Adult female. Above olive-green, but not nearly so bright as in the male, less of a yellow and somewhat of a grey shade in it. Below pale yellow, brighter on the chin, throat, and upper breast, tinged with olive on the flanks and shading off into yellowish white on the vent and abdomen. Bill shorter than in the male. Iris dark hazel; bill black, base brown; tarsi slaty grey; soles of feet pale orange.

Young birds are more like females, but more greyish above and less yellow below, where they are of a pale dirty yellowish colour. All the colours appear to be somewhat blurred through the presence of narrow and indistinct dusky edges to the feathers.

---

[1] This is also distinctly mentioned in Lafresnaye's description "la mandible inférieure est creusée dans toute sa longueur, comme de coutume, et recourbée comme la supérieure."

Iris dark brown; upper mandible brown with yellow markings at base; lower mandible whitish yellow with brown tip, and brown at each side of base; gape pale yellow; tarsi lighter than in adults.

Dimensions of perfect and sexed adult males and females.

Wing.	Tail.	Tarsus.	Culmen.	Length of lower mandible from the mental apex to the tip.
A. Males.				
in.	in.	in.	in.	in.
3·4	2·05	0·95	1·2	0·55
3·3	2	0·8	1·07	0·55
3·35	2	0·86	1·04	0·5
3·3	1·95	0·9	1·17	0·6
3·4	2	0·88	1·05	0·55
3·25	2·08	0·9	1·1	0·6
3·3	1·96	0·8	1·16	0·53
3·3	1·9	0·9	1·1	0·56
3·35	2	0·95	1·12	0·5
3·3	2·09	1	1·07	0·57
3·3	1·98	0·9	1·1	0·55
3·05	1·9	1	1·1	0·5
B. Females.				
3·15	1·8	0·83	0·9	0·5
3·2	1·85	0·84	0·95	0·45
3·15	1·8	0·84	0·91	0·5
3·21	1·8	0·83	0·9	0·51
3·15	1·9	0·87	1	0·45
3·2	1·8	0·88	0·9	0·5
3·15	1·9	0·86	0·92	0·5

Hab. Island of Hawaii.

PALMER found this bird very common in Kona district on the south-west side of the island, at heights of 2000 to 4000 feet, and a few were seen at higher elevations. They were also common above Hilo, north-east, at 1400 to 3000 feet. Numbers were seen in the district of Hamakua, north-east side, at from 2000 to 5000 or 6000 feet. Palmer says: "I believe them to inhabit all the upland forests."

Palmer and Perkins have watched and described the process of feeding in this peculiar bird, and how it uses its bill. It taps the branches and bark with its under mandible, while holding the bill wide open. The sound of the tapping is distinctly heard at a distance in the quiet forests. After the bird had tapped vigorously for some time he was seen either going on again or stopping to take something out of the bark. He hooked the insects from holes or crevices. The tapping of the bark and branches, Palmer thinks, was done to find out whether they were sound or insect-eaten, and also to lay open the holes of larvæ, &c., for the under mandible of this bird is very strong.

Wilson says :—" In the Island of Hawaii, to which, as far as we know at present, it is peculiar, this bird is decidedly rare, and I obtained only three specimens during a stay of some five weeks in June in Kona, where it frequents the Koa trees alone, running up their great smooth trunks and along their limbs in search of insects. In the mamáne woods near Mana I subsequently found it in considerable numbers in the month of January, when these trees are in full flower, resembling laburnums with their golden clusters. Its movements are very rapid, and the quickness with which it slips from one side of a limb to the other is surprising."

Perkins's description of the habits of this bird is published in the 'Ibis,' 1893, pp. 106, 107. His observations agree mainly with Palmer's, and I think it therefore better to give Perkins's words almost in full. He writes:—"It is a common bird from rather below 4000 feet to some hundreds of feet above that altitude, and most probably much higher still. It is most partial to the larger acacias, running up and down the limbs with equal ease, and also both on the upper and lower surfaces of the branches. It was on the 11th of July that I first saw one, a fine bright male, feeding. When I first caught sight of it, it was some ten yards off; but I easily got closer without scaring it in the slightest. Being bare-footed and bare-legged at the time, and the ground being overgrown with a prickly introduced thistle, after following it for half an hour I found my feet somewhat painful. Meanwhile the bird kept straying over the fallen trunks turning its head, now right, now left, in its desire for food. In this manner it searched both sides of the tree in one journey without retracing its steps. And this is how it used its bill :—The upper mandible it plunges into the small holes or cracks of the wood, while the lower presses on the surface of the bark. By this means, I imagine, it gets a considerable leverage to help it in opening out the burrows of the insects. In the same way it thrusts its upper bill under the loose bark, resting the lower one on the surface, and in this way strips the bark off. The upper mandible, though so thin, is very strong and somewhat flexible; while the curve of the bill follows the curve of the burrow, for insects nearly always burrow more or less in a curve. Should the curve of the burrow not agree with the curve of the bill, the difficulty is overcome both by the slight flexibility of the beak and by the wonderful flexibility of the bird's neck, which it twists round so as to bring the curve of the bill to follow that of the burrow. In this manner it gets out its prey, being largely aided by the long tongue, which is as long as the upper beak. Every now and then it gives several blows to the trunk, the sound of which may be heard at a considerable distance, sometimes, I think, to frighten out its prey to the entrance of the burrow, sometimes for the purpose of actually breaking the wood. . . . . The blows that it gives to the trunk and branches are dealt with great vigour and with the beak wide agape, so that the points of both mandibles come in contact with the surface. . . . . Into these blows it throws its whole weight, swinging backwards from the thighs to renew each stroke. In some cases at least these blows are for the purpose of driving out insects, or at any rate have that result; for several times I saw the bird after a stroke make a sudden dart, sometimes even taking an insect on the wing, and, after swallowing it, return again to its labour. Its song is short but rather pleasing, and, as one would expect from its habits, full of life and energy."

P

13. HETERORHYNCHUS HANAPEPE (*Wilson*).

NUKUPUU.

Heniignathus hanapepe, Wilson, Ann. & Mag. Nat. Hist. ser. 6, iv. p. 401 (1889) ; id. Ibis, 1890, p. 192,
pl. vi. fig. 1 ; id. & Evans, Aves Hawaiienses, pt. iii. text & pl. (1892).

This rare species was discovered and described by Mr. Scott Wilson ; it was obtained at an
altitude of some 3000 feet, in the forest-region of Kauai. It is nearest to my *H. affinis*
from Maui, but can easily be distinguished from that species by the much yellower colour
of the back, which merges gradually into the yellow of the head. From *H. lucidus* of
Oahu it differs chiefly in the yellow head and the want of a well-marked superciliary
stripe, and from *H. wilsoni* of Hawaii it differs most obviously in its curved under
mandible.

Adult male. Above, head dark gamboge-yellow, almost golden yellow, brightest on the
forehead, gradually passing into the yellowish olive-green colour of the back, rump,
upper tail-coverts, and wing-coverts. Quills and tail-feathers edged with the colour of
the back on the outer webs. Lores black, as well as a very narrow line above the bill
and a tiny spot on the upper chin. Throat, sides of the head, and breast bright gamboge-
yellow, passing into whitish on the abdomen ; flanks and sides of the breast slightly
tinged with olive-green ; under tail-coverts dusky white. Under wing-coverts and
axillaries yellowish white. Upper mandible deep brown, paler at base, more black
towards the tip ; lower mandible ashy grey at base, blackish towards the tip. Tarsi and
toes slate-blue ; soles of feet whitish flesh-colour.

Adult female. Above olive-grey, slightly washed with green on the forehead, lower rump,
and upper tail-coverts ; the upper wing-coverts, quills, and tail-feathers bordered with
olive-green. Below dusky white, more greyish on the sides of the body, and washed
with yellow on the chin and upper throat ; in dimensions smaller than the male, and
with a shorter bill.

Young birds are similar to the female, but the males soon become brighter yellow on the
throat and sides of the head, and show yellowish feathers here and there above and
below.

P 2

Dimensions of fully adult and sexed specimens.

	Wing.	Tail.	Tarsus.	Culmen.	Under mandible from base to tip, measured in a straight line.
	in.	in.	in.	in.	in.
♂	3·16	2·05	0·95	1·3	0·62
♂	3·15	2	0·9	1·2	0·65
♂	3·15	2	0·9	1·25	0·65
♀	3	1·96	0·8	1·08	0·55
♀	2·95	1·96	0·8	1·1	0·55
♀	2·95	1·97	0·8	1·09	0·55

(Imperfect, young, and doubtfully sexed specimens were not measured.)

PALMER found this species very scarce. All he procured were collected on the southern side of Kauai—two above Makaweli at a height of about 1000 feet, the rest at elevations of 3000 feet or so, in the district of Waimea, all within a range of fifteen miles. None were seen elsewhere. They kept singly or in pairs, and were almost invariably seen in company with *Oreomyza bairdi* (Akikiki).

These birds were never observed among the foliage, but were always busy on the bark of the trees, in the crevices of which they obtained their food. They were pretty lively in their movements, but not at all shy, in fact less so than many other birds. Their food consists of beetles and other insects, caterpillars, grubs, &c., which are usually hooked out of crevices and holes in the bark and decaying trunks and branches.

The song is a short but rather sweet warble, consisting of some half a dozen notes.

When shot, these birds had a strong smell, like all the other species of *Heterorhynchus* and *Hemignathus*, but less so than *Hemignathus procerus*. This peculiar smell, unlike any other known to him, but certainly disagreeable, was also noticed by Perkins in other species, such as "*Loxioides, Psittacirostra, Chloridops,* and *Rhodacanthis,*" but most strongly in *Himatione*. If this is the case, then the peculiar odour of the *Hemignathi* would not help to confirm the theory that these birds breed in hollow trees : but why is it that on these islands birds of different families should possess such an odour ?

14. HETERORHYNCHUS AFFINIS (*Rothsch.*).

Hemignathus affinis, Rothsch. Ibis, 1893, p. 112 (Maui).

Adult male. Top of the head bright gamboge-yellow, almost golden yellow, passing into yellowish green on the nape and hind neck, which colour is not gradually, but somewhat abruptly, terminated by that of the back, which is greyish olive-green without yellow, becoming brighter and greener on the upper tail-coverts. Quills and rectrices deep brown, bordered with bright olive-green. Lores black, connected by a black band across the forehead, which is distinctly broader than in *H. hanapepe*, Wils. A very tiny black chin-spot. Throat bright yellow, gradually passing into a sulphur-yellow on the abdomen and pale yellow on the vent and under tail-coverts. Sides of the body tinged with greenish olive. Iris dark brown. Bill blackish brown, slate-colour at base. Tarsi and feet dark grey with a yellowish tinge.

Adult female. Quite different from the corresponding sex of *H. hanapepe*, Wils. Above olive-green with a greyish shade, brighter and more green on the upper tail-coverts; on the forehead from the base of the upper mandible runs a bright yellow line to just above the eyes. Lores dusky. Sides of the head yellowish. Chin and throat bright yellow ; middle of breast and abdomen pale yellow, much like " Naple's yellow " (Ridgw. Nomencl. Colors, pl. vi. fig. 18); sides of breast and flanks washed with olive, under tail-coverts slightly so. Under wing-coverts yellowish white.

Dimensions of adult and properly sexed birds of **H.** affinis.

	Wing.	Tail.	Tarsus.	Culmen.	Under mandible from the base to the tip.
	in.	in.	in.	in.	in.
♂	3	2	0·75	1·10	0·66
♂	3	1·66	0·85	1·1	0·66
♂	3	No tail.	0·8	1·24	0·68
♂	3·05	1·9	0·9	1·25	0·65
♂	Imperfect.	1·9	0·85	1·15	0·7
♂	3	1·9	0·8	1·1	0·63
♂	3·05	2	0·9	1·15	0·66
♂	2·9	1·8	0·86	1·26	0·66
♂	3·06	1·95	0·86	1·2	0·69
♂	3·05	1·95	0·83	Destroyed by shot.	0·64
♂	3·05	1·87	0·8	1·22	0·64
♀	2·8	1·65	0·8	1	0·55
♀	2·75	1·75	0·7	0·9	0·5

The adult male of this species closely resembles the corresponding sex of *H. hanapepe*, but can always be distinguished by the colour of the back, the much less whitish lower abdomen, and by having some more blackish feathers on the forehead, besides other slight differences. The females differ much more from each other.

One specimen, marked "young male" by the collector, closely resembles the only two females he obtained.

Hab. Island of Maui.

THIS species was discovered by my collector in 1892. It inhabits the forest-region of the little-known and unexplored island of Maui. Although not at all plentiful it was less rare than *H. hanapepe* on Kauai, to which it is very similar in habits. It is not shy, but somewhat scarce.

The stomachs contained small beetles, skins of large grubs, and other insects.

Sometimes this species gives vent to a prolonged song, almost similar to that of the introduced American *Carpodacus*, (though considerably lower in tone. Its call-note is like that of *H. wilsoni* from Hawaii.

As regards its mode of feeding, Palmer says :—" It uses its lower mandible to hammer away the bark and then hooks the insects out with its long upper mandible. The hammering sound can be heard at a short distance. It also pulls insects out of the moss with the upper mandible and then, to kill them, shakes them like a dog does a rat."

15. HETERORHYNCHUS LUCIDUS (Licht.).

Hemignathus lucidus, Licht. Abh. d. königl. Akad. Wissensch. Berlin, 1838, p. 451, pl. v. figs. 2 ♂, 3 ♀
(1839) (Oahu); Bp. Consp. i. p. 404 (1850); Peale, U.S. Expl. Exp. p. 153 (1848) (rightly confounded
with *H. olivaceus*); Reichb. Handb. Scans. p. 313, Taf. dxci. figs. 1012-13 (1853); Licht. Nomencl. p. 55
(1854); ? Cass. U.S. Expl. Exp. p. 178 (1858) (no minute description nor exact locality given, therefore
it is impossible to say whether this species is intended or not); Prévost & des Murs, Voy. 'Venus,'
pp. 191, 192; iid. Atlas, Ois. pl. i. figs. 1 & 2, ♂ ♀ (sub nomine *H. olivaceus* errore, cf. p. 192) (1855);
Dole, Proc. Bost. Soc. Nat. Hist. xii. p. 298 (1869); Scl. Ibis, 1871, p. 359; Sharpe, Cat. B. x. p. 5
(1885); Wils. Ann. & Mag. Nat. Hist. ser. 6, iv. p. 400 (1889); id. Ibis, 1890, p. 192; Hartert,
Katalog. Vögels. Mus. Senckenb. p. 28 (1891).
Drepanis lucida, Gray, Gen. B. i. p. 96 (1847).
Drepanis (Hemignathus) lucida, Gray, Cat. B. Trop. Isl. Pacific Oc. p. 9 (1859) (Oahu); id. Hand-l. B. i.
p. 113 (1869).
Heterorhynchus olivaceus, Lafr. Mag. de Zool. 1839, pl. x. and text (Oct. 1839).
Hemignathus olivaceus, Reichb. Handb. Scans. p. 313, Taf. dxci. figs. 4010-11 (1853); ? Cass. U.S. Expl. Exp.
p. 179 (1858) (uncertain: see synonymy of *H. wilsoni*); Dole, Proc. Bost. Soc. Nat. Hist. xii. p. 298
(1869) (exclus. syn. *Certhia olivacea*, Gm. Syst. Nat. i. p. 473, quid *Zosterops*!); id. Hawaiian
Almanac, 1879, p. 45; Scl. Ibis, 1871, p. 359; id. Ibis, 1879, p. 92; Sharpe, Cat. B. x. p. 4 (1885).
Hémignathe brilliant, Prévost & des Murs, Voy. 'Venus,' pp. 192, 193 (very good synonymy and description).
Vestiaria heterorhynchus, Less. Rev. Zool. 1840, p. 209.

Adult male (in the Paris Museum). Above olive-green, darker and more olive on the back,
lighter and more green on the head, wing, and tail-coverts. Lores and a line behind the
eye brownish black; across the forehead a narrow and above the eyes a conspicuous
orange-yellow superciliary stripe. Quills deep brown, outer webs edged with greenish
yellow. Chin, throat, and upper breast bright orange-yellow; abdomen yellow and
fading into pale greenish grey on the vent and under tail-coverts.

Adult female or immature male (in the Senckenbergische Museum at Frankfort). Above dull
brownish olive, tinged with greenish on the top of the head, rump, and upper tail-coverts,
and on the edges of the quills and tail-feathers. Lores dusky; a somewhat ill-defined
but distinct superciliary stripe; sides of head and throat yellowish. Rest of underparts
light greyish brown, brownish olive along the sides of the body and flanks.

An *immature male* in the Leiden Museum is similar to the specimen in the Frankfort Museum,
but much more strongly tinged with yellow on the throat and sides of the head, the
edges of the quills more yellowish.

A skin in the Museum of the University of Cambridge, apparently a *female*, is more uniform below, where it is of a dull yellowish white.

The single specimen in the British Museum seems also to be a female of this species and it is much like the one at Cambridge, and does not clearly show the superciliary stripe, which is also absent in the latter. The British Museum specimen has some whitish tips to the wing-coverts, apparently indicating its immaturity. It has been minutely described in the ' Catalogue of Birds,' vol. x. p. 5.

Lichtenstein, in the original description of *II. lucidus*, says of the male that " the olive-green colour of the upper parts merges into grass-green on the top of the head and on the outer edges of the quills " ; that the sides of the head and throat are pure orange ("rein orangefarben "), shading into deep yellow (" dottergelb "), and becoming paler and duller on the lower abdomen ; vent and under tail-coverts greenish grey.

Lafresnaye gives a very recognizable figure and a short description of the colours, which runs as follows :—" All above of a dull olive, with the top of the head yellowish ; the eyebrows, fore part of the neck, and breast lively yellow."

Prévost and des Murs give a Latin diagnosis : " Supra olivaceus, capite dilutius ; superciliis et corpore infero lucide flavis, pectore fere junquillaceo ; striga olivacea a commissura usque ad aures transeunte" ; and a French description, of which I give an extract, translated :—" All above of a deep olive, clearer on the top of the head ; forehead, eyebrows, cheeks, throat, fore neck, and breast bright yellow, nearly jonquil-colour on the latter ('presque jonquille ') ; a dark streak through the eye. The secondaries are bordered with olive-green ; the primaries and their greater coverts with yellowish. Bill and feet horn-bluish." The female has some yellow only on the eyebrows, cheeks, and throat.

The Frankfort specimen measures :—Total length 5½ inches, culmen 1·1, length of under mandible 0·7, wing 2·95, tarsus 0·76, tail 2·9.

According to Lafresnaye the culmen measures 1·18 inch ; Sharpe (Cat. B.) gives total length 5·2 inches, culmen 1, wing 2·8, tail 1·75, tarsus 0·8. The culmen in adult males is much longer than in females or immature birds.

I intend to give a Plate of this species in Part III.

THIS species was found by Deppe on Oahu in January 1838, and was seen in great numbers flying round the flowers of the banana-plantations.

Palmer was told that it used to be not uncommon just above Honolulu, but no recent collector has been able to find this bird.

It is evidently extinct and one of the rarest birds in collections.

GENUS VIRIDONIA.

Viridonia, gen. nov., Rothsch. Ann. & Mag. Nat. Hist. 1892, x. p. 112.

Bill straight or very slightly curved, high and stout at base, attenuating towards the tip, which is sharply pointed. Nostrils protected by an upper operculum, only at base a little overhung by short feathers. Wing rather broad; first primary entirely rudimentary; fourth and fifth about equal and longest, gradually becoming shorter towards both sides; second slightly shorter than the seventh, and about equal to the eighth. Tail somewhat short, nearly even at tip. Legs and feet strong. Plumage rich and soft.

Sexes similar.

One species only known from Hawaii.

This genus is nearest to *Oreomyza*, and especially to the strong-billed *Oreomyza maculata* (Cab.), but cannot be included in it.

16. VIRIDONIA SAGITTIROSTRIS, *Rothsch.*

Viridonia sagittirostris, Rothsch. Ann. & Mag. Nat. Hist. 1892, x. p. 112.

Adult male. Upper parts bright olive-green, rather brighter and more yellowish on the forehead, sides of the head, and upper tail-coverts. Underparts more yellowish olive-green; wings blackish brown, the primaries narrowly and the secondaries more broadly margined with yellowish green; tail blackish brown, with yellowish-green margins. Under surface of the wings dark ashy, the quills margined with dull white on their basal half; margin of the wing yellowish; under wing-coverts dull white, tinged with greenish yellow. Iris hazel, with reddish tinge; upper mandible black, under mandible black with a brown tinge, both bluish slaty at base; tarsi bright blackish grey; soles flesh-colour with a yellow tinge.

Adult female. The first female sent by Palmer, the type of the original description, is rather duller in tinge of colour both on the upper and under parts, and appears to be a younger specimen, for another female is entirely similar in colour to the male. The iris of the former was " dark hazel," much darker than in the adult male.

♂ ad. (type). Wing 3·25 inches, tail 2·1, tarsus 0·9, culmen 0·9.

♂ ad. (figured on Plate). Wing 3·2 inches, tail 2·1, tarsus 0·86, culmen 0·89.

♀ ad. Wing 3·05 inches, tail 2·03, tarsus 0·85, culmen 0·86.

♀ (type) (figured on Plate). Wing 2·96 inches, tail 1·8, culmen 0·9, tarsus 0·86.

Total length, measured by Palmer in the flesh, 6¼ to 6½ inches.

Hab. Island of Hawaii.

PALMER describes the discovery of this rare and singular bird as follows :—"*April 23rd*, 1892. When marching along the slopes of Mauna Kea, above Hilo, and not far from Wailuku, through old forest of Ohia, here and there intermixed with the Koa, and much enjoying the beautiful contrast of the thousands of bright crimson-coloured flowers of the lofty Ohias, which were frequented by numerous birds of the common Hawaiian species, I heard a strange note coming from an Ohia tree that I had just passed. At the first moment I thought it might be an Akialoa's note, but on hearing it again was convinced I had never heard it before. I soon saw the bird among the leaves, but before I could fire it

Q 2

flew into the tree just above my head, where it began to call again, as if calling its mate, until I shot it. On picking it up I saw it was quite new to me, and I hope it may be so to science. Its call-note might be described as a high clear 'chirrup' uttered three or four times in quick succession at short intervals. After waiting a few minutes its mate came up, but was shy and escaped me. I waited on the spot for about an hour, but none came back: on returning to my camp, however, I was fortunate enough to shoot another specimen, to which I was attracted by its cry, and I think I heard two or three more which I could not get; but so many birds of various kinds were singing among the flowering trees that it was very difficult to distinguish their various notes from a distance."

Palmer obtained altogether four specimens only of this apparently rare bird.

They are obviously forest-birds, but not inhabitants of high elevations, for they were only observed at heights of about 500 to 1500 feet in Hilo district and all within a radius of a few miles. They were mostly seen in the higher trees, but one was once observed on a banana-leaf not ten feet from the ground. It is pretty active in its movements and not very shy, as a rule, but difficult to catch sight of in the dense foliage of the trees.

Its song is a whistling regular " trill," much like that of *Himatione virens*, but louder, with the exception of the last two notes, which sound much lower. It consists of six or seven notes, clear and distinct.

The stomachs of all contained insects.

GENUS OREOMYZA.

Under mandible straight or at least not perceptibly curved. Nostrils covered with an upper operculum; nasal fossæ quite bare or only very slightly concealed by a few tiny feathers at base. Tarsus covered in front with four, five, or six distinct scales and laterally also with scales.

The tip of the wing is formed by the third, fourth, and fifth, or by the fourth, fifth, and sixth primaries; the second primary is more or less shorter than the sixth, but always so and mostly very much; first entirely rudimentary.

Plumage rich and soft, red or greenish; females and young birds mostly much duller in colour.

Their food consists of insects, and they are arboreal in their habits.

THIS genus is only known from the Sandwich Islands, where nearly every island has a different representative form, while two entirely different ones are known from Kauai. The type of the genus is *Oreomyza bairdi*, Stejn.

The genus *Oreomyza* is most nearly allied to *Himatione*, from which its members can always be distinguished by the straight bill, which is perceptibly curved in *Himatione*. Moreover, the wing-formula is more or less perceptibly different from *Himatione*, and they have different habits. Palmer observed them most frequently on the branches and trunks of trees, while the *Himationes* were more often seen amongst the foliage.

The short-billed species, *O. bairdi* and *O. mana*, may be regarded as the most typical members of the genus *Oreomyza*, while *O. maculata* tends somewhat to *Viridonia* with its great bill, and *O. parva* comes nearest to *Himatione*.

The genera *Himatione* and *Oreomyza* are very easy to define, and it is most convenient to separate them.

Key to the adult birds of the Genus Oreomyza.

A. Plumage red . . . . *O. flammea.*
B. Plumage not red.
   a. Bill short and somewhat stout, tail shorter.
      a^1. Beneath whitish, ashy on the flanks, no green on breast . . . . *O. bairdi.*
      b^1. Beneath greenish . . . . *O. mana.*
   b. Bill longer and slenderer, tail longer.
      c^1. Yellowish green above.
         a^2. Head uniform yellowish, not perceptibly different from the back . . . . *O. parva.*
         b^2. Forehead yellow, and quite different from the greenish back . . . . . . *O. montana.*
      d^1. Olive-green above.
         c^2. Broad dusky loral mark; bill stronger; colour deeper olive above, more
           golden beneath . . . . *O. maculata.*
         d^2. Loral mark smaller; bill more slender; more greenish above, paler yellow
           beneath . . . . *O. newtoni.*

Immature birds of these species are extremely alike, and in some cases not distinguishable, or hardly so, without knowing the locality whence they came, although the males are always easy to distinguish. The young of *O. mana* and *O. bairdi* much resemble each other, but are very different from those of the other species; the young of *O. parva* are the smallest, those of *O. maculata* have a more conical bill, and those of *O. flammea* have generally a somewhat different and more brownish colour than all the others.

17. OREOMYZA MACULATA (Cab.).

Himatione maculata, Cab. Mus. Hein. i. p. 100 (♂ and ♀ jun., Oahu) (1850); Wils. Ibis, 1890, p. 186 ("fairly common in the district of Halemauu, Oahu"); Stejn. Proc. U.S. Nat Mus. 1887, p. 94 ("apparently quite distinct from both *H. virens* and *H. chloris*").
Included in the synonymy of *Himatione sanguinea* as "Juv." by Gray, Cat. B. Trop. Is. Pac. Oc. p. 9, and Hand-l. i. p. 113, 1869; and in the synonymy of *Himatione virens* by Sharpe, Cat. B. x. p. 9, 1885.)
Viridonia maculata, Rothsch. B. O. C. i. p. lvii (1893); id. Ibis, 1893, p. 571 (reprint). (When I first received a few specimens of this bird the powerful beak of the adult males induced me to include it in my genus *Viridonia*; but as Palmer has since brought home a fine series in all stages of plumage, I feel sure that it cannot possibly be removed from the straight-billed *Himationes*, which I have united under Stejneger's name *Oreomyza*, and that it is not a *Viridonia*.)

Adult male. Forehead, a very conspicuous superciliary stripe, chin, cheeks, ear-coverts, throat, and breast golden yellow. Rest of the upper parts olive-green. Wings and tail-feathers deep blackish brown, with olive-green edges. Big loral spot dusky; broad line from the eye to the hind neck dusky olive-green, making the superciliary line very conspicuous. Underparts from the breast downwards yellow, more whitish on the middle of the lower abdomen; sides of breast and flanks strongly washed with olive-green. Under wing-coverts whitish, washed with olive-yellow. "Iris dark brown. Upper mandible blackish brown, tip horny grey; lower mandible, base whitish, greyish brown towards the tip. Toes and tarsi grey; soles flesh with yellow tinge."

Adult female. Very different from the male. Forehead, superciliary stripe, and underparts yellowish white; sides of breast and flanks washed with olive-grey. Above olive, the greater wing-coverts with large greenish-white tips. "Iris dark brown. Upper mandible dark brown, with a tinge of grey on the sides of the base; lower mandible greyish brown, pearly white at base. Tarsi and toes grey, toes a little lighter; soles flesh-colour and orange."

Quite young birds and nestlings of both sexes are much like the females, but are browner above and of a mottled appearance, owing to the dusky edges of the feathers above and below. The superciliary line is already marked in the youngest specimens. The wing-spots in the young birds are more buffy than in adult females.

Wilson ('Ibis,' 1890, p. 186) remarks that Cabanis's name "*maculata*" is inappropriate, because the old birds were not spotted. However, *O. maculata* (Cab.) is the only species of the genus *Oreomyza* in which the adult female has spotted wing-coverts. No mistake is possible, since I have before me twenty-six carefully sexed specimens, and Palmer's observations (who often saw them building nests and feeding their young) leave no doubt that his "adult females" are fully aged birds. Therefore Cabanis's name is most characteristic.

Total length about 4½ to 5 inches; wing in males 2·7 to 2·81, in females 2·6 to 2·72; tail 1·85 to 2; tarsus 0·8 to 0·86; culmen in males 0·6 to 0·65, in females 0·56 to 0·65, mostly shorter than in males.

Hab. Island of Oahu.

PALMER found this bird only in the upland region in the district of Wailua, on the south-west side of the mountains, at elevations of 1500 feet to the top of the range. In habits it much resembled the species of the genus found on the other islands. It was not at all rare, and I received a fine series in all stages of plumage.

Wilson says:—"This species is peculiar to the island of Oahu, and is fairly common in the district of Halemanu (house of the birds), where there is still some forest remaining."

Wilson obtained only immature birds (see 'Ibis,' 1890, p. 186).

18. OREOMYZA NEWTONI (*Rothsch.*).

Himatione newtoni, Rothsch. Bull. B. O. C. i. p. xlii (1893) (Maui) ; id. Ibis, 1893, p. 443 (reprint).

Adult male. Forehead, line above the eye, cheeks, chin, and throat bright yellow. Entire upper parts, except the forehead, olive-green, a shade brighter and lighter on the rump and upper tail-coverts and on the top of the head. Below from the throat to the under tail-coverts lemon-yellow, strongly washed with olive-green along the sides of the body. Quills and tail-feathers deep brown, bordered with the colour of the rump; inner wing-lining whitish. Thighs brownish olive. " Iris dark hazel; bill brown, base of lower mandible light grey, with a yellow tinge ; legs and feet silvery grey, with a pinkish tinge." (*H. C. P.*)

Adult female. Similar to the male, but duller above and below.

Young birds have light spots to the tips of the wing-coverts, are dull brownish olive above and along the sides of the body, light brownish buff along the middle of the underparts.

The tip of the wing is formed by the third, fourth, fifth, and sixth primaries, of which the third is the shortest ; the seventh is shorter, and the second is much shorter than the sixth and seventh; the first is quite rudimentary.

Total length about 4½ to 5 inches, wing 2·5 to 2·6 (female 2·17), culmen 0·5 to 0·6, tarsus 0·79 to 0·85, tail 1·8 to 1·9. (Measured 5 adult males and 1 female.)

Hab. Island of Maui.

THIS species was discovered by Palmer on the 16th of July, 1892, in the thick forest on the slopes of Mount Haleakala, in the district of Makawao. It was rather rare and extremely local, being found only on that one side of the mountain.

It was gregarious and often seen in company with *Loxops ochracea*, Rothsch. In its habits it reminded Palmer much of *Oreomyza mana* (Wils.), from Hawaii, but differed from the latter in the colour of the legs.

Like the other members of the genus *Oreomyza* it was fairly active, and invariably seen running up the smaller limbs and branches of various trees in search of food.

The food consists of insects and grubs, which they extract from the moss and bark on the branches.

II

The note is a sharp and short sort of " twit."

These birds are very difficult to obtain, as they live in the thick wood, where progression is very tedious and slow, on account of the dense and high ferns, which—to make things more unpleasant—mostly drip from the continuous rain that so often prevails upon those hills.

I named this bird in honour of Professor Alfred Newton, of Cambridge, whose active interest in the birds of the Sandwich Islands has brought forward so much to increase our knowledge of the Hawaiian avifauna.

19. OREOMYZA MONTANA (*Wilson*).

ALAUHIO.

Himatione montana, Wils. P. Z. S. 1889, p. 446; id. Ibis, 1890, p. 186; Wils. & Evans, Aves Hawaiienses, pt. iii. text & pl. (1892) (Lanaihale, Lanai).

Adult male. Forehead, line above the eyes, and cheeks bright lemon-yellow, approaching gamboge-yellow; rest of upper parts light yellowish green, more greenish yellow on the rump and upper tail-coverts. Quills and rectrices bordered with the colour of the back on the outer webs; quills lined with whitish inside. Below from the throat downwards bright lemon-yellow, in some specimens shading off into whitish yellow or buffy whitish on the lower abdomen. Thighs brownish buff.

Adult female. Similar to the adult male, but more greenish and duller above, less bright yellow below, and distinctly tinged with yellowish olive along the sides of the body.

Young. Greyish olive above, becoming more greenish as the bird advances in age. Below greyish white, subsequently becoming more yellowish; wing-coverts with large buffy-whitish tips.

The tip of the wing is formed by the fourth, fifth, and sixth primaries, which are almost equal in length. The third and seventh are slightly shorter, and the second is much shorter than the sixth and seventh; first rudimentary.

Total length about 5 inches, wing 2·35 to 2·4, tail 2, culmen 0·55 to 0·59, tarsus 0·8 to 0·82. Females have the wing generally a trifle shorter than adult males.

Hab. Island of Lanai.

PALMER describes the soft parts of several specimens as follows :—" Iris dark brown. Upper mandible brown, shading to almost black at tip; lower mandible brownish grey at tip, lighter and with a tinge of yellow at base. Tarsi grey with a brown tinge; toes light greyish; soles flesh-colour." Again :—" Iris hazel. Upper mandible dark brown; lower mandible brown, pale yellowish at base. Tarsi and toes greyish brown; soles flesh-colour, with a yellow tinge."

" The toes and feet of the ' Alauhiio ' vary somewhat in colour and are much darker in

R 2

some individuals. Wilson described the bill and feet as light pinkish, but I never saw one with pink bill or feet, but merely with a faint pinkish tinge in some specimens." (*H. C. P.*)

From Palmer's notes it is evident that the bill is somewhat light and perhaps not very hard during life, and probably changes to a kind of pinkish colour after death, as is the case with many birds. Palmer's notes, however, are almost invariably taken on the spot, at the moment when the bird is shot.

Palmer found this species on the south-west side of Lanai at elevations of about 1500 feet to the uppermost peaks.

The call-note is a short and plain "tweet."

Wilson obtained a male and a female at a spot called Lanaihale (the house of Lanai), at a height of about 3000 feet, and subsequently two immature examples in a gulch at a much lower elevation.

It is perhaps not so very rare, for Palmer sent me a fine series, but the dense bush and the broken country make it very difficult to obtain. In reference to this Mr. Wilson writes :—"No words of mine can convey an idea of the difficulty and danger of collecting in the mountains of Lanai; this is due to the almost impenetrable bush which covers the upland plateau, to the fogs which render riding extremely dangerous, and to the rains which make the nearly perpendicular mountain-trails treacherous even to the sure-footed Lanai horse; indeed, inured as I was to 'steep bits' in my island travels, I must confess that the first trip we made into these solitudes surprised me."

20. OREOMYZA PARVA (*Stejn.*).

ANAUANII.

Himatione parva, Stejn. Proc. U.S. Nat. Mus. 1887, p. 94 (Kauai); Wilson & Evans, Aves Hawaiienses, pt. iii. text & pl. (1892).

Adult male. Above almost uniform greenish yellow, a little brighter on the head and upper tail-coverts. Quills and tail-feathers deep brown, outwardly edged with the colour of the back, inwardly (except at the tips of the feathers) edged with white. Below all over yellow, with a greenish tint, buffy or whitish on the vent. Thighs creamy white. "Iris brown; bill brown, paler at base, pearly grey on basal half of lower mandible; tarsi and feet brownish grey."

Adult female. Similar to the male, but much darker and more greenish above, less bright yellowish below.

Younger birds are similar to the female, but much duller in colours.

In this species the tip of the wing is formed by the third, fourth, and fifth primaries, while the sixth is only slightly shorter than the fifth and the second slightly shorter than the sixth; first primary quite rudimentary.

Oreomyza parva (Stejn.) can always be easily distinguished from all other members of the genus by its small size and the almost uniform yellowish colour of its upper parts. Its not perceptibly curved, but nearly quite straight mandible, as well as its wing-formula (the second primary not longer, but a little shorter, than the sixth), unite it much better with *Oreomyza* than with the very different genus *Himatione*, although it is not the most pronounced form of *Oreomyza*.

Total length nearly 4½ inches; adult males have the wing 2·3 to 2·4, mostly 2·35, the tail 1·6 to 1·65, tarsus 0·73, culmen 0·5 to 0·53; adult females are smaller, having the wing 2·28 to 2·3, the tail 1·48 to 1·51, the culmen 0·5.

This pretty little *Oreomyza* inhabits the hill-forests of Kauai. Stejneger first described it from specimens sent by Mr. Knudsen, and Wilson found it also in several places. The latter author describes its call-note as a low "chirp," but he did not hear it sing. He says that the range seems to extend throughout the forest-region, as it was found by Mr. Francis Gay towards the summit of Waialeale (4000 feet), the highest point of Kauai.

According to Palmer this species generally goes about in small flocks and often in company with other small birds, and it is not very rare. Palmer sent me a very fine series.

Small insects were found in the stomach of this bird. According to Messrs. Wilson and Knudsen it feeds on insects and also occasionally on the honey of flowers.

The nearest ally to *Oreomyza parva* is, in my opinion, *Oreomyza montana*, which, however, is a larger species.

21. OREOMYZA FLAMMEA (*Wilson*).

KAKAWAHIE (*Wilson*).

— · ·   · — — ·

Loxops flammea, Wilson, P. Z. S. 1889, p. 445 (Kalae, Molokai); id. & Evans, Aves Hawaiienses, pt. i. text & pl. (1890).

Adult male. All over coloured with such brilliant scarlet that the artist had immense difficulty in copying the natural tint in the figure on the Plate, and, in my opinion, has not perfectly succeeded. In Wilson's figure the head, neck, and breast are much too dull [1]. Above the colour is a deep scarlet-vermilion, lighter and more flame-scarlet (Ridgw. Nomencl. Col. pl. vii. fig. 14) on the forehead, rump, and upper tail-coverts. Below altogether of a brilliant flame-scarlet. Quills and tail-feathers blackish brown, edged with brownish scarlet on the outer webs. Wing-lining pale scarlet; quills inwardly edged with reddish buffy white. Directly after being shot the iris is hazel, the upper bill light grey at tip and base, darker in centre; lower mandible light grey at tip, shading to pale yellow at base; tarsi and feet greyish, with a pink tinge; soles flesh-colour.

Young males have a more or less strong mixture of deep ferruginous brown or rufous brown above, washed with deep brown along the sides of the body, more obvious in younger individuals.

Quite young males are deep greyish olive, with a slight brownish tint in the plumage, which generally distinguishes them at once from the young of *Oreomyza montana* (Wils.), to which they bear a very close resemblance. Wing-covets largely tipped with buffish white. Below dusky white, tinged with yellowish and strongly washed with olive along the sides of the body.

All sorts of intermediate coloration between the brilliant scarlet adult birds and the dull greyish-olive young are met with, and my large series of these birds is the most complete that could be desired to show the changes of plumage. It would be most interesting for an ornithological student resident on Molokai to observe how long a period is required to complete this entire change of plumage, for the many parti-coloured feathers of most of the intermediately coloured specimens show that the red plumage cannot be acquired suddenly by one moult. Very likely these birds do not assume their final plumage until they are several years old.

[1] The bird described and figured by Wilson as an adult female seems to be an immature male.

Young birds have the iris dark brown; upper mandible brown with the tip light, lower mandible deep salmon-colour; tarsi and feet grey; soles flesh-colour.

Adult female. Above dark olive-brown, washed with orange all over, somewhat redder on the head and on the rump and upper tail-coverts. Below cadmium-orange or orpiment-orange (Ridgw. Nomencl. Col. pl. vi. figs. 1 & 2), deepest on the chin, throat, and breast, lighter and paler on the abdomen, washed with olive on the flanks.

Very few specimens show this final brilliant plumage; in most examples, no doubt younger individuals, the abdomen is strongly mixed with buffy white, the breast is more of a yellow colour, and the upper parts are more olive, even with a greenish-olive mixture.

Quite young females are like the youngest males, but even more greenish above and more tinged with yellowish on the abdomen, so that they are scarcely distinguishable from the young of *Oreomyza montana* (Wils.) from Lanai.

Every intermediate stage is also met with in the females, but they are distinguishable from immature males by having the gay mixture in the plumage of a yellowish or rather orange colour instead of a more or less scarlet shade.

Total length about 4¾ inches, and not above 5 inches, in skins; but, as measured in the flesh, Palmer gives it always above 5 inches. The wing varies in over twenty adult males only from 2·5 to 2·63 inches, the tail from 2 to 2·2, tarsus 1·85 to 1·9, culmen 0·55 to 0·62. The few fully adult females which I have are a little smaller—wing 2·3 to 2·45 inches, culmen 0·5 to 0·55, tail and tarsus about the same as in male.

Nothing but the red colour could have induced Mr. Scott Wilson to call this bird *Loxops flammea*. The colour alone, however, cannot decide a bird's systematic position. In the Sandwich Islands especially we have a number of bright red birds belonging to different genera—the *Himatione sanguinea*, the *Vestiaria coccinea*, the *Loxops coccinea*, and the present species. Not only the unfeathered operculum of the nostrils and the straight elongated bill, but also the different wing-formula, the much stronger feet and legs, and the habits altogether remove this bird from the genus *Loxops*. Wilson, after uniting *Oreomyza mana* (which is most nearly allied to *Oreomyza bairdi*) with *Himatione*, would have been more correct in classifying the bird under consideration with *Himatione* than with *Loxops*.

The colours of the soft parts given above are those of the first adult and the first young bird shot by Palmer, but he subsequently gives the following note regarding these birds in his diary :—"The colour of the bill and feet varies much (probably according to age), from brown to grey on the upper mandible, from creamy yellow to deep rose-colour on the lower mandible, and from brown to light grey with a pink tinge on the feet."

Hab. Island of Molokai.

My collector found this remarkably beautiful bird quite common in all the upper regions of Molokai, chiefly in the districts of Pukoo, Alawa, and Kalae.

The birds were met with almost everywhere in the forests, from the deeper valleys up to the highest peaks, and were not at all rare, but were never noticed among the Kukui trees at the bottom of the ravines. They are fairly active and seemed to have no preference for any particular kind of tree, but were perhaps more often seen on the Ohia trees.

They were not at all shy, but, on the contrary, rather inquisitive, and could without difficulty be attracted when their call-note was properly imitated. The latter is a short and plain " tweet," not to be distinguished from that of *Oreomyza montana* (Wils.) from Lanai.

Their food consists of insects.

Palmer found that the fully coloured bright red birds were much less numerous than the brownish and dull specimens. The dense undergrowth in the forests made it often very difficult to discover the bird, even when its note was heard close by, and it was equally hard to find a bird when shot, but his good dog retrieved every one.

Wilson, who first discovered this beautiful bird, obtained only three specimens, all at Kalae. He justly remarks :—" It may not, however, be safe to consider it rare, as my host easily obtained the native name for me, thus showing the bird to be known to the aborigines." He further says :—" We had lost the way, and while wandering about and searching for the trail, I heard a curious sound—a continued ' chip, chip, chip,' not unlike the sound of chopping wood when heard at a distance—which at first I did not think could belong to a bird ; soon, however, I was undeceived, as a flash of brilliant orange colour passed us in the fog ; when, on trying to follow it up, the continuous metallic note enabled me to get within range and I fired, bringing down two birds, which proved to be male and female.

" The name applied to this bird in the Hawaiian language means firewood ; but whether this is given to it from the note, which, as remarked above, resembles the sound of chopping wood, or from the brilliant flame-colour of its plumage, I am unable to say."

s

22. OREOMYZA BAIRDI, *Stejn.*

AKIKIKI.

Oreomyza bairdi, Stejn. Proc. U.S. Nat. Mus. 1887, p. 99 (Kauai, Knudsen coll.); id. Proc. U.S. Nat. Mus. xii. 1889, p. 385; Wils. Ibis, 1890, p. 193; Gadow, Aves Hawaiienses, pt. ii. p. 14 (structure); Wils. & Evans, Aves Hawaiienses, pt. ii. text & pl. (1891).
Oreomyza wilsoni, Stejn. Proc. U.S. Nat. Mus. xii. 1889, p. 386 (Kauai).

Adult male. Above clear olive-grey, tinged with olive-green on the rump and margins of the quills and rectrices. Beneath pale olive-buff, nearly white on chin, throat, under wing-coverts, and middle of abdomen, where also a more or less distinct sulphur-yellow tinge is obvious; strongly washed with olive-grey along the sides of the body. Bill light brown, with a pinkish tinge; the feet light pinkish brown; iris dark hazel.

Adult female. Exactly similar to the male, except that it may be a shade greener and duller.

Younger birds are similar to adults, but have the lores, forehead, a short superciliary stripe, and the space round the eyes dull white, or rather white suffused with a tinge of palest olive-buff.

Such specimens are very numerous, and I believe them to be younger birds, especially since the closely allied species of Hawaii, *Oreomyza mana* (Wils.), has the forehead also whitish when younger, and in fact much resembles the white-fronted specimens of *O. bairdi*, Stejn.

Palmer found very few pale-fronted birds in the higher hill-parts, while lower down the majority had this character, and often more developed than all those shot at higher elevations.

Stejneger has founded another species on these pale-fronted specimens, which he named after Mr. Scott B. Wilson, but it has no specific value. Not only have I birds in every possible intermediate phase of coloration in my collection, but there is no constant difference in the dimensions or in the wing-formula. It is not surprising that Palmer found more fully adult specimens in one locality, while more immature birds were procured in another, but he also obtained them in both plumages in each place on the same day. There is, however, no certain proof that the white-fronted birds are really the youngest.

Quite young birds appear to have whitish tips to the wing-coverts, as indicated in several of my specimens.

The fourth, fifth, and sixth primaries form the tip of the wing, the fourth and fifth being longest; the third is only very little shorter, but the second is much shorter, than the sixth and seventh; the first is quite rudimentary.

Total length nearly 4½ inches, wing 2·6 to 2·77, tail 1·65 to 1·75, culmen 0·46 to 0·6, tarsus 0·7 to 0·8. (Measured 20 specimens.)

Hab. Island of Kauai.

THIS species was first described by Dr. Stejneger from specimens sent by Mr. Knudsen, of Kauai. Afterwards it was collected and observed by Mr. Scott Wilson, who gave a good plate and descriptions in the 'Aves Hawaiienses.' I extract the following notes from the latter work:—

"It is usually met with in small flocks of from eight to twelve, and is a particularly active bird, continually running up and down the limbs and trunks of the high trees in search of insects; it is, in fact, the most energetic bird of the Hawaiian forests. Its short tail, in Dr. Stejneger's opinion, indicates terrestrial habits, but I only observed it at some considerable height from the ground, in the lofty Ohia and Koa trees, for the dead branches of which it evinces a decided preference. The note is a simple 'twit, twit, twit,' repeated constantly. Its range seems to reach an elevation of 3000 feet."

Palmer's notes are as follows:—

"The 'Akikiki' (this is its proper native name) is undoubtedly found all over the island at altitudes from a thousand feet probably to the uppermost peaks. They were common on the south side of the island between 1000 and 3000 feet; the same may be said of the west side; while they were rare on the north and north-east side at about the same altitudes. They feed on insects taken from the wood and branches. They do not keep so much to the higher trees as many other members of its family, for Palmer and Munro saw them several times very low and flying from low ferns if not from the ground.

"They go about in small flocks and very often in company with other birds. Its note can easily be distinguished from that of *Loxops cæruleirostris* (Wilson)[1], which inhabits the same localities, it being a much shorter note and sounding more like 'twit,' while that of the Ou-holowai, *Loxops cæruleirostris* (Wils.), is deeper and more prolonged, somewhat like 'tweet.'

"No nest was found, but on March the 5th a female was shot with the ovaries greatly enlarged."

[1] *Chrysomitridops cæruleirostris*, Wilson, but a true *Loxops* in my opinion, as will be conclusively shown in Part III. of the present work.

Tⁿⁿ Author's collectors have been actively engaged for some years past in exploring the Islands of the Hawaiian Archipelago, and many species of birds, new to science, have been discovered by them; these, with others, will be figured in a series of about 46 Hand-coloured Plates,—most of which will be delineated by that master-hand Mr. J. G. KEULEMANS, and others, including tinted Plates, are entrusted to Mr. F. W. FROHAWK's careful treatment.

In addition to the above, a most interesting series of Collotype Photographs, showing various phases of bird-life and landscape, will be included in the Volume.

The size of the book will be imperial 4to, and will be issued in 3 Parts, price £3 3s. each, net.

As no separate Parts will be sold, Subscribers are expected to continue their Subscriptions until the work is completed.

The Edition is limited to 250 Copies.

CONTENTS OF PART II.

History of the Birds of the Hawaiian Islands.

PART III.]   [DECEMBER 1900.

16/20

THE

AVIFAUNA OF LAYSAN

AND THE

NEIGHBOURING ISLANDS;

WITH A COMPLETE HISTORY TO DATE OF THE

BIRDS OF THE HAWAIIAN POSSESSIONS.

BY

The Hon. WALTER ROTHSCHILD.

Illustrated with 25 Coloured and 2 Black Plates by Messrs. Keulemans and Frohawk.

LONDON:
R. H. PORTER, 7 PRINCES STREET, CAVENDISH SQUARE, W.
1900.

23. OREOMYZA MANA (*Wils.*).

HAWAIIAN AKIKIKI.

Himatione mana, Wilson, Ann. & Mag. Nat. Hist. ser. 6, vii. p. 460 (1891); id. Aves Hawaiienses, pt. iv. (1893) (Hawaii); Perkins, Ibis, 1893, p. 105.

Adult male. Above of a light and bright olive-green, shading into ashy green on the hind-neck and top of the head, a little paler on the forehead. Wings and tail greyish brown, outwardly edged with olive-green. A dusky spot before the eyes. Beneath dull greenish buff, more yellow in the middle of the abdomen, more green on the sides of the body, shading into whitish greenish buff on the throat and chin. Iris dark hazel. Upper mandible brown, lighter at base; lower mandible dark grey, lighter at base and tip. Tarsi and toes greyish brown; soles pale yellow.

Adult female. Similar to the male, but perhaps a little more olive above.

Immature birds have a buffy-white forehead and are more olive above and paler below. In this plumage they closely resemble the white-fronted specimens of *Oreomyza bairdi*, but they are greener above and below.

Total length about 4·75 inches, wing 2·55 to 2·76 in adult birds, tail 1·6 to 1·8, culmen in adult birds 0·5 to 0·55, tarsus 0·7 to 0·75.

Palmer states that the birds with the light mark on the forehead are the young. In August 1893, when in the district of Kona, he found all the adult birds in a poor state of plumage, mostly without tails. They were evidently moulting after the breeding-season, and Palmer procured young birds only at this time, with a few exceptions.

This bird was discovered by Mr. Scott B. Wilson on the island of Hawaii, where he procured three specimens. Its affinity to *Oreomyza bairdi* is striking, which was at once noticed by Count Salvadori (see Wilson, Aves Haw. pt. iv.).

Both Perkins ('Ibis,' *l. c.*) and Henry Palmer found this bird abundant in Kona, though not quite so numerous as *Chlorodrepanis virens*. Perkins says he "failed to notice *H. mana* in the lower forest, though it was common enough at 4000 feet, nor did the male appear to assume such a bright yellow plumage as did the curved-billed species."

Palmer made the following observations :—

"This species is quite numerous in the upper forest of Kona, and is generally met in

T

small flocks of from two to six individuals. It is also not uncommon, though less numerous, in the districts of Hamakua and Hilo on the N.E. side, and in Puna on the east side.

"It was never seen very low down, and inhabits the forests of from about 2500 to 5000 or 6000 feet. While *Chlorodrepanis virens* generally searches for food among the leaves, this bird hunts for its food on the trunks and branches of the trees.

"It is rather active, and has a clear, short, and somewhat sharp call-note, very much like that of *O. bairdi*, somewhat like a shrill 'chip.' It is easily distinguished from the notes of *Chlorodrepanis virens*."

According to Mr. Perkins, the notes of *O. mana* and *O. bairdi* are rather less sharp than in the other species of the genus.

It inhabits only the island of Hawaii.

Mr. Henshaw writes me that he found it rare in Olaa, even in winter, occurring at altitudes of about 1000 feet up to an indeterminate height.

NOTE ON OREOMYZA PARVA.

IN Part II. of this work I have included the *Himatione parva* of Wilson in the genus *Oreomyza*, together with *O. maculata, newtoni, montana, flammea*, and *bairdi*. Formerly *O. maculata, newtoni,* and *montana* were placed in the genus *Himatione, O. flammea* most erroneously in *Loxops*, and *O. bairdi* alone was placed in the genus *Oreomyza*. I am glad to see that I am followed by Perkins, Wilson, and other authorities on Sandwich-Islands birds in my arrangement, except that Perkins has created a new genus for *Oreomyza parva*, which he calls *Rothschildia* ('Aves Hawaiienses,' part vii.). Apart from the fact that the generic term *Rothschildia* has already been employed for a genus of moths, I cannot see the necessity for this genus. The bill of *Oreomyza parva* is so little curved and so much straighter than in *Himatione* and *Chlorodrepanis* that it will at once be placed with *Oreomyza*. The nostrils are not more bare than in several *Oreomyzæ*, and the structure of the wing seems rather to point to *Oreomyza* than to *Himatione*. Thus there are no outward characters to separate it generically ; and there remains only the structure of the tongue, alluded to by Mr. Perkins, and the fact that *O. parva* is, according to Mr. Perkins, more a honey-sucker than *Oreomyza*. If all these observations are correct, they would, I should say, tend more to uniting the two genera *Chlorodrepanis* and *Oreomyza* than to the erection of a new genus.

[*Oreomyza newtoni (antea*, Pt. II. p. 115) is described and figured, under the name of *Himatione newtoni*, in part iv. (1896) of the 'Aves Hawaiienses.']

24. OREOMYZA PERKINSI, sp. nov.

Adult male. Above light olive-green (Ridgway, Nom. Colours, pl. x. no. 18), brighter on the rump. Quills black, edged with oil-green; tail-feathers dark brown, edged with oil-green. Below olive-yellow; vent greenish white. Thighs dirty white. Under wing-coverts white, with a yellow tinge; lores black. Iris dark brown. Legs and feet greyish brown. Soles of feet yellowish flesh-colour. Upper mandible dark brown, with paler base; lower mandible grey. Total length about 5·5 inches, wing 2·6, tail 1·7, tarsus 0·85, culmen 0·63.

One male, Puulehua, Hawaii, September 25th, 1891.

This remarkable specimen has a long but straight bill; the nostrils are covered by an operculum, which, however, leaves a minute space at the lower margin open; the second primary is about one-tenth of an inch shorter than the third; the third, fourth, and fifth are nearly equal, but the fourth is a trifle longer. The coloration is that of *Chlorodrepanis virens.* The bird thus occupies a somewhat intermediate position between *Oreomyza* and *Chlorodrepanis*, and it might be a hybrid between *Oreomyza mana* and *Chlorodrepanis virens*; but, as *Oreomyza parva* of Kauai occupies a similarly intermediate position between the genera *Oreomyza* and *Chlorodrepanis*, I think it is quite possible that it is a good species, and I have much pleasure in naming it after Mr. Perkins, who has done such very good work on the Hawaiian Islands in furthering our knowledge of their biology.

THE GENUS CHLORODREPANIS.

THIS comprises the green forms of the old genus *Himatione*. They are all greenish and very closely allied to each other. They differ from *Himatione* in having a softer and somewhat more fluffy plumage generally, in having no elongated, narrow, and stiff feathers on the head, and in having the primaries more pointed, not at all truncate at the tips. In the wing-formula, shape of tongue and nasal operculum, and manner of nesting, they agree with *Himatione*. The bill is more curved than in *Himatione*.

The differences from *Oreomyza* are stated under the genus *Himatione*.

Both *Himatione* and *Chlorodrepanis* live more among the foliage, while the species of *Oreomyza* are more particularly found frequenting the branches and trunks of trees and feeding among them.

Key to the Genera Himatione *and* Chlorodrepanis.

THIS key applies only to the adult males, because the young of several of the species are not always distinguishable. The species of *Chlorodrepanis* are all somewhat closely allied, and should be regarded as representative races of one species. The largest is *C. stejnegeri* from Kauai, characterized by its large and powerful bill. The other three green species are much more alike; but *C. virens* is considerably less yellow and not so bright as *C. chloris* and *C. wilsoni*, and of these latter two *C. chloris* is darkest about the head, brightest on the breast, and a dark shade runs right and left from the throat, thus making the throat very conspicuously yellow. In some species the females differ from the males, while in others they are almost if not quite similar. Thus the females of *C. chloris* and *C. wilsoni* are much alike and different from the males, while those of *C. virens* and *C. stejnegeri* are almost similar to their males.

A. Plumage chiefly red, feathers of the head harder, stiffer, and pointed. (*Himatione*.)
    a. Larger, wing more than 2·8 inches, of a much deeper blood-red colour . . . . . *H. sanguinea.*
    b. Smaller, wing less than 2·8 inches, of a lighter red colour . . . . *H. fraithi.*
B. Plumage chiefly green; feathers on the head rounded and not harder or stiffer than usual.
    (*Chlorodrepanis.*)
    c. Bill very broad and high at base . . .   . *C. stejnegeri.*
    d. Bill less powerful.
        a'. Breast and abdomen brightest, and almost pure dark yellow . . . . *C. chloris.*
        b'. Breast and abdomen with a light but distinct greenish tinge; bill larger . . . *C. wilsoni.*
        c'. Breast and abdomen with a much darker olive-green tinge; bill smaller . . . . *C. virens.*

                                                                                         u

25. CHLORODREPANIS VIRENS (*Gm*).

"AMAKIHI."

Olive-green Creeper, Lath. Gen. Syn. i. pt. 2, p. 740 (1782) (Sandwich Islands) (Leverian Museum).
Certhia virens, Gm. Syst. Nat. i. p. 479 (1788, ex Latham) ; Lath. Ind. Orn. p. 290 (1787) ; Tiedem. Anat. &
    Naturg. Vög. ii. p. 431 (1814) (Ins. Amicis ! !).
L'Héorotaire vert olive, Audeb. & Vieill. Ois. Dor. pls. 67, 68 (1802).
Melitreptus virens, Vieillot, Nouv. Dict. d'Hist. Nat. xiv. p. 330 (1817) ; Bonn. & Vieill. Enc. Méth. ii.
    p. 607 (1823).
Nectarinia flava, Bloxam in Byron's Voy. ' Blonde,' p. 249 (1826). (Type in Brit. Mus. examined.)
Drepanis flava, J. E. Gray, Zool. Misc. p. 12 (1831) ; Dole, Proc. Boston Soc. Nat. Hist. xii. p. 298 (1869) ;
    Scl. P. Z. S. 1878, p. 348 ; id. Report Birds 'Challenger' Exped. p. 95 (1881) ; Dole, Haw. Altman.
    1879, p. 45.
Phyllornis tonganensis (sic !), Less. Rev. Zool. 1840, p. 165.
Himatione virens, Sharpe, Cat. B. x. p. 9 (1885) ; Scott Wilson, Ibis, 1890, p. 185 (peculiar to Hawaii) ;
    Perkins, Ibis, 1893, p. 105 (abundant in Kona, Hawaii), p. 106 (nests) ; Newton, P. Z. S. 1897, p. 893
    (eggs) ; Wilson & Evans, Av. Hawaii. pt. vi. text & plate (1896) ; iid. photograph of nest in pt. vii.
    op. cit. (1899).
[To this synonymy must also be referred a part of Gray's synonymy of his supposed female of *Drepanis
sanguinea*, Cat. B. Trop. Is. p. 8 (1859).]

Adult male. Above yellowish olive-green, lighter on the rump and upper tail-coverts, and
with a golden tinge on the forehead. Lores blackish. Quills blackish, bordered
outwardly with the colour of the back, inwardly lined with ashy. Rectrices blackish
brown, bordered with the colour of the back. Below green-yellow, with an olive-green
tinge. Under tail-coverts much paler. Under wing-coverts ashy white. Iris brown.
Maxilla dark horny brown, lighter at base. Mandible greyish, dark at tip. Feet brown.
Total length 4·75 to 5 inches, wing 2·6 to 2·7, tail 1·8 to 1·88, tarsus 0·85, culmen 0·6
to 0·64.

Adult female. Nearly a dozen females before me, shot in August 1893 at Kona, Hawaii, are
all much less brilliant in colour, having a dark olive and greyish tinge above, and the
underparts paler and somewhat cloudy ; but it is quite possible that not any of them
are quite adult, for one female (if rightly sexed) from the same place is exactly like the
adult males.

Young birds have the upperside with an olive and a grey tinge, the underparts clouded with
dusky. The wing has two, not always very regular, cross-bars, formed by the olive-buff
tips to the greater series of wing-coverts.

u 2

This bird has been known since the earliest days of Hawaiian ornithology, and for a long time was believed to be the female of the red *Himatione* (see synonymy).

It is restricted to the island of Hawaii, though very closely allied forms inhabit the other islands.

PERKINS ('Ibis,' 1893) says "*Himatione virens* is abundant in Kona, and particularly in the higher forest.   However, it ranges down in some numbers even as low as 1100 to 1500 feet.   They are chiefly insectivorous, feeding on lepidopterous larvæ and other insects, but are in places very partial to the Lehua-flowers.   On the rough lava, on which this tree grows abundantly, at the foot of the Mauna Hualalai, where the mountain rises suddenly from the high table-land, I frequently observed them sucking the honey of these blossoms, and, in company with *Himatione sanguinea*, on the same tree, certainly as high as 7000 feet up that mountain—at an altitude where in the morning the ground was covered with hoar-frost."

Palmer's notes declare that the Amakihi is found all over the island, but most numerous in the Kona districts, S.W. and W. of the island, in Hamakua, N.E., and in Hilo and Puna districts.   It ranges from 1000 feet above the sea to the uppermost forest regions, about 10,000 feet high.

It is generally seen searching for its food amongst the smaller branches and leaves, and sucking nectar from various flowers, and even feeding on "Cape gooseberries."

Palmer says it is one of the commonest birds of Hawaii and that he saw it quite low near the sea and up as high as he had been, *i. e.* about 9000 feet.   It has a chirping call-note, not unlike that of a Canary-bird, and rather a sweet kind of warbling song of several notes. Palmer calls it active in its movements, very swift in flight, and says it "seems to alight all of a sudden without lessening its speed beforehand."   On May the 5th one was seen carrying moss or some other green material as if for building a nest.   On May the 17th three nests were found, but all empty.   They were "much like the one received from Mr. Baldwin, and were outside built of sticks, then of moss, and lined with very fine roots, open above and about 2½ inches in diameter."

Palmer received a nest of *Chlorodrepanis virens* from Mr. Baldwin.   It is a roundish structure of grass, twigs, roots, and moss, lined with finer roots and hairs.   It seems to be somewhat flattened, and is certainly not a deep cup like that of *Chasiempis*; but its exact shape and size cannot be stated with certainty, as the nest is no longer on the branch (the only way nests should be collected if possible) and was foolishly pressed into a small box.

Perkins describes the nest as being lined with roots, and with fruit-capsules of the "poha," dry and more or less skeletonized, woven in the outside.

The two eggs sent by Palmer are ovate in shape and measure 19·5 × 14·5 millimetres each.   They are white, marked with brown blotches and spots, and some lilac-grey paler underlying patches, the brown as well as the greyish patches forming a wide ring near the broad end, more distinct in one than in the other.   If held against the light they shine through white with a faint tinge of greenish yellow.   The eggs sent by Palmer agree fully with those figured by Wilson, but less with the figures in the 'Proceedings' of the Zoological Society, which seem to be somewhat roughly executed.   Mr. Wilson also figured a nest.

26. CHLORODREPANIS CHLORIS (*Cab.*).

AMAKIHI OF OAHU.

Himatione chloris, Cab. Mus. Hein. i. p. 99 (1850) (Oahu); Wilson, P. Z. S. 1890, p. 446, and Ibis, 1890,
  p. 185 (partim) (Oahu, Lanai, Molokai, all called *chloris*, although some differences were stated between
  specimens from the different islands); Wilson & Evans in Aves Hawaiiensis, part vi. plate & text.
Himatione flava, Reichenb. (nec Bloxam!) Handb. Spec. Orn. Scans. p. 255; Pelzeln, Journ. f. Orn. 1872,
  p. 28 (Oahu).
[Sharpe, see Cat. B. x. p. 9, and others before him never tried to distinguish between *H. chloris* and
  H. virens.]
[Part of the synonymy of Gray's supposed females of *Drepanis* (*Himatione*) *sanguinea*, Cat. B. Trop. Is. p. 8,
  is referable to *H. chloris*.]

Adult male. Above bright yellow-green or green-yellow, a shade darker on the head and a
shade lighter on the rump and upper tail-coverts. Feathers round the bill, those
immediately above it on the forehead, as well as lores and chin blackish. A narrow
yellow line from the forehead to the upper part of the eye, but not produced through the
eye. Quills deep brown, outwardly margined with yellowish green, inwardly with pale
ashy. Under wing-coverts white, washed with yellow on the margins. Underparts
almost pure dark yellow, brightest on the breast, paler on the middle of the lower
abdomen, vent, and under tail-coverts. Sides of the head dark yellow, with a strong
wash of greenish olive. A dark shade runs right and left along the throat, thus leaving
the middle of the throat bright yellow by contrast. Sides of the body washed with
olive. Rectrices dark brown, outwardly bordered with yellowish green. Bill: maxilla
blackish brown, lighter at base; mandible lighter, only deep brown towards the tip.
Tarsi and toes greyish brown; soles flesh-colour. The irides are marked as "brown"
and "hazel." Total length marked about 5½ inches by the collector, as measured in the
flesh, but not exceeding, if reaching, 4½ inches in skins; wing 2·58 to 2·65, tail 1·75,
tarsus 0·8 to 0·85, culmen 0·55 to 0·65.

Adult female. Quite different from the adult male. Above olive-grey with hardly any
greenish tint, washed with greenish olive on the rump and upper tail-coverts. A whitish
line from base of bill to eye. Below pale whitish ashy, very slightly washed with olive-
greenish. Quills and rectrices bordered with olive-greenish. Two rows of creamy-white
spots on the upper wing-coverts. Under wing-coverts whitish. Total length in the
flesh (collector's notes) about 4½ and 4¾, in skins about 4¼ inches; wing 2·4 to 2·45,
tail about 1·5, tarsus 0·75 to 0·8, culmen 0·55 to 0·6.

Immature males are like the females, but as they advance in age a greenish tint is soon recognizable and increases the more the bird advances in age.

The *young* in first plumage have two broad buff bars across the wings, formed by large tips to the greater series of wing-coverts. The secondaries are also broadly tipped with brownish buff.

C. chloris inhabits Oahu. Its nearest ally, *C. wilsoni*, Rothsch., is found on the islands of Maui, Molokai, and Lanai.

C. CHLORIS is generally distributed over the hilly parts of Oahu. It frequents all kinds of trees, but is more often seen on flowering Koa than anywhere else, probably because these flowers seem to be more frequented by insects than others. It is generally seen flitting about very actively amongst the leaves of the trees, and is often rather shy. Its song is a short warble of a few notes, very much like that of other species of the group. Its call-note is a kind of " sweet," whistled through the teeth.

The food consists of both insects and nectar.

A specimen obtained by Townsend is in the Liverpool Museum.

Mr. Hartert has also examined the type of Professor Cabanis' description.

27. CHLORODREPANIS WILSONI (*Rothsch.*).

INTERMEDIATE or WILSON'S AMAKIHI.

— · —   —   —   —

Hemignathus obscurus (sic—errore, nec Gmelin), Fiusch, Ibis, 1880, p. 80 (Maui—"the commonest species").
  (Specimens with the name of *Hemignathus obscurus* in Dr. Finsch's handwriting on the label in the British
  Museum examined.)

Himatione chloris, Scott Wilson (errore, nec Cabanis), P. Z. S. 1889, p. 447 (Lanai and Molokai—differences
  from typical *H. chloris* from Oahu stated); id. Ibis, 1890, p. 185 ("Molokai, Lanai, and, I believe, Maui").

Himatione wilsoni, Rothsch. Bull. B. O. C. vol. i. p. xlii (April 1893); id. Ibis, 1893, p. 113 (reprinted) (Maui);
  Wilson & Evans, Aves Hawaiienses, part vi. plate & text (1896) ("peculiar to Maui").

Himatione ch'oridoides and *kalaana*, Wilson & Evans, ibidem.

Himatione steynegeri (sic, errore, non *H. steinegeri*, Wilson), Schauinsland, Abh. nat. Ver. Bremen, xvi. 3
  (1900) (Molokai).

Adult male. In general aspect and colour similar to the male of *Chlorodrepanis chloris* (Cab.),
from Oahu, but the breast and abdomen are equally tinged with olive-green, and the
throat and sides of the head are of the same tint. As a rule, the bill is larger than that
of *C. chloris*. It is easily distinguished from *C. virens* (Gm.) of Hawaii by the larger
bill, while the underside, although being less bright than in *C. chloris*, is more yellow.
Iris dark brown or hazel; maxilla deep brown, with lighter base and edges; mandible
brown, greyish towards the base. Wing 2·45 to 2·55 inches, tail 1·65 to 1·78, culmen
0·65 to 0·7, tarsus 0·8.

Adult female. Above greyish olive-green; below pale yellowish, washed with greyish olive
along the sides of the body, more yellow on the breast. Wing-coverts with two rows of
light yellowish terminal spots. Smaller than the male; wing 2·2 to 2·4 inches.

Immature males are like the females, but they become more yellow as they advance in age.

Young birds have very conspicuous pale yellow cross-bars on the wings.

The differences between the adult males of *C. wilsoni* and *C. chloris* are stated in the "key."
The females of *C. chloris* are more greyish above, those of *C. wilsoni* more greenish.
Young birds of the two species are sometimes hardly or not distinguishable.

The types of *C. wilsoni* were shot by Palmer on the 14th July on Maui.

I HAVE now before me many specimens from Maui, Molokai, and Lanai. Mr. Wilson formerly
united the birds from these islands with *C. chloris* of Oahu, but now he has given names to
all of them, accepting my name *wilsoni* for the Maui form, calling the Lanai birds *chloridoides*,

and the Molokai ones *kalaana*. When first trying to diagnose my *wilsoni* from Maui, I compared it with *stejnegeri* of Kauai, not having true *chloris* before me, but *C. chloris* is no doubt the nearest ally. Mr. Wilson's views about the distinctness of these forms are apparently somewhat unsettled, but finally he accepted Mr. Perkins's view that they were all distinct. However, these forms cannot be distinguished at all. Mr. Wilson's differences do not exist in my series, nor in those collected by Perkins and now in the British Museum.

Mr. Wilson (P. Z. S. 1890) is certainly wrong in calling the underparts of *C. chloris* "whitish buff, tinged with yellow." He repeats this statement on p. 2 of his article on *C. chloris*, while on p. 1 he had described it correctly as "golden yellow." Also the statement that " *C. chloridoides* " has a "light lemon-yellow" underside "shading into buff' on the flanks " is not correct, the flanks being more olive than " buff." My Molokai specimens are just as bright as those from Maui and Lanai.

This is not the only instance in which islands of the central group are inhabited by the same form.

Mr. Wilson tells us that all the specimens he shot on Lanai were observed in some fine guavas, quite 30 feet in height. On Molokai he saw them searching for their insect-food among the low shrubs of " ohia " which cover the sunny slopes of the ravines.

I need not say that this species is named in honour of Mr. Scott B. Wilson, who was one of the first who brought specimens of it to Europe, and to whom Hawaiian ornithology owes so much of its recent progress.

28. CHLORODREPANIS STEJNEGERI (*Wils.*).

AMAKIHI OF KAUAI.

"? *Himatione chloris*, Cab.," Stejn. Proc. U.S. Nat. Mus. 1887, p. 96 (Kauai).
Himatione stejnegeri, Scott Wilson, P. Z. S. 1880, p. 446; id. & Evans, Aves Hawaiienses, pt. iii. text & plate (1892).

In general colour like *Chlorodrepanis virens* (Gm.), but always easily distinguishable by its large beak, which is very high and broad at base. Culmen 0·75 to 0·8 inch, wing 2·57 to 2·7, tail 1·65 to 1·72, tarsus 0·85 to 0·9.

The female seems to be similar to the male; and one young specimen, shot on March 24th, 1891, is marked female and is the brightest in plumage of all before me. Most of the females before me, however, are somewhat less greenish above and have shorter bills, but this is not constant in my series.

This species was first separated by Wilson, *l. c.* It seems to be generally distributed over the island of Kauai wherever there is thick bushwood or forest, but is perhaps less numerous than some of its allies on other islands.

It feeds on insects and nectar, but Palmer saw it once feeding on an over-ripe banana, from which it may have sucked the juice.

THE GENUS HIMATIONE.

This genus was first established by Cabanis (Museum Heineanum, i. p. 99), with *Himatione sanguinea* as the type, and including the green-coloured species. This custom has been followed to the present day, until Perkins proposed (in Wilson's 'Aves Hawaiienses,' pt. vii.) to separate the green forms as a new genus, *Chlorodrepanis*. This seems to be sufficiently founded in fact, and I therefore accept it. In *Himatione*, as restricted to the red species (*H. sanguinea* and *H. frailhi*), the adult bird is red, the feathers of the head are elongated, pointed, and somewhat stiff, the primaries are somewhat truncate at the tip, the nostrils covered with a bare operculum, at the base of which are some bristles; the third or fourth primary is longest, the second a little shorter, the first rudimentary; bill strong, curved, and with a sharp point; tongue brush-like and tubular. Open nests on trees; eggs spotted.

Both *Himatione* and *Chlorodrepanis* differ from *Oreomyza* (see p. 111) in having a distinctly curved, somewhat longer beak, in having a larger and better fitting operculum over the nostrils, and in having the second primary very little shorter than the third, which is generally equal to the fourth and forms the tip. In typical *Oreomyzæ* the second primary is, as a rule, much shorter than the third, which again is shorter than the fourth and does *not* form the tip.

29. HIMATIONE SANGUINEA (Gm.).

APAPANE.

Crimson Creeper, Lath. Gen. Syn. i. p. 739 (1782) (Sandwich Islands, Leverian Museum).
? Bird of a deep crimson colour, Cook, Voy. Pacific Ocean, ii. p. 227 (1784).
Certhia sanguinea, Gm. Syst. Nat. i. p. 479 (1788) (based on Latham's Crimson Creeper); Lath. Ind. Orn. i.
   p. 290; Donndorff, Orn. Beytr. i. p. 643 (1794); Steph. in Shaw's Zool. viii. p. 231 (1812); Tiedemann,
   Anat. Naturg. Vög. ii. p. 431 (1814).
L'Hérotaire cramoisi, Vieill. Ois. Dor. ii. p. 128, pl. lxvi. (1802).
Nectarinia sanguinea, Cuv. Règne Anim. i. p. 410 (1817).
Le Pichion cramoisi, Petrodroma sanguinea, Vieill. Nouv. Dict. xxvi. p. 108 (1818) (It is curious that Vieillot
   should give *Tunna* as the habitat—an island south of Erromanga, New Hebrides ! the quotations and
   description of Vieillot, however, make it clear that he referred to our present bird); id. Enc. Méth.
   p. 621 (1823).
Nectarinia byronensis, Bloxam, Voy. ' Blonde,' p. 249 (1826).
Drepanis byronensis, Gray in Griffith, Anim. Kingd. p. 390 & pl. (1829); id. Zool. Miscell. p. 12 (1831).
Myzomela? sanguinea, Gray, Gen. B. i. p. 118 (1846).
Drepanis sanguinea, Hartl. Syst. Verz. Mus. Bremen, p. 16 (1844) (teste Wils. & Evans); id. Arch. f. Naturg.
   1852, p. 131; Gray, Gen. B. i. p. 96 (1847) (excl. synonyms, in which is included *Certhia virens*, Gmel.,
   Nectarinia flava, Bloxam, and *Phyllornis longanensis*, Less.); Bp. Consp. i. p. 404 (excl. synonyms)
   (1850); Licht. Nomencl. p. 55 (1854); Cass. U.S. Expl. Exp. Mamm. & Orn. p. 439 (1858); Dole,
   Proc. Boston N. H. Soc. xii. p. 297 (1869) (Hawaii, Kauai); id. Hawaiian Almanac, 1879, p. 44; Scl.
   P. Z. S. 1878, p. 347; id. Ibis, 1879, p. 92; Finsch, Ibis, 1880, pp. 79, 80 (Maui : sexes alike).
Himatione sanguinea, Cab. Mus. Hein. i. p. 99 (1851); Reichb. Handb. spec. Orn. p. 255, partim, pl. 612.
   fig. 3834 (1853); Scl. Ibis, 1871, p. 360; Pelz. J. f. O. 1872, p. 27 (spec. from Enero, Oahu; descript.
   & measurements: from the size and form of bill, von Pelzeln was inclined to consider *Certhia virens*
   [*Chlorodrepanis virens* of modern orn.] the young bird of *H. sanguinea*); Sundev. Tent. p. 48 (1872);
   Sharpe, Cat. B. x. p. 8 (1885); Stejn. Proc. U.S. Nat. Mus. 1887, p. 95 (Kauai); Wilson, Ibis, 1890,
   p. 183; id. & Evans, Aves Hawaiienses, pt. ii., plate & text (1891); iid. in Introd. p. xxi, fig. of nest
   (1899); Gadow, Aves Hawaiienses, pt. ii. p. 13, pl. iii. figs. 40, 41 (1891); Rothsch. Avif. Laysan etc.,
   pt. i. pl. 1 (August 1893).
Drepanis (Himatione) sanguinea, Gray, Cat. B. Trop. Is. p. 8 (1859) (excl. about half of the synonyms :
   Gray erroneously supposed that the green *Chlorodrepanes* were the females of the red *Himatione*);
   id. Hand-l. i. p. 113 (1869).

Adult male. Above of a dark blood-red crimson, lighter and much brighter on the head.
   Underparts deep crimson, shading into brownish white or white on the lower abdomen
   and under tail-coverts. Thighs brownish black. Primaries black, the longer ones with
   a narrow but distinct outer edge of white or whitish. Secondaries black, most of

them with broad dark crimson outer edges. All the remiges lined with pale ashy inside.
Rectrices black. Under wing-coverts ashy. All the feathers of the body are dark grey
at their bases, with more or less developed whitish shaft-lines; those of the head and
hind-neck, throat, and upper breast have also whitish subterminal spots, which are
easily seen showing through on the hind-neck. Iris dark brown; bill and feet black.

Adult female. The differences in colour stated by Scott B. Wilson in ' Aves Hawaiienses ' do
not exist. On the contrary, in most of my females the secondaries are not at all edged
as Wilson says (*i. e.* with the colour of the back), but with orange; this, however, is not a
sexual character, as some females have not got it, while it is, on the other hand, visible in
some males also. The females are entirely similar to the males in colour, but generally
smaller.

Measurements of 10 *adult males* :—Total length about 5 inches, wing 2·8 to 2·95 (average 2·9),
tail 2 to 2·2 (never 2·90 as given by Scott B. Wilson), tarsus 0·9 to 0·96, culmen 0·65
to 0·75.

Measurements of 10 *adult females* :—Total length about 4·75 to 4·8 inches, wing 2·65 to 2·85,
average 2·7 (2·9 in one specimen, which probably is wrongly sexed), tail 1·8 to 1·95,
tarsus 0·78 to 0·9, culmen 0·55 to 0·7.

The quite young bird has no crimson colour in its plumage at all. It is brown above, the
bases of the feathers being ashy grey. Wing and tail black ; primaries edged narrowly
with white outside ; secondaries broadly edged with buffy brown. Wing-lining inside
ashy. Beneath brownish buff, shading into whitish buff on the abdomen ; throat washed
with orange. When advancing in age the head becomes first almost blackish, intermixed
with orange; crimson feathers appear here and there, apparently first on the back and
abdomen, as a rule. Young birds have the maxilla brown, with the utmost tip yellowish ;
the mandible brown, with the base yellowish ; feet brown.

This species is distributed over the whole Hawaiian archipelago. Palmer procured good
series on Hawaii, Kauai, and Oahu. He records it as being most numerous on Molokai and
on Maui. The bird, an adult male, from Maui is as pale on the top of the head as the
Laysan species, *Himatione fraithi*, Rothsch., but otherwise agrees with those from the other
islands, and is biggest of all, the wing measuring at least 3 inches (slightly damaged), the
tail 2·3. This specimen is in very worn plumage, and probably not specifically distinct.
Scott B. Wilson obtained *H. sanguinea* also on Lanai.

The native name of *Himatione sanguinea* is on all the islands "Apapane," but in the
Hilo district on Hawaii it is often called "Akakane." It is a common bird in the hills and
reaches up to a considerable elevation. It has often been observed by Wilson, Perkins, and
Palmer sucking honey from flowers, especially on the Ohia, Mamane, and Lehua plants.
Its food, however, consists both of nectar and insects. Scott Wilson gives the following
particulars as to its habits :—"The note of the Apapane is a feeble though clear 'tweet'
twice repeated, but it also has a pretty simple song, generally heard soon after sunrise or
towards sunset. In its flight the white under tail-coverts are very conspicuous and serve to
easily determine it on the wing."

Palmer writes as follows :—" On November 23rd a pair were observed building their nest, but it was as yet only half built. It was at a great height, almost at the top of a tree, in a fork, well concealed from below, and built of moss." Mr. Scott Wilson writes :— " Although I did not find a nest of the Apapane, I shot a female on the 24th of May, 1887, at Kaáwaloa in the district of Kona, in the ovary of which was an egg almost ready for exclusion, a circumstance which enables me to fix approximately its breeding-time, which seems to be later than that of the Iwi, for I had shot several of the latter before the above date."

A nest, an open flat cup-shaped structure, is figured by Wilson (part vii. 1899).

The only very close ally of *H. sanguinea* is *H. fraithi* of Laysan, which differs in being smaller, with a smaller bill, and in being generally of a much paler red. The fine series collected by Professor Schauinsland, however, shows that freshly-moulted individuals have a deeper red and are not easy to distinguish from *H. sanguinea*, but the latter does not fade to such a pale red as *H. fraithi*.

GENUS PALMERIA.

Bill strong and round, about as long as the head, resembling that of *Viridonia*, but somewhat slenderer and slightly curved. First primary quite rudimentary; third, fourth, and fifth nearly equal and longest; second and sixth only about one-tenth or one-eighth of an inch shorter than the longest three. The long scutes of the metatarsus more or less fused. Tail almost square, the central and lateral rectrices very little shorter.

The most remarkable peculiarity of this genus among the *Drepanidæ* is the structure and coloration of the plumage. The feathers of the forehead are erect, elongated, and narrow, those of the crown narrow but not erect, those of the underside all more or less pointed. The grey and scarlet tints of the plumage are very singular.

The sexes do not differ in colour; but the young birds are almost deficient of all scarlet and silky-grey colours, and the feathers of the head are much less modified, although the frontal crest is just indicated. In the structure of the feathers this genus resembles *Himatione* more than any other.

WHEN I first described the genus in 1893 I hastily concluded, from some superficial characters of the plumage, that it belonged to the *Meliphagidæ*; but in 1894 I noticed my error, and corrected it in a preliminary note in the Bull. Brit. Orn. Club in February. This note was published over two months before part v. of the 'Aves Hawaiienses' appeared, in which Dr. Gadow corrected my original mistake of the position of *Palmeria*.

30. PALMERIA DOLEI (*Wils.*).

MAMO OF THE NATIVES ON MOLOKAI.

Himatione dolei, Wilson, P. Z. S. 1891, p. 166 (Maui : short description of a very young bird which did not show the striking peculiarities of the adult bird).

Palmeria mirabilis, Rothschild, Ibis, 1893, p. 113 (Maui : genus *Palmeria* established) ; id. Bull. B. O. C. i. p. xvi (Dec. 1892) (specimens exhibited).

Palmeria dolei, Rothschild, Bull. B. O. C. iii. p. ix (Nov. 1893) (*P. mirabilis* identified with *H. dolei*) ; id. Bull. B. O. C. iii. p. xxv (Feb. 1894) (corrected the previous mistake that *P. dolei* was a *Melphagine* bird, concluding that it belongs to the *Drepanidæ* : see reprints, Ibis, 1893, p. 249 ; 1894, pp. 121, 301) ; Wilson & Evans, Aves Hawaii. part v. (1894) (plate and text, and note on systematic position by Dr. Gadow).

Adult male. Forehead with a heavy crest of greyish-white feathers, curled forward. Top of head with crest-like narrow and pointed feathers of a dark grey colour and with narrow and sometimes obsolete red tips, these feathers being grey along the middle, black on their edges and tipped with red. The vermilion or bright flame-scarlet tips increase in extent till on the neck they occupy half the feathers, a broad bright red band being thus formed across the hind-neck. Feathers of the back blackish, with a narrow whitish-grey shaft-line, which at the tip of the feather forms a roundish subterminal spot, followed by a terminal spot of a dull yellowish pink. Feathers of the rump and upper tail-coverts black, with the base greyish and a broad yellowish-red tip, like Ridgway's orange-chrome (Nomencl. Col. pl. vii. fig. 13). Primary-coverts and greater wing-coverts black, with small white tips. Smaller wing-coverts ashy white, broadly tipped with red. Quills black ; primaries outwardly obsoletely edged with dull red in the middle of the outer web, and at the tip with white. Secondaries black, broadly tipped with white. Tail-feathers black, tipped with white. Lores dull blackish. Eyelids and a stripe from above the eye to the nape dull scarlet. Ear-coverts, sides of the head and neck, throat and breast leaden grey ; on the breast small dull-red tips to the feathers. Abdomen like the rump. Thighs yellowish vermilion. Vent and under tail-coverts yellowish grey. Under wing-coverts ashy, washed with red near the outer margin of the wing. Iris hazel. Bill and feet black.

Adult female. Exactly similar in colour to the male.

There are some slight variations in colour. They do not seem due to sex or age, nor is there any constant difference between specimens from Maui and Molokai ; but some specimens have no white tips, or only obsolete ones, to the rectrices. Sometimes all the secondaries, sometimes some of them only, have white tips, and the red spots of the back and rump are in a few specimens less in extent.

The *young* differ very much from the adult birds, but the curled feathers on the forehead and the peculiar plumage is recognizable even in the youngest specimens. The head is greyish brown, slightly tinged with red on the sides of the crown. The frontal crest is brownish grey. Back and rump dark dusky brown, washed with buffish olive. Below buffy olive, more grey towards the throat, and passing into pale grey on the under tail-coverts. Feathers of the shoulders with dull red tips. Greater wing-coverts with brownish-pink edges. Tail black.

All intermediate plumages between the young and old are represented in my collection. On the Plate I have given figures of a fully adult male of the type of Wilson's *Himatione dolei*, kindly lent me by the author, and of an intermediately coloured bird.

The length of the adult bird, as measured in the flesh, is given by Palmer as about 7½ inches, but skins are only about 6½ inches or little more. The wing measures 3·5 to 3·7 inches (fifteen adult specimens measured), the tail 2·7 to 2·95, the tarsus 1·1 to 1·25, the culmen 0·75 to 0·84.

THE specific name of this bird was first given by Mr. Scott Wilson, who referred his very young example to the genus *Himatione*. I received at first some perfectly adult males and females, and, comparing their description with that of "*Himatione dolei*"[1], no one could tell that they refer to the same bird. Mr. Wilson, however, most kindly lent me the type specimen, which at once revealed the truth, when being compared with the adult birds.

Palmer obtained a large series on Maui and Molokai, and afterwards Mr. Perkins collected many more specimens. They were found by Palmer only at considerable altitudes, but were said to shift their quarters with the flowering of the Ohia-trees, on the flowers of which they feed. The food consists of nectar and insects.

Palmer described the call-note on Maui as a hoarse "O-o-o," while on Molokai, he says, it was a clear flute-like whistle, commencing shrill and becoming lower towards the end. On Maui he heard also another note, somewhat like a long "tee-ee," and the males very often produced a kind of chuckle, which may be their song. I am, of course, not able to find out whether Palmer's memorandum about the different call-notes is correct or not. The alarm-note is, according to Palmer, a shrill whistle.

[1] The original correct spelling of the name *dolei*—named in honour of Judge Dole—was afterwards altered by its author to "*dolii*."

THE GENUS VESTIARIA.

The genus *Vestiaria* was first founded in John Fleming's 'Philosophy of Zoology,' ii. (1822) p. 240, with Latham's *Certhia vestiaria* as the type.

Bill with a rounded ridge, nearly semicircularly bent; mandible fitting into the premaxilla. The tip of the latter projects about 2 or 3 millimetres over the mandible. Cutting-edges perfectly smooth, without any serration. Nostrils near the base, bare, covered by a somewhat swollen operculum.

Tail when opened square, when closed slightly forked; all the rectrices obliquely pointed. First primary rudimentary, concealed by its coverts, not visible from below; third, fourth, and fifth about equal and longest. Plumage of the body full and soft; feathers of the throat narrowed and slightly stiffened.

Sexes alike in coloration; young rather different from adult.

Tongue, pterylosis, and alimentary canal described at length by Gadow in part ii. of the 'Aves Hawaiienses.'

Only one species known.

31. VESTIARIA COCCINEA (*Forster*).

"IIWI."

Carmosinrother Baumläufer (Certhia coccinea), G. Forster, Goetting. Magaz. Wiss. i. p. 347 (1780).
Certhia coccinea, Blumenb. Handb. der Nat. Gesch. 2. Aufl. p. 190 (1782) ; id. Abbild. naturh. Gegenst. ii.
　pl. 16 (1797) ; Shaw, Nat. Miscell. iii. pl. 75 (1791) ; Donndorff, Handb. Thiergesch. p. 251 (1793) ; id.
　Orn. Beytr. i. p. 621 (1794) ; Tiedemann, Anat. Naturg. Vög. ii. p. 430 (1814).
Hook-billed Red Creeper, Lath. Gen. Syn. i. p. 704 (1782) ; id. Suppl. p. 127 (1787).
Polytmus .... *flavo-aurantius*, Märter, Phys. Arb. Wien, i. p. 76, pl. 2 (♂ & ♀) (1783).
Mellisuga coccinea, Merrem, Av. Rar. Descr. et Icon. p. 14, pl. iv. (1786).
Eee-eve, King, Voy. Pacif. Oc. iii. p. 119 (1784).
Certhia vestiaria, Lath. Ind. Orn. i. p. 282 (1790) ; Shaw, Gen. Zool. viii. p. 229 (1812).
L'Hérotaire, Vieill. Ois. Dor. ii. p. 109, pl. 52 (1802).
Nectarinia coccinea, Tiedemann, Anat. Naturg. Vög. ii. p. 431 (1814) ; Bloxam, Voy. 'Blonde,' p. 247 (1826).
Melithreptus vestiarius, Vieill. Nouv. Dict. xiv. p. 322 (1817) ; id. Enc. Méth. p. 601 (1823) ; id. Galerie
　d'Ois. pl. 181 (1825) ; Less. Traité d'Orn. p. 300 (1831).
Drepanis vestiaria, Temm. Man. d'Orn. i. p. lxxxvi (1820) ; Hartl. Syst. Verz. Mus. Bremen, p. 16
　(1844).
Vestiaria evi, Less. Rev. Zool. 1840, p. 268.
"*Le Vestiaire*," Léclancher, Rev. Zool. 1840, p. 322.
Drepanis coccinea, Gray, Gen. B. i. p. 96 (♂) ; Bp. Consp. i. p. 404 (1850) ; Cab. Mus. Hein. i. p. 99 (1851) ;
　Hartl. Archiv f. Naturg. 1852, i. p. 131 ; id. J. f. O. 1854, p. 170 ; Licht. Nomencl. p. 55 (1854) ; Cass.
　U.S. Expl. Exp., Mamm. & Orn. p. 177 (1858) ; Dole, Proc. Bost. Soc. Nat. Hist. xii. p. 297 (1869)
　(Hawaii) ; id. Hawaiian Alman. 1879, p. 44 ; Gray, Cat. B. Trop. Is. p. 8 partim (♂) (1859) (Hawaii,
　Maui, Oahu) ; id. Hand-l. i. p. 113, no 1405 (1869) ; Scl. Ibis, 1871, p. 358 ; id. P. Z. S. 1878, p. 347 ;
　id. Report 'Challenger,' Birds, p. 95 (1881) ; Sundev. Tentamen, p. 48 (1872) ; Pelz. J. f. O. 1872, p. 26 ;
　Finsch, Ibis, 1880, p. 79 (Maui).
Vestiaria coccinea, Reichenb. Spec. Orn., Seans. p. 254, pl. 562. figs. 3830–3832 (1853) ; Sharpe, Cat.
　B. x. p. 6 (1885) ; Scott Wilson, Ibis, 1890, p. 181 (generally distributed through the Archipelago) ; id.
　& Evans, Aves Hawaiienses, pt. i. (1890), text, and two plates showing adult and immature plumages ;
　Gadow, in Aves Hawaiienses, pt. ii. p. 12 (1891) (excellent description of structure) ; Stejn. Proc. U.S.
　Nat. Mus. 1887, p. 97 (Kauai: good synonymy, exact measurements, &c.) ; id. Proc. U.S. Nat. Mus.
　1890, p. 385 (Kauai) ; Perkins, Ibis, 1893, p. 106 (common, Kona, Hawaii: a flattened parasitical fly,
　superficially the same sort of thing as the smaller of the two on the owl).
Drepanis rosea, Dole, Hawaiian Alman. 1879, p. 44.
Loxops rosea, Sharpe, Cat. B. x. p. 509 (1886).
[The unfortunate and rather unfounded idea that the females of this species were not red has greatly
encumbered the synonymy given by Gray, *l. c.* The same is the case with *Himatione sanguinea*, several
names pertaining to species of our present genus *Chlorodrepanis* having formerly been quoted as referring
to the female of *Him. sanguinea*.]

Adult male. Above and below generally bright vermilion, deepest on the breast and abdomen, distinctly lighter on the top of the head and on the throat. The bases of the feathers on the top of the head are nearly quite, those of the feathers on the throat pure white, while those of the other parts have blackish-grey bases and a lighter whitish shade along the shaft and before the red tip. Wings and tail black, in freshly-moulted specimens a narrow brown margin is obvious on some of the primaries. Primary-coverts black ; smaller wing-coverts red; greater black, with deep red edgings near the tip on the outer web, not obvious if the birds are in abraded plumage. The innermost (smallest) secondary is white, the next one blackish, with the outer web ashy grey or ashy brown. Under wing-coverts and axillaries white, with a faint rosy tint, which is sometimes nearly absent. Iris dark hazel. Bill red, a little darker on the upperside. Feet vermilion-red in life, whitish in skin. Total length nearly 6 inches, wing 3·2 to 3·4, tail 2·25 to 2·3, culmen 1·15 to 1·25, tarsus 1, bill from gape to tip 0·95 to 1.

Adult female. The female does not (according to the series examined by me) differ from the male in colour in any constant or obvious way, but it is distinctly smaller, the wing measuring only 2·98 to 3·08, the culmen about 1 to 1·15 inches.

The above-stated facts and measurements are based on a material of over thirty sexed adult birds, less than one-third of which are females. The average of the length of the wing in the male is 3·25, and only one male from Molokai measures 3·4 inches. The birds from the various islands are not different.

Young birds are quite different in colour. The feathers above and below are brownish orange or olivaceous orange, those above with blackish spots at the tip, those on the throat and breast similar, while those of the abdomen are not so distinctly spotted at the tip. The red feathers appear here and there generally, if not always, first on the head and breast, and get more and more plentiful as the moult advances ; but there are also, in several of my specimens, orange-brown feathers that are washed with scarlet-vermilion, as if the red colour was overspreading the old feathers together with the moult. A series of very instructive figures of immature birds are given in Mr. Scott Wilson's book.

In the young bird the iris is hazel ; the upper mandible is brown on the culmen, rose-coloured on the edges ; the lower mandible tipped with brown, reddish at base ; tarsi and toes brown, soles orange.

This species, which was first described by Professor Forster of Cassel, is generally distributed over the various Hawaiian islands, and no constant differences can be traced between the birds from the different islands. They inhabit all wooded regions, from the lower hills up to 8000 feet or more. Their food consists of nectar, taken from various flowers, and in the higher situations almost entirely from the Mamane, and also of insects. They have a great variety of notes, their call-note being a clear, powerful, and flute-like whistle, besides which they have a warbling song of many notes, and when Palmer saw one chase an Amakihi (*Chlorodrepanis*) it uttered a hissing sound.

Hardly anything is known about the breeding-habits. Wilson found a nest, which *probably* belonged to this bird, and which he describes as a round and shallow cup, composed of mosses and dry bents. Palmer, in spite of much attention, did not succeed in discovering the eggs. From Hawaii he reports having seen these birds collecting twigs on November 9th and on November 18th. They were then in splendid plumage, always going about in pairs and being very plentiful at elevations of from 5000 to 8000 feet and so on. On November 24th Palmer makes the following note :—"Their breeding-season is, I believe, now on, for they are always seen in pairs, and several were seen carrying moss in their bills. The *Himatione sanguinea* and *Phæornis obscura* are now also, I believe, nesting."

The following notes seem to contradict the former, except we believe that these birds breed both in spring and autumn. In the first week of May, on Hawaii, Palmer says he saw them carrying moss in their bills, and saw a pair *in copula*, and on June 2nd Palmer found a nest on Mauna Kea, on which the bird was still building. It was almost on the top of a tall Ohia-tree. Palmer says :—"These birds place their nests on the thin branches out of reach from the stem, so that it is *impossible* (sic!) to take eggs from them. One would have to climb the tall Ohias and to cut the branches off and let them fall."

It is to be deeply regretted that none of the collectors showed enough pluck to procure the eggs of *Vestiaria*, for it is not very likely that any such efforts to explore the ornithological treasures of the Sandwich Islands as were made within the last decade will be made in the near future.

In part vii. of his work, however, Mr. Wilson gives a photographic plate of the nest of *Vestiaria coccinea* ; it is an open structure, like those of *Himatione* and *Chlorodrepanis*.

THE GENUS DREPANIS.

— —

THIS genus, from which the characteristic family of *Passeres* only known from the Sandwich Islands received its name *Drepanidæ*, was created by Temminck in 1820 (Man. d'Orn. i. p. lxxxvi).

It is characterized by its long, curved, non-serrated bill, in which the upper mandible is a few millimetres longer than the lower. It has a striking black and yellow coloration, and the somewhat loose-webbed elongated under tail-coverts cover about three-quarters of the tail. The nostrils are large, but almost entirely covered by an operculum. The first primary is quite rudimentary, hidden by its covert and is white like the latter. The wing reaches to about 25 mm. from the end of the tail. The third and fourth primaries form the tip of the wing. The tips of these primaries are rather obtuse, not much pointed. The secondaries are about 28 mm. shorter than the longest primaries and rather square at the tips. Some of the axillaries are silky, long, and fluffy. The tail is slightly rounded, the distance between the shortest lateral and longest central rectrices being about 5 to 7 mm. The metatarsus is covered in front with about 5 or 6 scutes, which are somewhat fused : in the Plate they are shown too distinctly. The sides of the metatarsus are covered each with a long unbroken scute.

Only one species is known. Inhabits Hawaii (the locality Kauai is probably erroneous).

32. DREPANIS PACIFICA (*Gm.*).

"MAMO."

Great Hook-billed Creeper, Lath. Gen. Syn. i. p. 703 (1782) (types in the Leverian Museum : "Friendly Isles in South Seas") ; id. Suppl. p. 126 (1787) (" Common at Owhyhee ").
? *Hoohoo*, King, Voy. Pacif. Oc. iii. p. 119, partim (1784).
Certhia pacifica, Gm. Syst. Nat. i. p. 470 (1788) (ex Latham) ; Lath. Ind. Orn. i. p. 281 (1790) ; Donndorff, Orn. Beytr. i. p. 621 (1794) ; Steph. in Shaw's Gen. Zool. viii. p. 227 (1812) ; Tiedemann, Anat. Naturg. Vög. iv. p. 431 (1814) ; Peale, U.S. Expl. Exp., B. p. 149 (1848) ; Hartl. Arch. f. Naturg. 1852, i. p. 100.
Le Hoho, Vieill. Ois. Dor. ii. p. 124, pl. lxiii. (1802) ; Less. Compl. Buff. ix. p. 156 (1837).
Le Mérops jaunoir, Levaill. Hist. Promérops et des Guêpiers, p. 45, pl. xix. (1807).
Melithreptus pacificus, Vieill. Nouv. Dict. xiv. p. 323 (1817) ; id. Enc. Méth. p. 602 (1823) ; Cuv. Règne An., ed. II. i. p. 433 (1829).
Drepanis pacifica, Temm. Man. d'Orn. i. p. lxxxvi (1820) ; Gray, Gen. B. i. p. 96 (1847) ; id. Cat. B. Trop. Is. pp. 7 & 8 (1859) (" Friendly Islands [*ex* Lath.] [?] ; Sandwich Islands (Owhyhee, Atoui or Atooi or Atowi or Towi or Attoway or Tanai or Kauai)" [sic]) ; id. Hand-l. i. p. 113 (" Friendly Islands ") (1869) ; Bp. Consp. Av. i. p. 403 (1850) ; Hartl. Arch. f. Naturg. i. p. 131 ; Reichb. Handb. spec. Orn. p. 253, pl. 611. figs. 3828 & 3829 (1853) ; Dole, Proc. Bost. Soc. N. H. xii. p. 297 (1869) (" Hawaii, Kauai ") ; id. Hawaiian Alman. 1879, p. 45 ; Scl. Ibis, 1871, p. 368 ; id. Ibis, 1879, p. 92 ; Sund. Tent. p. 48 (1872) ; Pelz. J. f. O. 1872, p. 26 (the two types of Latham from the Leverian Museum, now in the Vienna Museum, are from Owaihi, marked as ♂ & ♀ , show no differences in size or colour !) ; id. Ibis, 1873, p. 21 (same specimens mentioned ; they are probably from Cook's voyage) ; Sharpe, Cat. B. x. p. 5 (1885) ; Lucas, Rep. U.S. Nat. Mus. 1888–89 ; Newton, Nature, 1892, p. 469 ; Wilson, Ibis, 1890, p. 178 (state-robes of feathers) ; id. & Evans, Aves Hawaiiens. pt. ii. plate & text (Sept. 1891) (very complete history, etc. etc.) ; Gadow in Aves Hawaiiens. pt. ii. p. 10 (1891) ; Hartl. Abh. nat. Ver. Bremen, xiv. p. 19 (1895) ; id. Beitr. Gesch. ausgest. Vög. (2ᵗᵉ Ausg.) p. 25 (1896).
Vestiaria hoho, Less. Rev. Zool. 1840, p. 269.

Description of Drepanis pacifica, *taken from the two skins in my collection.*

Adult male. Black with a slight gloss. Rump and upper tail-coverts, the shoulders, the ridge of the wing, the outer part of the under wing-coverts, thighs, vent, and under tail-coverts, all of a beautiful rich yellow, but paler on the ridge of the wing. The large primary wing-coverts white, blackish brown at base. Under wing-coverts white, washed with yellow on the outside. Primaries brownish black, outer web of the second (outermost) washed with ashy, the apical halves of the outer webs of the rest of the primaries washed with ashy, more white towards the tips. Rectrices brownish black, the outer ones with more or less indistinct whitish shaft-spots near the tip. Iris dark hazel. Maxilla black. Mandible black, dark brown at base. Tarsi and toes black, soles yellow [1]. Total length measured in the flesh as $9\frac{3}{4}$ inches, but only about 8 inches in the skin; wing 4 to 4·25; tail 2·9 to 3; bill along the culmen 1·75 to 1·85, in a straight line from the gape to the tip 1·5 to 1·6; metatarsus 1·1 to 1·2, middle toe with claw 0·85, hind toe with claw 0·68 to 0·88.

The measurements of the tail and metatarsus given by Mr. Wilson disagree not very materially with mine, but that of the tail must be wrong, as he gives it as only 2·5 inches. The adult male obtained by Palmer is slightly larger than the specimen (sex unknown) which I got from Mr. Wilson.

WHEN in 1891 Mr. Scott Wilson wrote his article in the second part of his ' Aves Hawaii- enses,' he called the " Mamo " an " extremely rare and apparently extinct species." And, indeed, it is rare in collections, for I know only of stuffed specimens in the Museums of Vienna (where Latham's types are preserved), of Leyden, Paris, and Honolulu; and Mr. Wilson brought the first two to England, one of which is now, after being " beautifully remounted " !!, in the Museum of the University of Cambridge, and the other in my Museum at Tring. These two latter were procured (about 1859 as Mr. Wilson tells us) by Mr. J. Mills, of Hilo in Hawaii. No specimens are reported to have been killed since that time; but Mr. Scott Wilson, "while staying at Olaa in the district of Puna in Hawaii, where Mr. Mills secured his specimens, was assured by the natives that the bird still existed, and at the time of his visit (October) had, together with the O-o, migrated to the mountains," but, he adds, " which is barely possible."

I am glad to be able to state that the " Mamo " is not yet quite extinct, because Palmer sent me a skin with the body in spirits.

As I mentioned in Henry Palmer's diary, he was severely kicked by a horse, and therefore, not being able to do long walks in the woods, sent the old bird-catcher Ahulan with several other natives, together with his "assistant," into the forests on the Mauna Loa above Hilo. I give herewith the history of that little expedition as it is written down in Palmer's diary from the notes of his assistant, with as few alterations of style etc. as possible :—

" *Monday, April 12th*, 1892.—Left Hilo at 8.30 A.M., and arrived at our first camping- ground at 1.30, where we pitched the tent and built a cook-house. This camp was on the same stream which forms the Rainbow Falls, and there was also a fair size fall near the camp. While half of the natives were busy with the camp the others went on to cut the trail through the forest. The bush up to this place is mostly Ohia with big tree-ferns and thick under- growth. The rain is coming down steadily, and a wet night it is sure to be. I saw near the camp many Iiwi, Apapane, Amakihi, Elepaio, and other birds.

" *April 13th*.—We left camp at 8 A.M. and marched on as far as the trail was cut, to a patch of thick bananas, where, according to the natives, Mr. Hitchcock stayed. We reached this spot at 9.15 and pitched the tent, four of the natives going on to cut the trail. I searched

---

[1] These colours were taken by Palmer from the live bird.

the forest around, but could find nothing I had not seen before except one bird, which, however, the natives said was an Ou—so I did not shoot it, as I wished to keep all the ammunition till we reached the place where the Mamo dwells and the Ulaaiwahane flies undisturbed. [I don't understand this note, for I suppose they did not see an 'Ulaaiwahane,' which is *Ciridops anna*.—W. R.] We passed several of the plants which the Mamo feeds on. They are called Haha by the natives. They were not in flower. The natives demonstrated the use of their snares, which are simple and clever. The bush still consists of big and lofty Ohias with a sprinkling of equally large Koa. We killed a pig to-day. Iiwi, Apapane, Akakane, Ou, Elepaio. Amakihi, etc., were pretty numerous here.

"*April 14th.*—Left camp at 7.30 A.M. The morning was fine, and we got to the end of the trail and camped here at 12 o'clock. The natives take things leisurely, and it is a good job they are not hired by the day. The Ou seems very plentiful here, but I do not shoot any here. I also heard an O-o and went after him, but failed to get a shot. Two natives have gone on cutting the pass, and they expect to reach the palm-region to-morrow, when I hope I shall have something to report. The dog killed another pig.

"*April 15th.*—Struck camp at 7 A.M. and marched along the trail, where we soon came across two Lolu-palms. I watched them close on an hour, but saw no signs of Ulaaiwahane. At 3 P.M. we struck the big gulch again and followed it till 5 P.M. There is here a beautiful waterfall, but I don't appreciate it much as I am very hungry. Not seeing a good camping place, we camped here in a very narrow and uncomfortable place.

"*April 16th.*—Broke the camp up at 6 A.M. and pushed on till 4 P.M. The old bird-catcher Ahulan was leading to cut the trail, whilst Holi and myself came next, followed by the others, who were a long way behind. We had not gone more than three miles, when I heard a call from the other side of the gulch, and thought it was a native calling, but immediately afterwards a bird flew across, and I saw in a moment it was the bird we were after. I was going to follow it up to shoot it, but Ahulan begged me not to shoot as it would scare the other away, which I had heard calling a little way off. Ahulan fixed the snares and bird-lime on a haha, which growed out on a tree-fern, and which has flowers somewhat like those of a fuchsia. Ahulan fulfilled his promise and caught the Mamo! He is a beauty, and takes sugar and water eagerly and roosts on a stick in the tent. I now feel as proud as if someone had sent me two bottles of whisky up."

In spite of two or three days' more work, no second specimen was caught or even noticed, and at last the natives declared that Palmer's "assistant" was mistaken about having seen two, and that there was no other specimen about there. When their food was finished and they had no hope of getting any more "Mamos," the whole party descended to where Palmer stayed. On April 21st Palmer killed and skinned the Mamo.

Now, whether the men saw two or only the one which was caught, it is evident that in April 1892 the species was not quite extinct, and it is a most hazardous assumption that the men in Palmer's employ got the only specimen then living in those wide and almost pathless woods. Moreover, Mr. H. W. Henshaw, the distinguished American ornithologist, who now resides in Hawaii, writes to me (dated October 9th, 1899) as follows:—

"*Drepanis pacifica* is still a living species, though unquestionably very rare. No doubt it is on the verge of speedy extinction. About a year ago last July I found what, no doubt, was a family of Mamos in the woods above Kaumana. I am sure that I saw at least three individuals, possibly four or five. They were flitting silently from the top of one tall Ohia-tree to another, apparently feeding upon insects. The locality was a thick tangle, and a momentary glimpse of a slim, trim body as it threaded its way through the leafy tree-tops was all I could obtain. After about two or three hours I succeeded in getting a shot at one bird in the very top of a tall Ohia-tree. It was desperately wounded, and clung for a time to the branch, head downwards, when I saw the rich yellow rump most plainly. Finally, it fell six or eight feet, recovered itself, flew round to the other side of the tree, where it was joined by a second bird, perhaps a parent or its mate, and in a moment was lost to view. I need not speak of my disappointment, which was bitter enough, for I had looked upon that bird as absolutely mine own. Of the others I saw no more, though I have repeatedly visited the locality again."

In former times the Mamo was probably a common bird. Its rich orange-yellow feathers were in great demand for certain feather-robes, capes, and wreaths. Most of these feather-robes are made of the feathers of *Moho nobilis*, but those of the Mamo were also plentifully used, while one cape examined by Mr. Scott Wilson consisted entirely of Mamo-feathers.

According to Palmer, the Mamo was very partial to the berries of the "Ilaha" or Hawaiian mistletoe, a parasitic plant with long oval leaves and bell-shaped purple flowers. Out of the juice of this same plant the old bird-catchers of the kings made a very good bird-lime.

It is hardly possible to say at present why this bird has become so rare. If its rarity is due to the value of its feathers, then one naturally wonders why the *Moho nobilis*, which has equally been sought, is still so (comparatively) numerous. In any case the Mamo is one of the rarest birds in collections, being represented, so far as known, only in six museums by eleven or twelve examples.

The upper figure on the coloured Plate is taken from one of Mills's specimens now in my Museum, the lower figure from the one sent home by Palmer. On the second Plate, with bills etc., fig. 36 shows the bill of the natural size and double size, fig. 56 the entire tongue, 56 A the tip of the tongue ten times magnified, and figs. 57 & 58 the sternum in two views.

THE GENUS DREPANORHAMPHUS.

If the genera *Drepanis* and *Hemignathus*, *Vestiaria*, *Himatione*, and *Chlorodrepanis*, *Loxops* and *Oreomyza* are separated, then it is necessary also to separate the wonderful " *Drepanis funerea* " of Professor Newton from *Drepanis*.

Drepanorhamphus, as I propose to call this genus, wants the elongated and loosely-webbed under tail-coverts, which, owing to their bright yellow colour, are so conspicuous in *Drepanis pacifica*. The bill is actually and comparatively longer, more strongly curved, thicker and higher at base. The culmen is about equal in length to the tail, while it is much shorter than the tail in *Drepanis*; the operculum of the nostrils is much larger and more elongated than in *Drepanis*. The upper tail-coverts are shorter and not so stiff and loosely webbed as in *Drepanis*.

One species known.

2 A

33. DREPANORHAMPHUS FUNEREUS (*Newton*).

Drepanis funerea, A. Newton, Proc. Zool. Soc. Lond. 1893, p. 690 ; Wilson & Evans, Aves Hawaiiens. pt. v.
text & plate (1894).

Adult male. Entirely dull black above and below, the feathers blackish grey at base. Primaries
sooty black, outer webs of a buffy or dirty white, passing into blackish at base. Feet
black, soles yellowish (in skin). Bill black, nasal operculum and base of upper bill
yellowish (in skin). This colour is undoubtedly quite natural and must have been
brighter in life. Total length about 8 inches ; culmen 2·6 ; bill from base to tip of
upper mandible 2, under mandible from base to tip 1·7, thus being 0·3 shorter than
the upper ; height of bill at nostrils 0·3 ; wing nearly 3·9 inches (damaged at tip), tail 2·8,
tarsus 1·15.

Female similar to the male, but smaller, bill shorter.

THIS peculiar bird is described from one of the original specimens discovered by Mr. R. C.
L. Perkins on Molokai. It was shot on June 18th, 1893, on Molokai, at about 5000 feet.
Its sex is not stated, but on the label is inscribed : " The body of this bird in spirits, evidently
male by the bill."

I am much obliged to Professor Newton and the " Royal Society and British Association
Joint Committee for Zoology of the Sandwich Islands " for their kindness in lending me
the specimen for figuring and describing. I have now, however, one in my own collection,
received in exchange from the British Museum. The Museum in Bremen has received this
bird from the sons of the late Mr. R. W. Meyer.

The bird has a very strong smell, somewhat like that of a *Thalassidroma*. It is
so strong that the box still smells in which it was packed coming to Tring from
London.

2 A 2

THE GENUS LOXOPS.

THIS genus was first established by Cabanis in the 'Archiv für Naturgeschichte,' 1847, p. 330. Three years later it received the name of *Hypoloxias* from Bonaparte (Conspectus Avium, i. p. 518); and a third name "*Byrseus*" was bestowed on it in the same year (1850) by Reichenbach. *Chrysomitridops* is clearly synonymous.

As the names alone would seem to indicate, the bill, in which, as in *Loxia*, the edges of the maxilla and mandible are not parallel but divergent laterally, was considered a chief characteristic of the genus [1]. It seems that recent authors have missed this very remarkable feature, as nobody even mentioned it, only in 1899 Gadow refers to it again. It is clearly obvious in all specimens, except in badly skinned and distorted ones and in one or two very young ones, but sometimes is more developed, sometimes less. It is most strongly marked in some specimens of *Loxops cœruleirostris* (Wils.).

The short and stout bill, reminding one of those of some of the *Fringillidæ* with slender and pointed bills, like *Chrysomitris*, is another character distinguishing the genus *Loxops* from all the other genera of *Drepanidæ*. In the wing, the first primary is absent, the third, fourth, and fifth are about equal and longest, the fourth being hardly distinctly longer than its neighbours. Tail deeply emarginate; tail-feathers sharply pointed. The bill is strong, conical, and pointed: the nostrils are near the base, covered on the upper part with a strong membrane, partly, but slightly, covered with tiny bristles.

The genus *Loxops* belongs undoubtedly to the *Drepanidæ*. (See also Gadow, in 'Aves Hawaiienses,' pt. ii.)

Loxops cœruleirostris (Wils.) differs in no character whatever from the other members of the genus, excepting the colour, and females and young birds may be mistaken for those of other species, if not labelled and without material to compare.

---

[1] Cabanis, Arch. f. Naturg. 1847, p. 330, expressly says as follows:—

" Loxors, nov. gen.

" Die Gattung *Loxops* (von λοξός, schief, seitwärts gebogen, und ὄψ, Gesicht) ist in Färbung und Schnabelbildung eine Wiederholung der Gattung *Loxia* im Kleinen. Die Schnabelspitzen sind indess weniger stark gekrümmt.

" Typus : *L. coccinea.*

" *Fringilla coccinea* Gm. Lath., von den Sandwichs Inseln."

Key to the adult males of the Genus Loxops.

a. Above green, below greenish yellow, lores black . . . .   *L. cæruleirostris* (Kauai).
b. Above and below orange-scarlet or scarlet-orange; wing about 2·5 inches   *L. coccinea* (Hawaii).
c. Above and below rufous orange with a brownish wash; wing about
  2·25 inches . . . .   *L. rufa* (Oahu).
d. Above and below orange ochraceous; wing about 2·5 inches . . . .   *L. ochracea* (Maui).

This key applies only to adult males, the females and young of the different species varying sometimes so little from each other, and passing through so many different shades of coloration, that a precise, short, and satisfactory key could not be given for them. The three reddish forms are hardly more than subspecies.

34. LOXOPS COCCINEA.

THE SCARLET AKAKANE.

Scarlet Finch, Lath. Gen. Syn. ii. p. 270 (1783) (Sandwich Islands, Læver. Museum).

Fringilla coccinea, Gmel. Syst. Nat. i. p. 921 (1788) (ex Latham!); Lath. Ind. Orn. i. p. 444 (1790); Donndorff, Orn. Beytr. ii. p. 541 (1795); Tiedemann, Anat. Naturg. Vög. ii. p. 433 (1814); Steph. in Shaw's Gen. Zool. ix. p. 454 (1815); Cuvier, Règne Anim. i. p. 387 (1817); Vieill. Nouv. Dict. xii. p. 167 (1817); id. Enc. Méth. p. 983 (1823); Gray, Griffith's Anim. Kingd., Aves, ii. p. 140 (1829); Gray, Gen. B. ii. p. 371 (1845).

Chardonneret écarlate, Vieill. Ois. Chant. pl. 31 (1805).

Carduelis coccinea, Less. Compl. Buffon, viii. p. 281 (1837).

Loxops coccinea (partim!), Gray, Cat. B. Trop. Isls. p. 28 (1859); id. Hand-l. B. i. p. 114 (1869); Scl. Ibis, 1871, p. 360, 1879, p. 92; Scott Wilson & Evans, Aves Hawaiienses, pt. i. (1890) (text and plate of male); Gadow, in Aves Hawaiienses, pt. ii. p. 13 (structure) (see also *t. c.* p. 17).

Hypoloxias coccinea, Hp. Consp. i. p. 518 (1850); Hartl. Arch. f. Naturg. 1852, p. 133; Dole, Proc. Bost. Soc. N. H. 1869, p. 301; id. Hawaiian Alman. 1879, p. 49; Scl. Ibis, 1871, p. 359.

Drepanis aurea, Dole, Hawaiian Alman. 1879, p. 45.

Adult male. Upper and under parts bright orange-scarlet or scarlet-orange. Feathers dark grey at base, lighter just before the scarlet tips. Quills and rectrices brownish black, at the tips a little paler, outer webs bordered with orange, inner webs with an ashy border. Under wing-coverts whitish, scarlet-orange towards the tip. " Iris brown. Maxilla pale grey, dark brown towards the tip, edges and base paler with a more or less developed rosy tinge. Legs and feet black, soles flesh-colour." Total length given as from nearly 5 to a little over 5 inches by Palmer, as measured in the flesh, but only 4 to 4½ inches in the skins. Wing 2·45 to 2·6 (averaging 2·5), tail 1·95 to 2·1 (not 1·65 as given by Scott Wilson), culmen 0·4 to 0·45, tarsus 0·8 to 0·9.

Adult female. Above olive-green, greener towards the rump and upper tail-feathers and on the wing-coverts; forehead of a pale greyish yellow. Underparts buffy yellow, breast and throat with an orange wash, sides of the body washed with olive. Wings and tail dusky brown, quills edged with yellowish green on the outer webs and with pale ashy en the inner webs. Under wing-coverts pale buff washed with yellow. "Iris dark brown. Bill grey, darker towards the tip; edges and base of mandible tinged with yellow. Tarsi and toes blackish, soles pale flesh-colour." Wing 2·35 to 2·45 inches. One very old female has an almost yellow breast.

Young males resemble the female, but are less green, and a strong orange wash soon becomes visible, progressing more and more as the birds advance in age. All possible intermediate colours between the adult orange-red birds and the olive-grey young birds are represented in my collection.

Hab. Hawaii only.

Loxops coccinea is also one of the discoveries of the early travellers, and was first described by Latham. The first Latin name bestowed on it was *Fringilla coccinea*, given by Gmelin. Professor Cabanis first of all, after having studied specimens sent from Oahu (*Loxops rufa*), distinguished the genus *Loxops*. Up to quite recent times this bird was rare in collections, and even Scott Wilson only got males and considered it "one of the rarest of Hawaiian birds, and cannot, I think, be far from extinct." There is, however, no reason to fear that the bird will soon disappear, as it is still numerous in the higher regions of Hawaii, while another species of the genus inhabits the island of Maui, not to speak of the green *Loxops cœruleirostris* of Kauai, and of the rare, and perhaps nearly extinct, *L. rufa* of Oahu.

Perkins (Ibis, 1893, p. 105) says of this species:—"Of the *Drepanidæ* the rarest species in the Kona district was *Loxops coccinea*. I saw only the adult and young males, and these were mostly feeding on insects, either amongst the blossoms of the *alii*-tree or on the foliage of the acacias. Their habits seemed almost identical with those of the yellow species of *Himatione*. I never heard any proper song, nothing more than a squeaking like that of the female *Himatione*. On several occasions it was in company with the small, straight-billed *Himatione*; on two of these I saw it pursue the latter from tree to tree, and on another the *Himatione* was itself the aggressor. On one of these occasions I shot the green bird, and it was beyond doubt the *Himatione*, and not a green female of *Loxops*. *Loxops* is apparently confined to the *upper forest.*"

So little being known about this pretty little bird, I instructed Henry Palmer to do his very best to discover the female and to make the history of the species more known. Strange to say, it was a long time before the female was discovered; for the first time Palmer saw none but the red males, and not knowing whether the female would be red also or green, he shot many males, and in fact more than I should have desired. However the large series now before me proves that the bird is not so rare as it was believed to be, and it enabled me to give the measurements and descriptions with greater accuracy and completeness.

According to Palmer, this bird inhabits the higher regions from between 5000 to 7000 feet, although one was occasionally shot not higher than about 3000 feet. It frequents chiefly old lava-flows, where the Aaka-flowers are plentiful. Higher up the females were not rare, although less numerous than the males, while lower down Palmer searched for them in vain. He says in his diary:—"I cannot understand why I did not find any females at Pulehua, while on the other side of the island and at higher elevations I found females, and they were altogether more numerous on Mt. Hualalai. The bird I cannot say is exactly very rare, as one can see from the number I shot, and it was comparatively numerous in Kona, between 4000 and 7000 feet, common on the Kohala Mountains at 4000, several seen in Hamakua, Hilo, Puna. They can be distinguished from the *Himatione* from a wide distance when one has learned to watch their habits, which are different. They are not at all shy. They are slower in their movements, their note is shorter and more of a squeak. A 'chirruping' noise was once heard from a female."

The stomachs contained insects and caterpillars.

Nothing is known about the nesting.

Palmer in his diary calls the bird the Hawaiian Akakane, and adds that it is called "Akapalau" in Hilo. Wilson says that its proper name is "Akepenie."

35. LOXOPS OCHRACEA.

OCHRACEOUS AKAKANE.

Hypoloxias aurea (errore!, non *Drepanis aurea*, Dole), Finsch, Ibis, 1880, p. 80 (Maui; descr.).
Loxops aurea, Sharpe, Cat. B. Brit. Mus. x. p. 50 (1885) (copied from Finsch. Locality "Sandwich Islands").
Loxops ochracea, Rothschild, Ibis, 1893, pp. 112 (Maui), 281 (note on name and synonymy).
Himatione aurea (errore!, non *Drepania aurea*, Dole), Wilson & Evans, Aves Hawaiienses, pt. iv. (1893).
Loxops aurea, Perkins, Ibis, 1895, p. 121 (habits) ; Wilson & Evans, Aves Hawaiienses, pt. vi. (1896).

Adult male. Dark orange ochraceous above and below, brighter and more yellowish towards the rump and upper tail-coverts above and towards the vent and under tail-coverts below. Wing-quills blackish dusky, outer webs bordered with orange-yellow, inner webs with pale cinereous. Rectrices deep dusky brown, outer webs edged with orange-yellow. Under wing-coverts yellowish. Iris dark hazel : beak slate-blue, tip dark brown. Tarsi and toes greyish brown. Size not perceptibly different from that of *L. coccinea.* Wing 2·45 to 2·54 inches, tail 1·9 to 2·2, culmen 0·47 to 0·49, tarsus 0·8.

Other males shot at the same time of the year are wax-yellow or gallstone-yellow. I am inclined to think that they are not so old as the orange-ochraceous ones, but Mr. Perkins, although admitting " occasional intermediate forms," is of opinion that the male is dimorphic in colour.

Adult female. Above dark green, lighter and more yellowish green on the rump and upper tail-coverts. Wings and tail dusky blackish, edged with yellowish green on the outer webs of the feathers. Below dull dirty yellow, washed with olive along the sides of the body.

Young males resemble the female, but between the first olive-greenish plumage and the ochraceous garb of the adult males some apparently intermediate stages occur, some being wax-yellow (Ridgway, Nomencl. Col. pl. vi. fig. 7), others very nearly gallstone-yellow (pl. v. fig. 6). The adult male is either dimorphic, as explained by Mr. Perkins, or the plumage would seem *to change colour*, for if the colour of the adult male is assumed by moult alone, it would take several moults to complete it. There are,

2 D 2

however, instances enough where it takes birds several years to assume the perfect plumage of the adult male.

This species can be distinguished by the colour of the adult male, and none of the intermediate plumages are quite similar to any of the intermediate stages of *Loxops coccinea*. The adult female is a good deal darker than the female of the latter species. Structural differences there are none, and even the size is about the same.

PALMER discovered this species in the hills of Maui, whence he sent a fair series, though it was not very numerous.

He saw it mostly creeping about among the leaves of the Ohia-trees, feeding on small insects and larvæ. In its manners it was like its congener *L. coccinea* of Hawaii, but its note was lower and more plaintive. It was not at all shy. The stomachs were generally full of small beetles and other insects.

Mr. Scott Wilson [1], in part iv. of 'Aves Hawaiienses,' describes this species as "*Himatione aurea*." Mr. Wilson has thus placed the three forms of the genus *Loxops* known to him at the time under three different generic names—the fourth (*L. rufa* of Oahu) having been united by him with *L. coccinea*. Although such a mistake was at the time pardonable if descriptions only of them were at hand, it is less so when examining specimens. Yet Mr. Wilson had before him some *Loxops coccinea*, some *Loxops cœruleirostris* (for which he created the genus *Chrysomitridops*), and some Maui skins collected by Dr. Finsch. These were lent to him by the authorities of the Berlin Museum, and, after having examined them, he declared "that they undoubtedly belong to *Himatione*." Dr. Finsch quite properly referred the Maui birds to the genus *Loxops* (or, as he called it, *Hypoloxias*), but unfortunately employed Dole's name "*Drepanis aurea*" for them. This name (*aurea*) has been accepted by Wilson ; but this is an extraordinary nomenclatorial proceeding, as it is perfectly clear that Finsch did not mean to create the name *aurea*, but united his Maui birds with Dole's *aurea*, which was, however, given to Hawaiian specimens, and is thus to be placed as a synonym to *L. coccinea*. There is thus no "*Hypoloxias aurea*, Finsch," but only a "*Hypoloxias aurea* (Dole)." In 1893 I described the Maui form as *Loxops ochracea*. There is therefore no "nomenclatorial puzzle" at all about this bird, and, according to all ancient and modern rules of nomenclature, the name *ochracea* must be accepted for the Maui form. Yet Mr. Wilson accepts the name "*aurea*," writing as follows:—"This term originally appearing in connexion with *Drepanis* was a wholly inaccurate generic assignment, and justice to the perspicuity of the distinguished ornithologist (Finsch) demands that his name should not be set aside. So far as practice is concerned no confusion is likely to follow from maintaining the term *aurea* in Dr. Finsch's sense ; and, as Mr. Rothschild was neither the discoverer nor the first describer of the species, and could not have known, except from my work, what the '*Drepanis aurea*' really was, there seems no need to treat his name for the Maui bird otherwise than according to the strictest law, which to me does not appear to require the adoption of his subsequently conferred designation of *ochracea*." Mr. Wilson would thus justify the use of one and the same specific term twice in one genus, if for the

---

[1] I say purposely Mr. Scott Wilson, and not Messrs. Wilson and Evans, as the author always speaks of "I" and not of "we."

first time used in connection with a wrong generic name, or else an older synonymous name could in the same genus also stand as a specific non-synonymous name. With the same right he might then conclude that his *Himatione dolei*, being not recognized as belonging to another genus, must not be used in the new genus *Palmeria*. Mr. Wilson then raises the question of "justice" to "distinguished authors"; but surely it is not out of a sense of "justice," but for the reasons of practical usefulness that we have to consider nomenclatorial questions, and we are confident that Dr. Finsch himself would never have desired to make himself the author of Dole's "*aurea*." I was, of course, not aware of the fact that Finsch had described Maui specimens when adopting Dole's name *aurea* in the 'Ibis,' or I should have mentioned it; but it was before Wilson's examination of the "type" of Dole's "*aurea*" that Finsch said it was a *Loxops*. Evidently Finsch must have had some information about Dole's types, and, though Wilson informs us first that he has actually compared the specimen in the Mills collection, and found it to be a young *Loxops coccinea*, the fact that Dole gave "Hawaii" as its home was enough to reject it for the Maui representative form. The fact that I learnt from Mr. Wilson's work "what the *Drepanis aurea* of Dole really was" had nothing to do with my naming *Loxops ochracea*, and, thankful as we must be for the examination of the "*Drepanis aurea*," we cannot, on the whole, admire Wilson's original work in the genus *Loxops* much, since he distributed three species of it in three genera, and failed to recognize the fourth form, by which he misled me to rename it "*Loxops wolstenholmei*." All these mistakes are, however, duly corrected in the later parts of the 'Aves Hawaiienses.'

That excellent observer, Mr. Perkins, describes the call of *Loxops ochracea* and the other species of the genus as a plain "*keewit*," "uttered once or repeated, and constantly to be heard. They seek," he proceeds, "their food amongst the leaves, especially at the ends of the branches, more rarely on the limbs themselves. It consists largely of caterpillars and small spiders. They also suck the nectar of the ohia flowers (*Metrosideros*); this I saw them do but rarely, and only two of the species, *L. aurea* (sic!) and *L. cæruleirostris*. Most often, when seen amongst the blossoms, they were merely seeking insects, thereby attracted; but several times I shot specimens with the beak dripping, and on tasting the fluid found it to be, beyond doubt, the nectar of these flowers. From the other green birds, the green young and females of *Loxops* are readily distinguished, at any height, by their more forked tails, which, combined with their short, thick beaks, give them a very finch-like aspect." On one occasion Mr. Perkins saw "a pair of this bird building, high up in a tall ohia tree, toward the end of a branch. They came down to the ground for material, stripping off the brown down that covered the young fronds of some stunted 'pulu' ferns." On another occasion he "watched a pair sporting on the wing, now ascending, now descending, but gradually rising upwards till they became mere specks in the sky. It must have been several minutes before they finally alighted at no great distance from their starting-point. Both were splendid males."

With regard to the plates in Mr. Wilson's book, I must say that the figures on his first plate, which are taken from Dr. Finsch's Maui specimens in the Berlin Museum, are good, but the upper figure on his second plate, which is meant for *L. ochracea*, is too red and appears smaller than the lower figure, which is said to be *L. rufa*, while in fact the latter is smaller and more reddish. Possibly the lower figure should be *ochracea* and the upper *rufa*, but the lettering of the plate contradicts this, while in the text it is said that both forms of *L. ochracea*, the "red" one and the "orange" one, are figured on one plate, the lower figure being taken from the Berlin specimen. There is, however, one plate showing a yellowish male and a female, evidently from the Berlin birds, another meant to show (according to the lettering) the red form of *L. ochracea* and *L. rufa*. With regard to my own plates, I must confess that the lower figure of *L. ochracea* is not bright enough, and that Mr. Keulemans has not quite successfully copied the colour of the adult male of *L. coccinea*, while the figure of *L. rufa* is correct, the bird shot by Wolstenholme being a little brighter than those in the British Museum. Wilson's figure of *L. rufa* is also good in colour, and his adult *L. coccinea* (in part i. of his work) is by no means bad in colour.

36. LOXOPS RUFA (*Bloxam*).

AKAKANE OF OAHU.

Fringilla rufa, Bloxam, Voy. 'Blonde,' p. 250 (1826) (types in the British Museum examined) ; Gray, Zool. Misc. p. 11 (1831).

Linaria ? coccinea (errore !, non *Fringilla coccinea*, Gmelin, 1788), Gould, Voy. 'Sulphur,' Birds, p. 41, pl. xxii (1843).

Drepanis rufa, G. R. Gray, Gen. B. i. p. 96 (1847).

Loxops coccinea (errore, non *Fringilla c.*, Gmelin !), Cabanis, Archiv f. Naturg. 1847, p. 330 (genus *Loxops* established, on specimens collected in Oahu by Deppe : specimens in Berlin Museum); Reichenow, Vög. Zool. Garten, ii. p. 365 (1884) (" ♂ rothbraun ").

Hypoloxias coccinea, Lichtenstein, Nomencl. p. 48 (1854) (only Deppe's skins from Oahu in Berlin).

Loxops coccineus, Pelzeln, Journ. f. Orn. 1872, p. 29 (Enero, Oahu, Deppe).

Loxops coccineu, Sharpe, Cat. B. Brit. Mus. x. pp. 49 (fig.), 50 (synonymy partim !).

Loxops wolstenholmei, Rothschild, Bull. Brit. Orn. Club, vol. i. p. lvi (June 1893).

Loxops rufa, Scott Wilson, Aves Hawaiienses, part vi. (1896) (plate and text).

Adult male (shot on Oahu by Wolstenholme ; type of *L. wolstenholmei*). Above dark rufous orange with a brownish wash, becoming lighter and brighter on the lower rump and upper tail-coverts. Wings blackish brown, edged with greenish orange on the outer webs and with pale ashy on the inner webs. Tail-feathers paler than the wings, and edged with bright orange on the outer webs. Underparts reddish orange, changing into bright salmon-orange on the lower abdomen and under tail-coverts. Under wing-coverts whitish grey, slightly washed with rusty brown. Iris dark hazel ; maxilla whitish blue with dark brown tip, mandible the same ; feet and toes black, soles light flesh-colour. Palmer measured the type in the flesh as 4¾ inches ; in skin it measures nearly 4 inches. Wing 2·2, tail 1·7, tarsus 0·74, culmen 0·45.

Besides this perfectly adult male, in my museum at Tring, I am aware of specimens of adult males in the following collections :—

Three in the British Museum, namely Bloxam's types and one from the Gould Collection, probably the one figured in the 'Voyage of the Blonde.'

One in the Berlin Museum, collected by F. Deppe on Oahu.

One or more in Cassel, collected by Deppe on Oahu.

One in Count Berlepsch's collection, also from Deppe, collected on Oahu.

One in Vienna, collected also by Deppe on Oahu.

These males agree with the type in colour, but some are less bright, more brownish above and more orange below.   This may partly be due to the remote date of their capture and exposure to light, or they may be a little less adult.   Their wings are 2·26 to 2·3 inches, and their other measurements are the same as those of the type of *L. wolstenholmei.*

Some specimens of *L. rufa* no doubt exist in Philadelphia, U.S.A., for one of Mr. Townsend's specimens is now in the Museum at Liverpool.

There is one immature male in the Berlin Museum, also from Oahu, which is coloured thus:—Above greenish olive, brighter on the rump and upper tail-coverts, strongly washed with orange on the head and occiput.   Below orange, washed with deep orange-brown on the throat and breast and with olive on the flanks.   Size similar to that of the adult male. Another young male is in Vienna.

An immature male in the Liverpool Museum, which was received there in 1838 "from Mr. Townsend per Mr. Audubon," is above of a deep rufous brown with some blackish edges to the feathers, below very pale brownish rufous with some orange (new) feathers. The wing is 2·3 inches long.

There are females in the Museums of Berlin, Vienna, and Count Berlepsch.   They agree in colour with the females of *Loxops coccinea*, but they have no orange wash across the breast.   They are above olive greenish, brighter on the rump and upper tail-coverts and on the edges of the wings and tail.   Sides of head and entire underparts dull yellowish, washed with olive along the sides of the body; quills brown, bordered with greenish, tail similar.   Wing 2·2 and 2·18 inches, tail 1·6.

The Berlin specimens bear the Museum numbers 827, 828, and 829; the Berlepsch specimens 1414 and 1415.

THIS bird is distinguished from *Loxops coccinea* by the much more brownish colour of the male and the smaller size; and the female (if those known to me are perfectly adult) by the colour of the breast and forehead, the latter not being lighter than the crown.   It is likewise easily distinguished from *Loxops ochracea* by the much more rufous colour of the male and the much smaller size.

It inhabits Oahu and is a very distinct form.   It was, no doubt, more numerous in former days.   At present it must be decidedly rare, as only one male was obtained in the course of many months.

It was shot on April 20th in the mountains of the Wailua district, Oahu, by Palmer's assistant, Wolstenholme.   Mr. Perkins was present when he shot it, and neither he nor anyone else has succeeded in observing another example.

The specimens of Count Berlepsch and those belonging to the Liverpool Museum are before me while I am writing this article, and Mr. Hartert has examined the Berlin specimens with the type and Berlepsch's for comparison.   Needless to say, I have also, together with Mr. Hartert, examined the three males in the British Museum.

The Plate was lettered *L. wolstenholmei* before I discovered its identity with *L. rufa.* It is delineated from the specimen obtained by Palmer.

37. LOXOPS CÆRULEIROSTRIS (*Wilson*).

GREEN AKAKANE.

Chrysomitridops cæruleirostris, Scott Wilson, Proc. Zool. Soc. 1889, p. 445; id. & Evans, Aves Hawaiienses, part i. (Dec. 1890).

Adult male. Yellowish olive-green above, much lighter and more yellow on the rump; a more or less ill-defined gamboge-yellow patch on the crown; lores black, this colour meeting on the forehead, and very narrowly on the chin; wing-quills brownish black, margined with olive-green on the outer webs, and with greyish white on the inner webs. Under surface greenish yellow; rectrices brownish black, bordered with olive-green, and with a wash of that colour on the two middle ones (between Ridgway's gamboge-yellow and olive-yellow, on plate vi.), washed with greenish along the sides of the body. Under wing-coverts lemon-yellow; iris dark brown; maxilla and mandible greyish blue, tipped with brown; legs and feet blackish brown. Total length, as measured in the flesh by the collector, about 5 inches, and about 4½ as measured from the skins before me. Wing 2·5 to 2·6 inches, tail 2 to 2·15, tarsus 0·75, culmen 0·44 to 0·46.

Adult female. Most of the females before me (about a dozen) are less brightly coloured above, the yellow patch on the crown is not so well defined, and the wing is a few millimetres shorter; but some males, probably younger ones, are quite similar, and there is one adult female which cannot be distinguished from the males. About this latter specimen (no. 913) there seems to be no mistake, as Palmer expressly says in his diary :— "You will observe that this bird has a yellow head like the males, but my assistant dissected it carefully and found that the ovaries were enlarged, and there could not be any doubt whatever that it was a female." One or two other birds marked "female" are hardly different from that specimen.

Mr. SCOTT WILSON discovered this species in October 1888 in the district of Waimea, in Kauai, at an elevation of about 3000 feet, and he believes it to be a rare bird.

Palmer sent me about 30 specimens from the mountains of Kauai, but being there collecting chiefly from December to February it seems he arrived at an unfortunate season for these birds, as the majority are in much abraded plumage.

They were not met in the valleys, but only at a considerable height in the mountains; only once were they observed close to Makaweli at low elevations. They were generally

2 C

seen feeding amongst the foliage of the trees, slipping through the leaves exactly in the same way as the red *Loxops* of Hawaii; and they " differ not in any way in their habits from the red *Loxops*," although they were seen in rather larger flocks when accompanied by the young in June. The note is like that of the Hawaiian bird, only a little stronger; in its movements it is slower than the *Himationes*. They were always seen in flocks and were very tame, not even flying off at the report of a gun—a similar tameness being peculiar to most birds on Kauai.

THE GENUS CIRIDOPS.

The bill is strong, the culmen arched, the maxilla overlaps the mandible at base. Nostrils covered by a membrane; no rictal bristles, but a few hairy feathers on the chin. Feathers of crown and throat short, strong, and pointed. Wings large; first primary absent, third, fourth, and fifth about equal and longest, second one-eighth of an inch shorter than the third. Tail moderate, rectrices pointed and nearly equal in length. Feet rather strong; tarsus hardly longer than middle toe with claw. Feathers of abdomen and others somewhat decomposed.

The tongue seems to prove that this genus belongs to the *Drepanidæ* and not to the *Fringillidæ*, the only two families which would have any chance of claiming it.

The genus was founded by Scott Wilson in 'Nature,' 1892, p. 469.

It seems to stand nearest to *Loxops*, from which it is at once distinguished by the form of the bill, the pattern of coloration, the strong feet, and the peculiar plumage, which is somewhat stiff and scanty, while it is soft and very rich in *Loxops*.

Only one species known.

38. CIRIDOPS ANNA (*Dole*).

ULAAIHAWANE.

—

Fringilla anna, Dole, Hawaiian Alman. 1879, p. 49; id. Ibis, 1880, p. 241 (reprinted).
Ciridops anna, Scott Wilson, Nature, xlv. p. 469 (1892); id. & Evans, Aves Hawaiienses, pt. iv. (1893).

Adult. Lores and forehead velvety black, this colour gradually shading into the ashy grey of the crown, nape, and hind-neck, which colour again shades off into the dark sepia-brown colour of the back. Rump and upper tail-coverts dark glossy red. Tail-feathers uniform black. Primaries and secondaries black, only the outer webs of the last three secondaries earthy brownish buff (nearest to Ridgway's "clay-colour" on plate v. fig. 8); scapulars and tips of some of the greater wing-coverts of the same colour. Feathers on the sides of the head and neck, chin, and throat black with silvery-grey shaft-stripes. Breast down to the middle of the uppermost part of the abdomen black. Middle of abdomen, vent, and under tail-coverts tawny brown. Sides of abdomen largely glossy red. Under wing-coverts blackish. Wing 3 inches, tail 1·75, culmen 0·43, height of bill at base 0·2, tarsus 0·8, middle toe with claw 0·76, hind toe with claw 0·64.

THIS specimen is the one described by Scott Wilson in the 'Aves Hawaiienses,' which was procured by the late Mr. Mills of Hilo.

Another specimen now before me was shot by Palmer's men. It agrees with the above description, except that the beautiful red has faded away in the spirits and that apparently the head has been darker. The wing is a little longer, measuring 3·1; the other measurements are the same as those of the preceding specimen.

This peculiar bird is undoubtedly one of the rarest in the world. There is no other specimen of it in Europe, and only one is known to be in the Bishop Museum in Honolulu, this making a total of only three specimens known in collections. It was described by Judge Dole, fifteen years ago, as coming from Hawaii, where it was probably discovered by the late Mr. Mills of Hilo.

Mr. Scott Wilson did not find it, but he says:—"I used to hear repeatedly of the 'Ulaaihawane,' by which name it is well known to the natives, who told me that it feeds on the fruit of the Hawane palm, whence its name—Ula (red), ai (to eat), Hawane (the Hawane palm); and therefore I have little doubt that it will be found, perhaps in some numbers, in the upland region of the interior, which I was unable to explore. My friend Mr. Francis Spencer, writing to me quite recently, says that his natives had seen the bird in the swampy forest-region above Ookala on Hawaii, and his description leaves no doubt of its identity."

It was of course one of Palmer's principal tasks to rediscover the *Ciridops anna*, and he succeeded in getting one specimen, which was shot by a native on the 20th of February near the head-waters of Awini on Mt. Kohala, Hawaii. There, the natives sent him message, the "Ulaaiwhane," as Palmer says it is properly spelt, was still seen in numbers, but when Palmer arrived on the spot the former report was denied and they said the bird shot was the only one seen there! The "Loulu" palms, they said, were the trees formerly frequented by the "Ulaaiwhane," and that they were very scarce now, although in some deep gulches there might still be some more. Palmer, after staying a short time at the place and seeing no sign of the bird, was apparently much disappointed with the place, and he concludes with the following words:—"Undoubtedly there are more 'Ulaaiwhane' in the place where the one was shot, but if a bird is rare in such a country one might travel about for months and not find one. I therefore thought it wiser to offer the natives a fair price for specimens and to go personally to a more promising place. I have shown the natives at the place how to make a skin, and they have promised to do their best."

Palmer's hopes, however, were not realized, for no further specimen has turned up as yet, and I am afraid this bird is nearly extinct.

THE GENUS PSEUDONESTOR.

Pseudonestor, Rothsch. Bull. B. O. C. vol. i. p. xxxv (1893) ; id. Ibis, 1893, p. 438 (reprinted).

THIS most remarkable genus belongs to the *Drepanidæ* and not to the *Fringillidæ*, as I thought from a superficial comparison when first describing it, accepting at the time Dr. Gadow's theory that *Psittirostra* was a genus of the *Fringillidæ*. It does not differ in any structural characters from *Heterorhynchus*, except in the formation of the bill, with its enormously hooked maxilla and the very broad mandible. Also in colour it is strikingly like *Heterorhynchus lucidus*. The feather-tracts, too, are like those of *Heterorhynchus*, the plumage is just as soft and full.

The tongue is distinctly *Drepanine*.

The sexes are similarly coloured, but the female is much duller and considerably smaller, the bill especially is very much smaller and less hooked.

Only one species is known.

39. PSEUDONESTOR XANTHOPHRYS, *Rothsch.*

Pseudonestor xanthophrys, Rothsch. Bull. B. O. C. vol. i. pp. xxxv, xxxvi (March 28th, 1893) ; id. Ibis, 1893, p. 138 (reprint) ; Perkins, Ibis, 1895, p. 118 ; Wilson & Evans, Aves Hawaiienses, pt. vi. text & plate (1896).

Adult male. Above olive-green. A broad superciliary stripe from the nostril to behind the ear and a few feathers just above the base of the bill golden yellow; lores below the yellow line dark brown ; feathers behind the eye deep brownish green ; throat and breast yellow, strongly washed with olive on the sides of the body ; under tail-coverts yellow, under wing-coverts yellowish white ; wing-feathers and tail-feathers dark brown, each feather bordered with olive-green on the outer web, and with whitish towards the base on the inner webs. Iris dark hazel; maxilla dark grey, light at base ; mandible dull white, with a slaty tinge ; tarsi and toes slate-colour, soles pale orange. Total length about 5½ inches, wing 3 to 3·1, tail 1·9 to 2·1, culmen 1 to 1·1. upper bill from base to tip in a straight line 0·8, under mandible from the mental apex to the tip 0·55, thickness of upper mandible at nostrils only 0·2, lower mandible at base 0·5 thick !, tarsus 0·9, middle toe without claw 0·6.

Female. Resembles the male, but has all the colours much duller, and is considerably smaller. The back is somewhat tinged with greyish, the abdomen paler and slightly washed with olive. Wing 2·6 to 2·7 inches, tail 1·6 to 1·7, culmen 0·68 to 0·78, upper bill from base to tip 0·5 to 0·55 measured in a straight line, under mandible from the mental apex to the tip 0·4 to 0·41, tarsus 0·76 to 0·8.

Hab. Maui.

This most interesting form was shot by Palmer on Maui, between July and October, 1892. A small number of specimens only were procured, and the bird appeared to be rare. It was mostly seen among the Koa- and Ohia-trees, on the trunks and limbs. It was not very shy, but on account of the thick undergrowth it was sometimes difficult to get near enough to shoot. Its call-note varies from a sound like "Ilow-it" to a loud "Iloo-ah," and its song is a clear warble of several notes. The stomachs contained caterpillars, larvæ of beetles, and other insects. Both Palmer and his assistant saw it several times "push up the moss off a branch and pull out caterpillars or grubs with its upper mandible."

Not long after Palmer's discovery of the *Pseudonestor,* Mr. Perkins collected some specimens on Maui and observed it frequently.

Perkins makes the following remarks :—

"Of the Fringillidæ (nearly all of which are peculiar to the island of Hawaii) I have already given some account of the habits ; but there remains one—*Pseudonestor xanthophrys* —peculiar to the island of Maui, which is perhaps the most remarkable form of all. It is

2 D

local and rare, and seems to be confined to the highest forest on Haleakala, at an elevation of some 5000 feet above sea-level. Being very tame and apparently unwilling to fly far, I had on several occasions excellent opportunities to learn something of its habits, and especially of the use of its curiously formed and exceedingly powerful beak. The bird has an evident predilection for the koa trees (*Acacia falcata*), and it is from these that it mainly gets its food. This consists of the larvæ of a highly peculiar endemic genus of Longicorn beetles (*Clytarlus*), of which there are in the islands a considerable number of species, nearly all of them attached to the different species of acacias. The larger ones usually burrow in the main trunks, the smaller in the limbs and twigs above. It is on the larvæ of the latter that *Pseudonestor* feeds, and in procuring them has developed the large hooked beak, the powerful jaw-muscles, and heavy skull, which constitute its chief peculiarities. It may be observed that the twigs in which the *Clytarli* have their burrows are not generally rotten, but dry, and of excessive hardness, often surpassing in this respect the still living and unaffected branches. The bird is sluggish, in its movements parrot-like in the extreme, especially in the varied hanging attitudes that it assumes, while the similarity is still further increased by the shape of its beak.

"'Those that I saw in the act of feeding were generally clinging to the under sides of the thin branches or twigs, the head raised above the upper surface; the point of the curved maxilla was thrust into the burrow, the short mandible opposed thereto, and pressed against the side or under surface of the twig, and the burrow opened out by sheer strength. All that I shot contained larvæ of these beetles, as many as 20 or 30 being found in the stomach of a single bird. No less than four species of *Clytarlus* were found on the acacias in the actual haunts of *Pseudonestor*; these, too, like the bird, are all of species peculiar to the same island. When alarmed the bird gave utterance to a short squeaking cry; it has besides a decided song, which reminded me much of that of the green *Himatione*. Once I heard it sing on the wing as it crossed a gulch.

"'The unpleasant scent of *Pseudonestor*, like that of many Drepanididæ and other Hawaiian Finches, is very noticeable.

"Looking at the Hawaiian Finches as a whole, it may be noticed how wonderfully the structure of each of them has been specially developed according to the nature of its own particular and most important article of food. Thus *Pseudonestor*, as above mentioned, has an enormous development of beak and skull and muscles attached thereto, for splitting the koa twigs; *Chloridops* has a huge beak and still heavier skull and muscles, which enable it to crack the hard nuts of the bastard sandal (*Myoporum*); then there is the strong cutting-beak of *Rhodacanthis* for dividing up the koa beans, and a large development of the abdominal portion of the body, in accordance with the large fragments that it swallows; the shorter bill of *Loxoides*, which deftly cuts off the bean of the mamane acacia (*Sophora*), while the bird holding it in position with its foot opens the pod and devours the seeds; and, lastly, the hooked bill of *Psittacirostra*, with which it digs out the separate components of the fleshy inflorescence of the 'ieie' (*Freycinetia*), for this is certainly its natural food, though it has now come to feed largely on various introduced fruits—guavas, oranges, and the like. Besides their special foods all the Finches vary their diet at times with the larvæ of Lepidoptera."

These notes were of course written before it had been recognized that *Pseudonestor* was a *Drepanine* and not a *Fringilline* bird.

THE GENUS PSITTIROSTRA.

Psittirostra, Temm. Man. d'Orn. i. p. lxxi (1820). (Afterwards often corrected into *Psittacirostra*.)

Sittacodes, Gloger, Handb. p. 249 (1842).

Psittacopis, Nitzsch, teste Cab. Arch. f. Naturgesch. 1847, p. 330. (I am unable to find where Nitzsch published this name. Scudder quotes *Psittacopis*, Nitzsch, 182-.)

Psittacina, Licht. Nomencl. p. 48 (1854). (Lichtenstein quotes *Psittacina*, Temm.; but Temminck did not publish that name.)

THIS genus is characterized by its stout and curved maxilla, the entire bill being more like that of a *Fringilline* bird than that of a member of the family *Drepanidæ*. Nevertheless it belongs to the latter family. It is only natural that in former days, with skins only in hand, the genus was placed among the *Fringillidæ*. Even the study of the skins, however, induced more recent ornithologists to class this form (like *Loxops*) with the *Drepanidæ*, or, as this family was not then recognized, at least along with the allied genera *Hemignathus*, *Drepanis*, *Loxops* in the *Dicæidæ*, or whatever they called the assemblage [1]. It was Dr. Gadow, who quite recently, in his chapter on the "Structure of Hawaiian Birds" in Wilson's 'Aves Hawaiienses,' came back to the old theory of its being a Finch, and he derived his view from anatomical researches. Dr. Gadow's theory, however, was evidently wrong, and he has recently admitted that not only this but also all the other thick-billed *Passeres* of the Sandwich Islands belong to the *Drepanidæ*.

---

[1] See, for example: Sclater, Ibis, 1871, p. 360; Reichenow, Vögel der Zoologischen Gärten, ii. p. 365 (1884); Sharpe, Cat. B. x. p. 2 (1885), and others.

40. PSITTIROSTRA PSITTACEA (Gm.).

OU.

Parrot-billed Grosbeak, Lath. Gen. Syn. ii. p. 108, pl. xlii. (1783) ("Sandwich Islands").
"Bird with a yellow head," King, Voy. Pacific Ocean, iii. p. 119 (1784).
Loxia psittacea, Gm. Syst. Nat. i. p. 844 (1788) (ex Latham); Latham, Ind. Orn. i. p. 371 (1790);
   Donndorf, Orn. Beytr. ii. p. 343 (1795); Tiedemann, Anat. Naturg. Vög. ii. p. 433 (1814); Steph. in
   Shaw's Gen. Zool. ix. pt. ii. p. 268 (1816); Bloxam, Voy. ' Blonde,' p. 249 (1826).
Strobilophaga psittacea, Vieill. Nouv. Dict. ix. p. 609 (1817); id. Enc. Méth. p. 1021 (1823).
Psittirostra psittacea, Temm. Man. d'Orn. i. p. lxxi (1820); Swains. Class. B. ii. p. 205 (1837); Gray,
   Gen. B. ii. p. 389, pl. 94. fig. 2 (1845); Bp. Consp. i. p. 492 (1850); Hartl. Arch. f. Naturg. 1852, i.
   p. 133; Cass. U.S. Expl. Exp., Mamm. & Orn. p. 432 (1858); Dole, Proc. Bost. Soc. N. H. xii. p. 301
   (1869); id. Hawaiian Alman. 1879, p. 49; Scl. Ibis, 1871, p. 360 (has nothing to do with the Fringillidæ, being merely aberrant forms of the same type as Drepanis and Hemignathus); id. P. Z. S. 1878,
   p. 347 (Hilo, Hawaii); id. Ibis, 1879, p. 92; Pelz. J. f. O. 1872, p. 30 (compared type of Latham with
   specimens from Oahu); id. Ibis, 1873, p. 21 (same); Finsch, Ibis, 1880, p. 81 (Maui, shot but lost
   specimens); Sharpe, Cat. B. x. p. 51 (1885); Pelz. & Lorenz, Ann. k. k. naturh. Hofmus. 1886, p. 263
   (types in Vienna); Stejn. Proc. U.S. Nat. Mus. 1887, p. 389; 1889, p. 386 (Kauai); Wilson, Ibis,
   1890, p. 194 ("throughout the group"); id. & Evans, Aves Hawaiienses, pt. ii. text & pl. (obtained
   on all islands, except Oahu, where prob. extinct, and Maui); Gadow, in Wils. Aves Hawaiienses, pt. ii.
   pp. 6 & 7 (stated that it is a Fringilline bird!); A. B. Meyer, Vogelskelette, iii. p. 38, pl. cxlv. (1892);
   Perkins, Ibis, 1893, p. 103 (field-notes). (Some authors spelt the name Psittacirostra.)
"Raouhi," Quoy & Gaim. Voy. ' Uranie' & ' Phys.,' Zool. ii. p. 36 (1842).
Psittirostra sandvicensis, Steph. in Shaw's Gen. Zool. xiv. pt. i. p. 91 (1826).
Psittacirostra icterocephala, Temm. & Laugier, Pl. Col. 457, livr. 77 (1828) (♂ only figured, much too deep
   green; ♂ & ♀ described); Cuv. Règne Anim. éd. 2, i. p. 415 (1829).
Psittacopis psittacea, Cab., Ersch & Gruber's Allgem. Encycl. sect. i. p. 219 (1849); Sundev. Tentamen,
   p. 32 (1872).
Psittirostra icterocephala, Gray, Cat. B. Trop. Isls. p. 28 (1859).
Sittacodes psittacea, Reichenow, Vög. Zool. Gärt. ii. p. 365 (1884).
"Ou (Loxioides bailleui)," Perkins, Ibis, 1893, p. 103 (Perkins evidently confused the names of the two
   birds; see p. 197 l).

Adult male. Head all round and fore part of neck above golden yellow. Rest of upper
surface olive-green, with a yellow tinge, slightly duller on the hind-neck and inter-
scapulium, brighter on the rump and upper tail-coverts, the feathers grey at base.
Wings blackish brown; quills outside edged with the colour of the back, narrower on
the primaries, much wider on the inner secondaries. Inner webs with paler edges.

Rectrices coloured like the remiges. Chin and upper portion of throat golden yellow ; rest of under surface pale greyish olive-green, brightest and most greenish on the sides of the body, whitish in the middle of the abdomen and on the under tail-coverts. The feathers of the underside are grey at base and on the shaft, those of the chest more or less washed with grey. Under wing-coverts greyish white, washed with yellowish green towards the bend of the wing. Total length about 7 inches, wing 3·8-4·1, tail 2·4, culmen 0·75, bill from gape to tip about 0·75, tarsus 0·95.

Bill and feet pinkish or greyish white in life, light horn-coloured in collection. Iris brown.

Adult female. Head above coloured like the back, hind-neck with a distinct greyish tinge. Underside, including chin and throat, whitish grey ; sides widely washed with olive-green. Dimensions generally less than those of the males. Wing 3·6 to 3·8 inches.

Young males resemble the adult female, but there is a more or less strong yellow wash over the grey of the underside, which is generally absent in adult females. Young birds have often small buffish tips to the wing-coverts, and their beaks are a little less hooked than in adult birds.

I have not been able to trace any differences in specimens from the various islands, except in those from Oahu. I have examples from Hawaii, Kauai, Maui, Lanai, and Molokai. The wings of adult males from Hawaii measure 3·8 to 3·85 inches, females 3·6 to 3·75 ; males from Kauai 3·8 to 4·1, females 3·6 to 3·8 (one 3·9, but this is probably a young male ?) ; males from Maui 3·8 to 3·9, one female 3·7 ; one male from Lanai 3·85 ; males from Molokai 3·85 to 3·95, females 3·65 to 3·8 inches.

THIS bird inhabits now the lower portions of the mountain-forests of all the islands, except Oahu. It is commonest on Kauai and Hawaii, rarer on the other islands. Perkins has once seen it in Kona at about 4000, Palmer up to about 7000 feet above the sea. It is a very noticeable bird and not at all shy. Its song is pleasant, and when feeding on guavas they utter at intervals a note which sounds somewhat like the word " sweet." On Hawaii, Palmer saw them rise in a flock, and on Lanai a flock thus rising " began to ring out their call-notes." It often sings while on the wing.

They are principally fruit-eaters. On Hawaii, Perkins found them very partial to the " kukui " trees. They are very fond of the guavas and bananas. Upon the former they fall in large flocks, when their fruit is ripe. They bite open the ripe oranges, and smaller birds of the genera *Chlorodrepanis* and *Oreomyza* then feed away at them.

Nothing seems to be known about the breeding of this bird, but from the end of March to the middle of May females were procured with greatly enlarged ovaries.

These birds have a strong and peculiar musky smell, which they have not even lost in the drawers in the Museum.

They are generally known as the " Ou," but Mr. Gay informed Mr. Perkins that on Kauai the male is called " Ou poelapalapa," the female "Ou laevo," which means the yellow-headed and the green Ou.

41. PSITTIROSTRA OLIVACEA, *Rothsch.*

THE HONOLULU OU.

Psittina olivacea, Licht. Nomencl. p. 48 (1854) (nomen nudum !, "Sandwich Inseln " !). (It is only because the specimens in the Berlin Museum are still in existence and are marked as the types of *P. olivacea*, and they are from Oahu, that we know that this name refers to the Oahu-form of *Psittirostra*.)
Psittirostra psittacea, Pelz. J. f. O. 1872, p. 30, partim (mentioned the differences between a specimen from Oahu and Latham's type) ; id. Ibis, 1873, p. 21 (ditto).

VON PELZELN has noticed, in the 'Journal für Ornithologie,' 1872, p. 30, that a male from Oahu (collected by Deppe), differed from Latham's type in the Vienna Museum in *having the middle of the breast and belly, the feathers of the tibiæ, and the under tail-coverts whitish.* Now I have compared a series of *Psittirostra* from Oahu (collected by Behn), kindly lent me by the director of the Museum at Kiel, Herr Professor Dr. K. Brandt, and Mr. Hartert has studied the specimens in the Berlin Museum from Oahu, and it became obvious to us that all the adult males had the breast, abdomen, and under tail-coverts buffy whitish, the breast washed with olive, the sides of the body olive-green. Such a coloration of the underparts is found only in immature birds of *Psitt. psittacea*, while adult males of the latter with bright yellow heads have the underparts olive-green, merging into whitish only in the middle of the lower abdomen, and the breast mostly with a strong ashy wash. The figure in Wilson's book represents them very well. Besides this the wing of the Oahu form is shorter and the upperside has a more olive tint, but immature birds and females are very much like those of *P. psittacea*, except being more brownish olive above.

Now under these circumstances it is pretty clear that the form of Oahu belongs to a distinct species, or perhaps one should say subspecies, as there are specimens which in certain characters make a step to approach the Oahu form. Thus one bird, a male, from Molokai has more of a whitish shade on its abdomen, but the dirty yellow of the head shows it to be somewhat immature, and its wing is longer than those of the Oahu birds. It being separated on another island, and perfectly intergrading adult birds not being known to me, I keep it here as a species.

Lichtenstein's name is published without any description and without an exact locality, and such " nomina nuda" cannot be accepted, of course, if the forms to which they refer are not described. However, I use Lichtenstein's name, not wishing to encumber the synonymy by rejecting it, reluctant as I am to do so, because my name will have to stand as its author, as here it is first distinguished under that name.

It may be diagnosed as follows :—

Psittirostra similis speciei Ps. psittacea dictæ, differt abdomine maris adulti albido, colore supra paullo olivascentiore, alis brevioribus.

Hab. Insula Oahu dicta, olim.

The measurements of an adult pair, which I received in exchange from the Museum at Kiel, are : Wing 3·75 inches in the male, 3·64 in the female; tail 2·6 in both; culmen 0·71 in the male, 0·64 in the female; tarsus 0·85 in both.

This Ou formerly existed on the island of Oahu, and a few specimens collected by Deppe are in the Museums of Berlin and Vienna, while quite a series, collected on Oahu in October 1816 by Professor Behn during the voyage of the Danish ship 'Galathea,' is in the Museum at Kiel, of which I was fortunate to obtain a male and a female by exchange.

Neither Mr. Scott Wilson nor Messrs. Palmer and Perkins were able to discover a Psittirostra on Oahu, and as these birds are not easily overlooked, we may safely suppose that it is extinct now. Palmer was told by several persons that it was formerly not uncommon.

THE GENUS LOXIOIDES.

THIS genus was first established by Oustalet in 1877, and considered to be a Finch. Sclater, in 1879, pointed out its close affinity to *Psittacirostra*, while Gadow, in 1891, declared it to be "clearly Fringilline," and "to approach the genera *Loxia*, *Coccothraustes*, and *Pyrrhula*" (sic), but eight years later united it, together with all the rest of thick-billed Hawaiian *Passeres*, with the *Drepanidæ*.

The nearest relative to *Loxioides* is doubtless *Telespiza*, and not *Psittirostra*. From *Telespiza* it differs merely in having a considerably shorter and very strongly curved bill with a less protracted point, the width of lower mandible at base wider than the length of the gonys, the whole bill much shorter than the head. The wings and tail are a little longer than in *Telespiza*.

The nostrils are overhung by short feathers, open in the middle, but dorsally protected by a distinct operculum. The third and fourth primaries form the tip of the wing. The tail is emarginate, the central rectrices being shorter than the lateral.

Only one species is known.

2 B

42. LOXIOIDES BAILLEUI, *Oust.*

PALILA.

Loxioides bailleui, Oust. Bull. Soc. Philom. Paris, sér. 7, i. p. 100 (1877) ; Sharpe, Cat. B. x. p. 49 (1885) ; Scl. Ibis, 1879, pp. 90, 92, pl. ii. (erroneously spelt *" bailleni "*) ; Wilson & Evans, Aves Hawaiienses, pt. i. text & figure (1890) ; Gadow, op. cit. pt. ii. p. 5, pl. i. figs. 11–16 (1891).

" Ou-po-papale, *Psittacirostra psittacea"* (partim), Perkins (non Gm.), Ibis, 1893, p. 103.   (There is, in my opinion, no doubt that Mr. Perkins had mixed up the names of *Psittacirostra* and *Loxioides,* as the description of the habits of his supposed *Psittacirostra*—the highland distribution, its principal food the *mamane*—leaves no doubt as to what bird he meant.)

Adult male. Head, neck, throat, and breast gamboge-yellow : lores and eyelids blackish. Interscapular region, back, and rump ashy grey, much lighter on the rump ; upper tail-coverts ashy grey, more or less tinged with greenish.   Quills and tail-feathers blackish, edged with whitish on the inner and with olive-yellow on the outer webs. Wing-coverts olive-yellow, blackish at base ; primary-coverts blackish, with olive-yellow outer edges.   Abdomen and under tail-coverts ashy white.   Under wing-coverts white, with a yellow tinge.   " Iris dark brown.   Bill brown.   Feet blackish brown, soles flesh-colour with yellow tinge."   Total length in the flesh measured as about 7½ inches by Palmer, but skins only 6 to 6½ inches in length.   Wing 3·5 to 3·65, tail 2·6 to 2·7, tarsus 0·9 to 0·95, culmen 0·5.

Adult female. Similar to the male, but the yellow of the head washed with dark brown, the yellow below not so bright and not reaching so far down over the breast.   The wing 3·4 to 3·5 inches.

Hab. Higher elevations on Hawaii.

THE " Palila," as Palmer heard this bird called, was discovered by Monsieur Bailleu in the Kona district of Hawaii.   I have seen the type in the Paris Museum.   It is not figured in this work, as fairly good figures of it are given in 'The Ibis,' 1879, and in Scott Wilson's book.

Oustalet believed this bird to be a Finch, but Sclater in his paper (Ibis, *l. c.*) expressed the opinion that it was related to *Hemignathus,* and belonged, like this latter and its allies, to the *Diceidæ,* or rather *Drepanidæ.*   Sclater also pointed out the difference between the bills of *Loxioides* and *Psittirostra,* although he believed these two genera closely allied.

2 E 2

He says: "In the shape of the bill only there is a considerable divergence, that of *Loxioides* being considerably shorter and much more swollen laterally than that of *Psittirostra*. This, and the differences in the feet, may justify the separation of the two forms into two genera. . . . ."

Wilson, Perkins, and Palmer all observed that the Palila's principal food consisted of the seeds of the "Mamane" (*Sophora chrysophylla*), but other kinds of seed were also found in their stomachs, and often caterpillars and other larvæ. Palmer says he saw it also once eat green "Aaka" seed, and that it will occasionally eat various berries, such as Cape gooseberries.

According to Palmer's observations, *Loxioides* belongs entirely to the upper forest region, and it is seldom found off the "Mamane" trees. In the Kona and Hamakua districts, between elevations of 4000 and 6000 feet, they were frequently met with, and in the Hilo district, on the slopes of Mauna Kea, they were seen even higher, at about 7000 feet. They were often seen in flocks of about half a dozen individuals. Wilson writes:—"The 'Palila,' as far as I know, has no song, but merely a very clear whistle-like note, which, when often repeated, is held by the natives to be a sign of approaching rain. While at Waimea, a specimen of *Loxioides* was brought to me alive though injured; it lived a few days, during which it constantly uttered the clear whistle without giving evidence of any further powers. On June 14th I found a nest, from which I saw the bird fly; it was placed in the topmost branches of a 'Naio' tree (*Myoporum santalinum*), about 35 feet from the ground, but contained no eggs, and when we subsequently revisited it we found it deserted. It may be briefly described as cup-shaped, 4 inches in diameter, and very loosely constructed of dry grass, among which is interwoven a considerable quantity of grey lichen, the inside being composed of the same lichen, with a few slender rootlets added."

The eggs are not yet known.

Palmer says that the note of the Palila is not unlike that of the "Ou" (*Psittirostra*), and that it has a kind of warbling song of several low but clear and sweet notes. He also writes that it is very fearless, somewhat heavy in its flight, and rather slow in moving about amongst the trees.

THE GENUS TELESPIZA.

This genus has been established by Mr. Scott Wilson (Ibis, 1890, p. 341) for the thick-billed finch-like bird from Laysan. Two years later I described a second species from Laysan, and figured both in the first part of this work; but a large and fine series sent to me by Professor Schauinsland, who recently made large collections on Laysan, shows beyond doubt, and Schauinsland's observations confirm, that my " *Telespiza flavissima* " is not a distinct species, but only the perfectly adult male of *T. cantans*.

I was at one time inclined to unite my genus *Rhodacanthis* with *Telespiza*, and I do not consider the differences between the two genera of great importance; but if I do so I cannot consistently keep *Loxioides* apart, and I therefore keep the sections *Loxioides*, *Telespiza*, and *Rhodacanthis* generically separate. They form a continuous series, *Telespiza* being in the middle, *Rhodacanthis* with the longest bill on one side, *Loxioides* with the shortest beak on the other. *Telespiza*, which lives more on the ground, has shorter wings and tail than the other two genera. The nostril is protected by overhanging bristly feathers, and has only a small operculum on the dorsal margin, its median portion being open.

THE GENUS RHODACANTHIS.

I CREATED this genus in 1892 (Ann. & Mag. Nat. Hist. x. p. 111) for the two thick-billed birds from Hawaii which I called *Rh. palmeri* and *Rh. flaviceps*. When I established *Rhodacanthis* I could not compare it with Mr. Wilson's *Telespiza*, which I believed to be more distantly related, but compared it merely with *Psittirostra* and *Chloridops*. I found, however, afterwards that *Telespiza* was so closely allied that I was inclined to unite the two genera; but if I do so I cannot keep *Loxioides* separate, as *Rhodacanthis* differs from *Telespiza* about as much as *Loxioides* does from *Telespiza*.

The only two species of *Rhodacanthis* known are large birds, with wings and tail longer than in *Telespiza*, the bill more elongated, with the cutting-edges sharper and their line more curved, the nostrils more bare of feathers, with a distinct operculum over the upper half and somewhat more dorsally situated. In the wing the third, fourth, and fifth primaries are nearly equal and longest. The plumage is soft and full. Habits arboreal; nidification not yet known!

The two species differ as follows:—

A. Wing more than 4 inches.
    a. Head and throat orange-red . .   . *R. palmeri,* ♂ ad.
    b. Head and throat olive-greenish .   . *R. palmeri,* ♀ & juv.
B. Wing less than 4 inches.
    c. Head yellow . .   . . *R. flaviceps,* ♂ ad.
    d. Head olive-greenish .   . . . *R. flaviceps,* ♀ or juv.

43. RHODACANTHIS PALMERI, *Rothsch.*

Rhodacanthis palmeri, Rothsch. Ann. & Mag. Nat. Hist. (6) x. 1892, p. 111; Perkins, Ibis, 1893, p. 103;
Wilson & Evans. Aves Hawaiienses, pt. v. plate & text (1894).

Adult male. Head and throat rich reddish orange; back and upper wing-coverts dull
greenish olivaceous, brightening to a dark dull orange on the back and upper tail-
coverts. Wings and tail dark blackish brown, the feathers externally margined with
dark dull yellow, paler and more greyish towards the tips of the primaries. Upper part
of breast dull reddish orange, passing into dull orange-yellow on the abdomen, becoming
much paler on the lower abdomen and under tail-coverts, which are pale yellow. Under
wing-coverts and axillaries olive-greyish, washed with orange. Bill bluish grey, tip
brown; feet grey with a bluish tinge, soles pale yellow. Total length about 8¾ inches
according to Palmer's measurings in the flesh, but only 7 to 7·5 in skins; wing 4·1 to
4·3; tail 2·95 to 3·15; culmen 0·8 to 0·87; tarsus 1 to 1·1; length of bill from gape
to tip 0·75, height at base 0·6.

Adult female. Upper parts olive-green, washed with yellow on the forehead and brightening
into yellowish green on the rump and upper tail-coverts. Wing and tail blackish brown,
the feathers externally margined with dull yellowish green. Throat and sides of body
yellowish olive-green; breast and middle of the abdomen dull white, washed with green.
Size similar to that of the male.

Young males resemble the female, but the feathers of the throat, breast, and abdomen are
spotted and clouded with dusky, thus giving these parts a very different aspect. As they
get older the underparts become more yellow, the throat dull orange, and the reddish
orange of the head is first beginning on the forehead. A fine series in my museum
exhibits all intermediate stages.

This fine bird, which is by no means rare in the Kona and Hilo districts of Hawaii, was
discovered by my collector Palmer, in whose honour I have named it. Many specimens were
shortly afterwards collected by Mr. Perkins in the forests at an altitude of about 4000 feet.
Palmer describes these birds as by no means shy, but rather fearless and easy to approach.
He found them to feed on the Koa seeds, but in their stomachs he also found seeds of the
Alaii and other plants. In the foliage of the tall trees the females and young ones are
difficult to see. The call-note is described by Palmer as a low whistle, sounding somewhat
like a prolonged "week." It generally consists of two or three notes, beginning high and

2 P

descending towards the end : it sounds melancholy. By imitating the cry Palmer succeeded in luring them very closely towards himself. The natives on Hawaii called the bird "Ponpou" and "Kopue," but did not seem to be well acquainted with it.

Mr. Perkins says that it frequents the tallest and most leafy acacias, both when growing on the roughest lava-flows and in the grassy openings in the forest. He found it entirely in the upper forest, and most numerous at about 4000 feet. He says that the peculiar whistle can be heard for a considerable distance, and that it can be easily called by the imitation of its whistle. Besides its principal food, the Koa bean, Mr. Perkins found it to eat occasionally lepidopterous larvæ, like *Loxioides*. Mr. Perkins says also :—"The female I have heard to utter a rather deep single note when alarmed. On one occasion when I had shot a male I heard his mate repeatedly utter this note, and she continued to do so for some five minutes, but seemingly possessed some ventriloquial power—the sound seeming now in front, now behind, now near, now far; yet it was utterly impossible that the bird could have flown without my being aware of it."

The nidification is unknown.

44. RHODACANTHIS FLAVICEPS, *Rothsch.*

Rhodacanthis flaviceps, Rothsch. Ann. & Mag. Nat. Hist. (6) x. 1892, p. 111; Wilson & Evans, Aves Hawaiienses, pt. vii. (1899) (description copied).

Adult male. Head and neck dull golden yellow; back and upper wing-coverts greenish olive, shading into bright olive-green on the rump and upper tail-coverts; wings and tail dark-brown, the feathers externally margined with green. Chin, throat, and upper breast dark wax-yellow; abdomen and under tail-coverts yellowish green. Total length about 7·5 inches according to Palmer's measuring in the flesh, but only 6 inches in skin; wing 3·8 to 3·85, tail 2·5, culmen 0·7 to 0·72, tarsus 0·95 to 1.

Adult female. Underparts of a yellowish green, much paler than in the male; above similar to the male, but the head greenish olive like the back, only a yellow wash on the forehead. The female is also a little smaller than the male, the wing measures only 3·6 to 3·7 inches.

Young males closely resemble the adult females, but they are paler beneath and indistinctly spotted or clouded on the breast.

PALMER obtained a small series in the district of Kona, at the same places where *R. palmeri* was first collected. The smaller size and yellow head of the adult male serve to distinguish this species very easily from the much larger *R. palmeri* with its orange-red head in the adult male.

Neither Wilson nor Perkins met with this bird.

THE GENUS CHLORIDOPS.

Chloridops, Wilson, P. Z. S. 1888, p. 218.   Type *C. kona*.

This remarkable genus, characterized by its enormous bill, which reminds one inadvertently of the *Geospizæ* of the Galapagos Islands, was very justly created by Mr. Scott B. Wilson. The most singular fact in the bill of *Chloridops* is that the two mandibles do not quite fit together, but that a narrow open slit, through which one can look, remains near the base. The coloured Plate, as well as the bill on the Plate of bills, shows this form well.

In the moderately sized wing the first primary is quite minute; the tip of the wing is formed by the third, fourth, and fifth primaries, which are equal in length, the second is about equal to the sixth.   Neither the tail nor the feet show anything specially remarkable. The plumage is rather soft and full; the feathers have a soft, fluffy aftershaft.

The sexes are similar; the first plumage of the young is not known.

One species only is known.

45. CHLORIDOPS KONA, *Wilson.*

Chloridops kona, Wilson, P. Z. S. 1888, p. 218 ; id. & Evans, Aves Hawaiienses, pt. iv. plate & text (1893); Perkins, Ibis, 1893, p. 101.

Adult male. Bright olive-green above and below, paler and washed with buff on the vent, more olive on the under tail-coverts. Lores dusky. Quills dusky blackish, paler towards the base of the inner webs, margined with bright olive-green on the outer webs. Under wing-coverts brownish buff, washed with olive-greenish. Rectrices dusky brown, edged outwardly with olive-green. "Iris dark hazel. Maxilla horn-grey ; mandible grey, much lighter at base. Legs and toes dark brown, almost blackish, not pale as in the figure in 'Aves Hawaiienses'" (*H. C. Palmer*). Total length, as measured in the flesh by the collector, 6½ to 7¼ inches, but hardly 6 inches in skin; wing 3·45 to 3·55, tail 2·2 to 2·35, culmen 0·8 to 0·85, bill from gape to tip 0·75 to 0·8, height of bill at base 7·3 to 7·6, width of maxilla at base 3·56, width of mandible at base 0·57, tarsus 0·8 to 0·9.

Quite a number of specimens differ materially from the above described specimens, which undoubtedly are the adult males. They are smaller, the wing measuring only 3·25 to 3·4, and are much lighter and pale yellow on the abdomen. These birds are apparently younger specimens. They cannot belong to a different species, because there are specimens intermediate between them and the darker-bellied form, and as the majority of them are marked as females they probably are the females, and some that are marked males are evidently young. One of the supposed females is figured on my Plate, together with an adult male.

This remarkable form was discovered by Mr. Scott Wilson, who, however, shot a single specimen only at about 5000 feet elevation. During a stay of four weeks in the Kona district the discoverer saw only three of them.

Mr. Perkins (Ibis, 1893, p. 101) says that *Chloridops kona* is a singularly uninteresting bird in its habits. "It is a dull, sluggish, solitary bird, and very silent—its whole existence may be summed up in the words 'to eat.' Its food consists of the seeds of the fruit of the *aaka* (bastard sandal-tree, and probably at other seasons of those of the sandal-wood tree), and as these are very minute, its whole time seems to be taken up in cracking the extremely hard shells of this fruit, for which its extraordinarily powerful beak and heavy head have been developed. I think there must have been hundreds of the small white kernels in those that I examined. The incessant cracking of the fruits when one of these birds is feeding, the noise of which can be heard for a considerable distance, renders the bird much easier to get than it otherwise would be.   It is mostly found on the roughest lava, but also wanders into the open spaces in the forest."   Mr. Perkins never heard it sing.   "Only once did I see it display any real activity, when a male and female were in active pursuit of one another amongst the sandal-trees.   Its beak is nearly always very dirty, with a brown substance adherent to it, which must be derived form the sandal-nuts."

H. C. Palmer found *Chloridops kona* not rare at Pulchua, Nawina, and Honaunau, Kona, within a range of from 15 to 20 miles, and at all altitudes from 3500 to 5500 feet, but he did not meet with it elsewhere.

Palmer says it apparently wholly subsists on the kernels of the Aaka or Bastard Sandalwood tree (*Myoporum santalinum*), and is more frequently found on the rough old lava-flows than elsewhere, even if the Aaka is just as plentiful in other places.   Palmer says that most of the trees look fresher on these old lava-flows, though the Aaka looks the same everywhere, and that the berries of the latter are sweeter on the old flows.   Palmer says the noise of the cracking of the seeds of the Aaka first attracted his attention to these birds, and whenever he saw these birds one or another was cracking the Aaka-berries, and hundreds of seeds were found in their stomachs.   These notes of Palmer agree remarkably with those of Perkins, and they were written down long before Perkins came to the island of Hawaii.   Palmer, however, thinks the bird is much more active than Perkins described it, for he writes :—" When I saw one hopping about I was rather surprised by its activity.   From the look of the bird I should have judged it to be very slow in all its movements, but the one to-day convinced me that I was wrong."

Its call-note, Palmer says, or at least the one which is more often heard than any other note, is a low prolonged "cheep," not at all loud, and apparently not to be heard at any greater distance than the cracking of the berries.   Besides this a low chirping noise was heard when the bird was on the wing, and a real kind of song was heard on October the 19th, consisting of several whistling notes, not very loud but clear.   Another time Palmer mentions a "plaintive whistling sound of a few notes."   They are not shy and do not fly far if shot at, and sometimes when feeding they do not take notice of the firing of the gun.   Palmer once (October 12th) shot as many as a dozen on a big lava-flow, and six of them during five minutes within a distance of fifty yards.   On another lava-flow about 200 feet or so lower down Palmer did not obtain one, although the Aaka-trees were there just as plentiful and just as full of seed.

There seems to be no regular native name for this bird.   Scott Wilson calls it " Palila," but this name seems to apply more properly to *Loxioides bailleui*.   Several persons told Palmer its name was " Omao," which, however, is the name for *Phæornis obscura* ; according to Andrew's Dictionary, on the other hand, " Omao " means green or greenish, so that it is not impossible that this name was originally meant for the *Chloridops kona*.

Key to the Genera of the DREPANIDÆ.

A. Bill much longer than the head and very strongly curved.
  a. General colour greenish.
    a^1. Mandible of about half the length of the maxilla . . . . HETERORHYNCHUS.
    b^1. Mandible much longer than half the maxilla, nearly as long as the maxilla . . . . HEMIGNATHUS.
  b. General colour blackish.
    c^1. Bill more strongly curved ; under tail-coverts normal . . . . . . DREPANORHAMPHUS.
    d^1. Bill less curved ; under tail-coverts elongated and loose-webbed . DREPANIS.
  c. General colour red . . . . VESTIARIA.
B. Bill in length about equal to the head, straight or gently curved.
  d. A crest of curled feathers on the forehead . . . . . . PALMERIA.
  e. No crest on the forehead.
    e^1. Bill in appearance more that of an insect-eating bird, its height at base not more (mostly much less) than half the length of the mandible.
      a^2. Bill slender and perceptibly curved ; second primary longer than the sixth ; nostrils fully covered with a membrane.
        a^3. Plumage red . . . . HIMATIONE.
        b^3. Plumage greenish . . . . CHLORODREPANIS.
      b^2. Bill slender but straight ; second primary shorter than the sixth ; nostrils with an open slit . . . . OREOMYZA.
      c^2. Bill strong and conical.
        c^3. Culmen nearly an inch long . . . . VIRIDONIA.
        d^3. Culmen about half an inch long.
          a^4. Mandible curved to the right, so that its tip is visible laterally beyond the maxilla . . . . LOXOPS.
          b^4. Mandible following the maxilla . . . . CIRIDOPS.
    f^1. Bill finch-like in appearance, its height at base more than half the length of mandible.
      d^2. Bill as high as it is long.
        e^3. Bill less than 0·5 in. ; head yellow . . . . LOXIOIDES.
        f^3. Bill more than 0·5 in. ; whole upperside green . . . . . . CHLORIDOPS.
      e^2. Bill considerably longer than high.
        g^3. Maxilla considerably longer than mandible.
          c^4. Maxilla much curved and laterally much compressed, much narrower than mandible . . . . PSEUDONESTOR.
          d^4. Maxilla less curved, not compressed, about as wide as mandible . . . . PSITTIROSTRA.
        h^3. Maxilla very slightly longer than mandible.
          e^4. Operculum of nostril small ; wings shorter ; mandible at gonys distinctly convex, cutting-edges a little drawn in and less sharp . . . . TELESPIZA.
          f^4. Operculum of nostril overhanging nearly the whole nostril ; wings longer ; mandible at gonys almost flat, cutting-edges not drawn in and very sharp . . . . RHODACANTHIS.

2 G

THE GENUS CHÆTOPTILA.

(*Cf.* Gadow in Part ii. of 'Aves Hawaiienses.')

‑‑ ‑‑‑‑ ‑ ‑

THIS remarkable form is doubtless a member of the family *Meliphagidæ*, and its nearest ally, as far as the external structure goes, seems to be *Acanthochæra mellivora* (Lath.) of Australia.

Bill longer than the head, distinctly curved, the edges of mandible and premaxilla very distinctly serrated for about the anterior third; not at all quite smooth, as Gadow supposed, who saw only the Cambridge specimen, which has lost the tip of the bill. Nostrils as in *Acanthochæra* and other allied genera, situated in a large groove near the base of the bill, and covered with a large upper operculum.

The fifth and sixth primaries are longest; the first two are considerably shorter than the following ones. The tail is long and strongly graduated; all the rectrices obliquely pointed at their tips.

Plumage of the body very soft, that of the head, throat, and chest fluffy; that of the chin, throat, and forehead ending in hair-like bristles. Metatarsus covered in front with long, more or less fused scales. Feet as in *Acanthochæra*.

No trace of fleshy wattles anywhere.

The extinct *C. angustipluma* is the only species known.

46. CHÆTOPTILA ANGUSTIPLUMA (*Peale*).

Entomiza? *angustipluma*, Peale, U.S. Expl. Exp., Birds, p. 147, pl. xl. fig. 2 (1848) (Hawaii).
Anthochæra? angustipluma, Hartl. Arch. f. Naturg. 1852, i. p. 131; Gray, Cat. B. Trop. Isl. p. 13 ("Society Isl. [Hawaii or Owyhee]").
Mohoa angustipluma, Cass. Proc. Acad. Philad. 1855, p. 440.
Moho angustipluma, Cass. U.S. Expl. Exp., Mamm. & Orn. p. 168, pl. xi. fig. 1 (1858) (Hawaii: specimen in Mus. Washington); Dole, Proc. Bost. Soc. N. H. 1869 (xii.), p. 296 (wooded region, Hawaii: specimen in Smiths. Inst.); id. Hawaiian Alman. 1879, p. 47.
Chætoptila angustipluma, Gray, Hand-l. i. p. 159 (1869); Scl. Ibis, 1871, pp. 358, 360; id. Ibis, 1879, p. 92; Wilson & Evans, Aves Hawaiicns. pt. ii. (1891); Gadow, in Aves Hawaiicns. pt. ii. p. 10 (pterylosis etc.; is a *Meliphagine* bird).
(No mention whatever is made of this bird in Vol. ix. of the 'Catalogue of Birds.')

Top of the head dark brown, mottled with light grey and yellowish, the feathers being pale greyish and bordered and tipped with brown, also with some yellow feathers at the tip. Nape and hind-neck rich citron-yellow, streaked with dark brown and with lighter lines of a brownish grey, the yellow tips having become very large here. Interscapular region and upper back dark olive-brown, with yellowish-white elongated spots, each feather being deep brown with a greyish base; a lighter shaft-line and a raindrop-like yellowish spot at the tip. Lower back and rump olive-brown with a faint rufous tinge; upper tail-coverts olive-brown with a greenish tinge. Quills dark brown, with the outer webs margined with pale brown and with greenish towards the base; secondaries with the green margins reaching along the whole feather, and with very small pale tips. Wing-coverts dark brown. Rectrices dark olive-brown, broadly margined with yellowish green on the outer webs and with obsolete pale tips. Lores blackish. Eyelids, a broad stripe from the under mandible to the ear-coverts, ear-coverts, and feathers behind the eye black. A very narrow obsolete blackish line from the under mandible to the neck. A spot of feathers under the eye greyish. Chin and throat greyish yellow, each feather with the black shaft elongated into a long hair-like bristle. Breast greyish, striped with brown, each feather being greyish with brown borders. Abdomen of the same colour, but much more distinctly and boldly striped. Flanks and anal region washed with rufous. Under tail-coverts pale rufous, whitish along the shaft. Under wing-coverts pale brown. Bill blackish; feet and claws dark brown. Total length nearly 13 inches, culmen 1·55, wing 5·65, tail 6·3, lateral rectrices 3·7, tarsus 1·6, middle toe with claw 1·3, hind toe without claw 0·6, claw of hind toe 0·5.

(Description from a very good specimen in my Museum, which I received in exchange from the "Bernice Pauahi Bishop Museum, in Honolulu.")

Chætoptila angustipluma is evidently extinct. Nobody in the Sandwich Islands seems to know anything about it, and its former native name is unknown. Besides the specimen in my collection, we only know the type of the species in Washington, one in the Cambridge Museum, from the same source as my example, and one in the Honolulu Museum. All these specimens came from Hawaii. Dole quotes also Molokai as the habitat of this bird, but we do not know why. All these specimens (except the type) were procured by Mills, who, according to Palmer's enquiries, collected them between the lower Volcano House and the crater of Kilauea.

Peale says (*l. c.*):—"This rare species was obtained on the island of Hawaii. It is very active and graceful in its motions, frequents the woody districts, and is disposed to be musical, having most of the habits of a *Meliphaga*; they are generally found about those trees which are in flower."

Gray (Cat. B. Tropical Islands) quotes as a doubtful synonym *Moho atriceps*, Lesson ('Traité d'Orn. p. 646), but hardly a word of the description can be supposed to refer to *Chætoptila angustipluma*, and the locality is unknown.

THE GENUS MOHO.

Moho, Less. Traité d'Orn. i. p. 302 (1831).
Acrulocercus, Cab. Arch. f. Naturg. 1847, p. 327.
Mohoa, Reichenb. Syst. Av. pl. 41 (1850).

MR. SCOTT WILSON has adopted Cabanis's name of *Acrulocercus* for this genus, under the pretext that, as Lesson's name of *Moho* was based on an error, *Moho* being the native name of the Rail *Pennula ecaudata*, while that of the *M. nobilis* was O-o, it could not be used. He also says it was not used in a strictly generic sense by Lesson, but I find that Lesson on page 302 says: " § 4. Les Mohos; *Moho*." The §'s of Lesson are subgeneric groups, mostly of full generic value according to modern ornithological views, and the word *Moho* is spelt in italics, thus showing that it was meant to be a Latin term for "Les Mohos." Such subgeneric terms of Lesson are generally accepted by ornithologists as generic names, and so was the name of *Moho*; only Cabanis, in his usual way, altered the "barbaric" name into a proper classic word, thus creating the long name *Acrulocercus* to replace the short *Moho*. I cannot see the necessity of abandoning the well-known *Moho*, and shall use it, as nearly all ornithologists have done. *Mohoa*, of course, was only an amendment of *Moho*.

Moho is a genus of the *Meliphagidæ*.

The bill is a little longer than the head; the large nostrils are bare, but covered with a large operculum. Both maxilla and mandible serrated near the tip, the maxilla more so. Tongue very protractile.

The plumage is very soft and long, except on the head, where it is short and scaly.

The wing has ten primaries and nine secondaries. The first primary is about half to nearly two-thirds of the second, the second considerably shorter than the next. The fourth primary is longest, and but little longer than the third and fifth. The tail is much graduated, and varies according to the different species. Large yellow pectoral tufts are developed most conspicuously in most of the species. Plumage brown or black generally.

Tarsus and feet large and strong. The large scales in front of the tarsus more or less fused together. On the back of the tarsus is a prominent ridge, connected with the lateral sheathing by soft skin. (See Gadow, *l. c.*, for anatomical remarks and others.)

The sexes are similar in coloration, but differ in size, the males being larger.

Moho braccatus is less typical than the other species, and might, by a very rabid genus-splitter, be placed in a different genus, for which, however, I do not see any necessity.

Eggs and nidification unknown.

The genus contains, so far as is known, four species, three of which have been known for a considerable time, while one was described by me quite recently. The species from Oahu is evidently extinct.

A remarkable peculiarity of the species of *Moho* is the very strong somewhat musky smell, reminding one of that of many of the Petrels. This smell has been supposed to be a special feature of the *Drepanidæ*; but this idea cannot be upheld, as it is even stronger in the species of the genus *Moho*. It is, nevertheless, most singular that so many of the Hawaiian Passeres have this extraordinary smell.

Key to the Genus Moho.

A. Central tail-feathers greatly elongated; tail considerably longer than the wing; conspicuous yellow pectoral tufts.

   a. Yellow feathers along the flanks; all but the central pair of tail-feathers tipped with white . . . . *M. apicalis.*

   b. No yellow feathers along the flanks; only the two outer pairs of tail-feathers, or none, tipped with white.

      a^1. No yellow tufts on the sides of the head . . . . . . *M. nobilis.*

      b^1. Yellow tufts on the sides of the head . . . . *M. bishopi.*

B. Central tail-feathers less elongated; tail shorter than the wing or nearly equal (not longer); no yellow pectoral shafts . . . . *M. braccatus.*

47. MOHO NOBILIS (*Merrem*).

O-O.

Yellow-tufted Bee-eater, Lath. Gen. Syn. i. p. 683 (1782) ("Met with in great plenty at O-why-hee, *and others of the* Sandwich Isles, by our late voyagers ; . . . . . *great variety* of the specimens of which are to be seen in the Leverian Museum.") (The description undoubtedly applies to *M. nobilis*, but it is evident that Latham had also one, if not two, of the other species before him); id. Suppl. i. p. 120 (1787).

Mohò, Ellis, Narrat. Voy. Capt. Cook & Clerke, ii. p. 156 (1782).

Hoohoo, King, Voy. Pacific Ocean, iii. p. 119 (1784).

Gracula nobilis, Merrem, Beytr. bes. Gesch. d. Vögel, i. p. 8, pl. ii. (1784) (type specimen presented by the King of Great Britain to the Museum of Goettingen); id. Av. Rar. etc. icones et descr. i. p. 8, pl. ii. (1786) (the same in Latin).

Gracula longirostris β *Oriolus nobilis* (sic), Gm. Syst. Nat. i. p. 398 (1788).

Merops niger, Gm. Syst. Nat. i. p. 165 (1788) (ex Latham's Yellow-tufted Bee-eater) ; Tiedemann, Anat. & Naturg. Vög. ii. p. 431 (1814).

Merops fasciculatus, Lath. Ind. Orn. i. p. 275 (1790) ; Less. Traité d'Orn. p. 302 (1831) (placed under his genus *Moho*).

Philemon fasciculatus, Vieill. Nouv. Dict. xxvii. p. 428 (1818) ; id. Enc. Méth. p. 613 (1823).

Meliphaga fasciculata, Temm. & Laugier, Rec. d'Ois., livr. 79, pl. 471 (1829).

Acrulocercus niger, Cab. Arch. f. Naturg. 1847, p. 327 ; Sundev. Tentamen, p. 50 (1872).

Moho niger, Gray, Gen. B. i. p. 96 (1847) ; Bp. Consp. i. p. 394 (1850) ; Hartl. Arch. f. Naturg. 1852, i. p. 131 ; Dole, Proc. Bost. Soc. N. H. 1869, p. 296 (Hawaii).

Ptiloturus fasciculatus, Peale, U.S. Expl. Exp., Birds, p. 148 (1848).

Mohoa fasciculata, Reichenb. Handb. spec. Orn. p. 333, pl. 614. fig. 4098 (1853).

Mohoa nobilis, Cass. Proc. Ac. N. S. Philad. 1855, p. 439 ; Scl. Ibis, 1871, pp. 358, 360 ; id. P. Z. S. 1878, p. 347 ; Pelz. Journ. f. Orn. 1872, p. 25.

Moho nobilis, Cass. U.S. Expl. Exp., Mamm. & Orn. p. 170 (1858) ; Gray, Cat. B. Trop. Isl. p. 9 (1859) Dole, Hawaiian Alman. 1879, p. 46 ; Gadow, Cat. B. ix. p. 284 (partim ! the description is that of a *Moho nobilis*, but in the synonymy *M. braccatus* is included) ; Stejn. Proc. U.S. Nat. Mus. 1887, p. 101 (differences between *M. nobilis* and *M. braccatus*).

Acrulocercus nobilis, Reichenow, Vög. Zool. Gärt. ii. p. 359 (1884) ; Wilson, Ibis, 1890, p. 177 ; id. & Evans, Aves Hawaiienses, pt. i. text & plate of both sexes (1890) ; Gadow in Aves Hawaiienses, pt. ii. p. 9 (structure) ; Perkins, Ibis, 1893, p. 109 (common in Kona, Hawaii ; antipathy to the *Vestiaria*).

Acrulocercus braccatus, Gadow, in Aves Hawaiienses (? partim) (the large yellow pectoral tufts are, of course, not present in *braccatus*, but in *nobilis*).

Adult male. Deep glossy black, duller on the lower abdomen and upper tail-coverts. In most specimens there is a faint brownish-yellow wash on the ear-coverts. Feathers of the back, breast, and upper part of the abdomen dull at base, greyish along the basal part of the shaft, and with a more glossy border, thus producing a somewhat scaly appearance.

2 u

There is a big bunch of bright yellow axillary tufts, which in some specimens seems to part naturally into two tufts, as Mr. Keulemans has shown on the Plate [1]. Under tail-coverts bright yellow. Central pairs of rectrices black. Outermost pair of rectrices with the basal half of the inner webs black, the other half white; outer webs white, strongly washed with yellow, pure yellow at base; next pair similar, but the white tip smaller, and with a conspicuous broad black shaft-stripe near the tip, this latter character less distinct in the outermost pair. Next pair (third from outside) black, with a broad bright yellow edge to the basal part of the outer webs; rest uniform black. Central pair greatly elongated and twisted round at the tip. Iris reddish brown; bill and feet black. Total length 12 to 13 inches and even more, wing 4·8 to 5·1 (not 5·95, as Wilson says), tail 7·5 to 7·95, lateral tail-feather about 4·5 to 4·9 inches shorter, culmen 1·15 to 1·28, tarsus 1·4 to 1·6.

Adult female. Similar to the male in colour, but much smaller, and with the central pair of rectrices less elongated, less pointed, and not distinctly twisted. Cheeks and ear-coverts, or only the latter, more or less washed with pale golden yellow. Total length about 9 to 9·5 inches, wing 3·9 to 4·3, tail about 5·3 to 5·4, culmen 0·95 to 1·25.

The *young birds* have no yellow tufts beneath the wings.

The Plate of the male of this species in this book shows two separate yellow pectoral tufts. This is not quite correct, but perhaps less the fault of the artist than of myself, as the specimen from which the figure was delineated showed two distinct tufts. There are also several other specimens before me in which the yellow tufts part very naturally in two. I am also afraid that the bird in sitting quietly on a twig cannot show the tufts as they appear in the Plate. If this is a mistake, however, it originated from the wish of the artist to show all the prominent peculiarities of the species to perfection.

This beautiful bird is, as Mr. Scott Wilson justly says, "perhaps the best known species both to natives and white inhabitants of the Sandwich Islands; for it was principally from the yellow feathers that grow beneath its wings, together with the still more beautiful and similarly coloured under tail-coverts of the now extinct [2] *Drepanis pacifica*, that the state-robes of the princes were fabricated. It was the privilege of those classes alone to wear them; nor can it be denied that they formed a becoming and magnificent garb, as beautiful as anything that the triumphs of civilized art can now produce." For further most interesting accounts of the garbs made of the feathers of these birds, and their history, my readers may refer to Wilson's book, *l. c.*

Mr. Scott Wilson tells us that the O-o extracts the nectar with its long tubular tongue from the flowers of the Ohia or from the great tree-lobelia, and that he also saw it feeding on bananas. "Their cry," says Wilson, "is somewhat harsh, and resembles the sound of the letter O repeated twice, with a well-marked interval. The yellow axillary tufts are very conspicuous when this bird is on the wing, and its dipping mode of flight somewhat resembles

[1] These tufts consist of about 16 to 20 feathers, not only of 12.
[2] See my account of the Mamo.—W. R.

that of the Magpie. The O-o exhibits a decided preference for the extreme top of a tree, on which it alights, and when thus perched may be seen continually jerking its long tail up and down almost at right angles to its body, all the while uttering its harsh cry."

Wilson further says that the O-o is an extremely wary bird, and most difficult to approach when met with in the Ohia-forest.

The ordinary vertical range of this bird is, according to Wilson, from 1200 to 4000 feet, but he was told that at certain times it has been seen at or above 6000 feet.

Perkins (Ibis, 1893, p. 100) says that it was a "common bird in the lower forest, frequenting, as is well known, the lofty lehua-trees, especially when growing on the rough lava. Save its antipathy for the red birds (*Vestiaria*), its habits are difficult to observe, as it usually keeps very high up in the trees. Its peculiar cry, rather more like 'ow-ow' than 'o-o,' is very curious, and it would readily respond and even approach when I imitated its voice."

Palmer's notes on this bird are not very extensive, but on the whole agree with those of the other observers. As to the notes he says that, besides their usual loud call, they have a short song with but little melody, a "kind of squeaking noise," and a few other notes, which are but seldom heard. He says they possibly may not be their own specific sounds, but imitations of those of other birds.

Most of Palmer's specimens were obtained at elevations of about 2000 feet above the sea, but a Mr. Smith told him that they were sometimes seen much higher, and he shot one at an altitude of at least 4000 feet. At such elevations, however, they are probably not to be found regularly, as they were not seen there for a long time, although the Ohia-trees, on which they fed principally, were in full flower up to 5000 feet. Palmer calls it very active and "the shiest of the bush-birds." He saw them always moving about and keeping in the high Ohia-trees. They were difficult to shoot.

Nothing is known about the O-o's nesting. Perkins was told by a native and a white boy that they saw such a bird enter a hole in the trunk of a Lehua-tree, very high up, but that they were unable to climb the tree. This was about the middle of June, the same time when Mr. Perkins obtained a young bird which "certainly had not been long out of the nest."

This species inhabits Hawaii only.

Henshaw (*in litt.*) says :—" It is found in the forest near the Wailuku, but is rare ; and constant persecution on the part of the Kanakas for the yellow feathers has made the bird exceedingly shy. The day of doom is fast approaching. In neither Olaa or Puna is it found at all, and these districts were formerly its favourite haunts."

There is a large series in the Kiel Museum, collected by Behn during the Danish expedition of the ship ' Galathea.'

48. MOHO APICALIS, *Gould.*

Yellow-tufted Bee-eater (non Latham !), Dixon, Voy. round the World, plate p. 357 (1789).
Merops fasciculatus β, Lath. Ind. Orn. i. p. 275 (1790) (recognizable description, *ex* Dixon).
Yellow-tufted Bee-eater var. A, Lath. Suppl. ii. p. 149 (1801).
Moho apicalis, Gould, P. Z. S. 1860, p. 381 ; Dole, Proc. Bost. Soc. N. H. 1869, p. 297 ; Gray, Hand-l. i.
      p. 114 (1869) ; Gadow, Cat. B. ix. p. 285 (1884).
Mohoa apicalis, Scl. Ibis, 1871, p. 360, 1879, p. 92 ; Pelz. J. f. Orn. 1872, p. 26. (Two specimens, H. Deppe
      coll., Enero, Oahu, 1837, in Vienna Museum.)
Acrulocercus apicalis, Wilson & Evans, Aves Hawaiienses, pt. v. text & plate (1891).

Adult male. Black ; brownish and with distinct pale shaft-lines on the back and rump.
Rectrices with white tips for about 0·6 to 0·1 in., only the central pair without white
tips ; the central pair ending in a large narrow point, which is twisted upwards. A
broad line along the sides of the breast and flanks ; vent and under tail-coverts yellow.
Under wing-coverts, with the exception of the outer ones near the bend of the wing,
white. Bill and feet blackish. Wing 4·65 inches, tail 6·2, culmen 1·8, tarsus 1·5.

The *adult female* is similar to the male, but smaller.

THERE have been some doubts and mistakes about the true home of this species. Gould
said that " Dixon's bird was obtained at Owhyhee," and he believed that his specimens came
from the same island, but these statements have but little value compared with the fact that
Herr Deppe procured several specimens at Enero, Oahu, in 1837. Except those in Berlin
and Vienna procured by Deppe, and the two in the British and one in my own Museum, I
do not know of any specimens of this bird in collections. These lines may perhaps remind
the curators of Museums to give information of the existence of any specimens, though I
have little hope that any more are preserved.

Neither Palmer, Wilson, nor Perkins met with this bird, and if any of the Sandwich
Island birds are extinct, this one certainly is. Mr. Wilson is " of opinion that the bird still
exists," though there seems to be no special reason for it. This species of *Moho* is a large
and conspicuous bird with loud notes, and would certainly be less likely to be overlooked
than the inconspicuous *Hemignathus* and others.

There are a number of O-o in the Museum of Kiel, brought home by Prof. Behn, who
collected much on Oahu, but Mr. Hartert has seen the specimens, and found them to be
M. nobilis and properly labelled as coming from the island of Hawaii.

49. MOHO BISHOPI (*Rothsch.*).

Acrulocercus bishopi, Rothsch. Bull. B. O. C. vol. i. no. viii. p. xli (1893) (reprinted Ibis, 1893, p. 442);
Wilson & Evans, Aves Hawaiienses, pt. v. text & plate (1894) (description of male only).

Adult male. Head and occiput deep black with a very slight metallic gloss, the shafts of the
feathers a little paler. Neck, back, breast, and abdomen smoky black, with narrow
white shaft-lines to the feathers. Rump and upper tail-coverts black. Wings and
wing-coverts black. Tail black, with very narrow white fringes to the tip. Ear-coverts
with an elongated tuft of golden-yellow feathers, these feathers black at lowest base.
Axillary tufts bright yellow, shorter than in *Moho nobilis*. Under tail-coverts bright
yellow. "Iris dark brown. Bill and feet black. Soles dark flesh-colour with a yellow
tinge." Total length about 11 to 11¾ inches, wing 4·5 to 4·7, tail 6·3 to 6·6, lateral
tail-feathers 3·5 inches shorter, tarsus 1 to 1·5, culmen 1·4 to 1·45.

Adult female. Similar to the male in colour, but smaller. Wing 4 to 4·15 inches, tail 5 to
5·4, culmen 1·2, tarsus 1·35.

Hab. Molokai.

The middle pair of tail-feathers are greatly elongated and pointed as in *M. nobilis*, but only
turned upwards at the tips, not twisted round as in *M. nobilis*.

This fine new species, which I named after the Hon. C. R. Bishop of Honolulu, was
discovered by Palmer on the island of Molokai, and was afterwards also obtained there by
Mr. Perkins.

Palmer found this bird in small numbers on Molokai during the months December and
January. At this time of the year they were seen keeping very low, but Palmer was told
they would be high up in the Ohias when these were in blossom. Palmer writes :—"This
bird is rather shy when approached by men, but sometimes it is rather inquisitive, if one
keeps quiet, and it generally answers and comes up to the spot when its cry is imitated.
The greatest difficulty in procuring specimens is the dense undergrowth. It has a call-note
not unlike that of the Hawaiian O-o (*M. nobilis*), but neither so loud or muffled in tone,
somewhat like 'Ó-ó,' the accent being on both syllables. This call-note is not so full
and loud if uttered by the female. Quite a variety of other notes were observed : a kind
of a chuckling note, heard but once, and a high clear flute-like whistle as if the bird
was trying to sing. Another note is somewhat like a cat's cry, I should say like 'wháo.'

Another time again sounds were frequently repeated which sounded to me like 'Ponk,' finishing with the cat-like ' wháo.' "

Palmer saw these birds "sucking" the Ohia-flowers and found in the stomachs remains of small beetles, tiny land-shells, and some insects which he did not know.

Palmer got very few males, and it was a long time before he procured one. He thinks they kept quiet in the undergrowth, but it is possible that they were breeding at the time.

50. MOHO BRACCATUS, Cass.

Merops fasciculatus γ, Lath. Ind. Orn. i. p. 275 (1790).
Yellow-tufted Bee-eater var. B, Lath. Suppl. ii. p. 149 (1801).
? *Certhia pacifica*, Peale (nec Gmelin), U.S. Expl. Exp., Birds, p. 149 (1848).
Mohoa fasciculata ♀, Reichenb. Handb. spec. Orn. ii. p. 33, pl. 614. fig. 4099 (1853).
Moho[a] *braccata*, Cass. Proc. Ac. N. S. Philad. 1855, p. 440; id. U.S. Expl. Exp., Mamm. & Orn. p. 172
(1858); Gray, Cat. B. Trop. Isls. p. 9 (1859); id. Hand-l. i. p. 114 (1869); Dole, Proc. Bost. Soc. N. H.
1869, p. 296; id. Hawaiian Alman. 1879, p. 46; Sel. Ibis, 1871, pp. 358, 360; 1879, p. 92; Pelz.
J. f. O. 1872, p. 26; id. Ibis, 1873, p. 21; Wallace, Island Life, p. 297 (1881); Stejn. Proc. U.S. Nat.
Mus. 1887, p. 100. (Most authors spelt the name *Moho*, not *Mohoa*.)
Acrulocercus braccatus, Wilson & Evans, Aves Hawaiienses, pt. i. text & plate (1890); Gadow, *op. cit.*
pt. ii. p. 7 (structure) (1891) (partim) (Gadow, in the Catalogue of Birds in the British Museum, ix.
p. 284, included *M. braccatus* in the synonyms of *M. nobilis*); Meyer, Vogelskelet. iii. p. 37, pl. clxiv.
(1892).

Adult male. Head above black with a faint gloss, the feathers ashy grey along the shaft,
except at the tip, and a few bristly feathers right and left on the forehead ashy white.
Some small whitish bristly feathers over the eye like the remains of a superciliary
stripe; lores deep black, without any gloss. Feathers of the throat and fore-neck black,
with a whitish bar before the tip. Feathers of the back and abdomen dark smoky
brown, with narrow greyish shaft-lines, those of the vent and under tail-coverts rufous
brown; rump and upper tail-coverts uniform brown. Wings and tail-feathers uniform
black, the former inwardly bordered with ashy greyish. Edge of the wing and under
wing-coverts pure white. Axillary tufts not conspicuously developed, and of a brown
colour; tibiæ golden yellow. Iris yellowish white. Bill and feet black with a greyish
tinge. Soles pale yellow. (In young birds the iris is more greyish.) Total length
about 7·5 to 8·5 inches, wing 3·8 to 4·1, tail 3·5 to 4 (20 adult males measured),
culmen 1·05 to 1·16 (but none at all reaching up to 1·5, as Wilson gives), tarsus 1·05
to 1·14.

Adult female. Entirely similar to the male in colour (the difference of the colour of the
throat mentioned in Scott Wilson's book does not hold good), but a little smaller. Wing
3·5 to 3·7 inches.

Young birds differ from the adults in having the shaft-lines of the feathers of the back and
abdomen less developed and almost imperceptible, in having the abdomen greyish olive,
and *the tibiæ blackish instead of yellow.*

Hab. Kauai.

2 ι

THIS is one of the few birds known in early days which evidently was not from Hawaii. Latham had already recognized the differences, but in ignorance of the proper locality and with so very little knowledge and material of the other forms he carefully noticed the differences, at the same time putting it under the larger species as a variety or different sex of the same, leaving it to future researches to decide its proper value.

Cassin was the first to give this species a specific name.

Wilson says it seems to be found at all elevations throughout the forests, and is called " O-o A-a," the dwarf O-o, by the natives. The call-note is, according to Wilson, " somewhat similar to that of the larger O-o, though in a higher key; the bird has also a sweet song, some of its notes possessing a bell-like clearness." Its chief food, Wilson says, is nectar, but it also eats ripe bananas, which it hollows out.

Palmer found this bird common on Kauai. He saw it more often than its congeners on the trunks of trees, especially Ohias, peeking in crevices of the bark. It was *very numerous* on flowering Obias. On one single tree of that kind Palmer saw more than once about a dozen at the same time. He says in his diary :—"When I sat down under the tree and imitated its cry, half a dozen came right close to me; then seeing my dog move, gave out shrill squeaking notes which I think were alarm-notes, for instantly others answered, and though not much frightened, they were now more cautious in coming close to me. When walking along under the trees, where many dry leaves covered the soil, one made a rustling noise which often attracted the curiosity of these birds, so that they came flying down to see what was going on, and after satisfying themselves went up the tree again to continue their interrupted search for food." According to Palmer they feed on nectar out of the Ohia-flowers as well as on insects taken out of the same and of the bark and decayed trunks of trees.

These birds have a remarkable somewhat musk-like scent, even strongly perceptible in a box of skins. This same scent is present in the other species of *Moho*. See also ' Ibis,' 1893, pp. 108, 109, where Perkins speaks of a similar scent in some of the *Drepanidæ*, where, however, it is apparently much less strong, to judge from the skins.

51. CORVUS HAWAIIENSIS, *Peale*.

ALALA.

Raven, King, Voy. Pacif. Ocean, iii. pp. 119, 161 (1784) (Kakoon, Hawaii).
Corvus tropicus, Bloxam (nec Gmelin!), Voy. ' Blonde,' p. 250 (1826); Wilson & Evans, Aves Hawaiienses, pt. iv. plate & text (1893).
Corvus hawaiiensis, Peale, U.S. Expl. Exp., Birds, p. 106, pl. xxviii. (1848); Hartl. Arch. f. Naturg. 1852, pp. 102 & 133; Cass. U.S. Expl. Exp., Mamm. & Orn. p. 119, pl. vi. (Hawaii, Mus. Acad. Philad.—
Corvus tropicus, Gm., based on Latham's Tropic Crow, is not applicable to the present bird) (1858); Gray, Cat. B. Trop. Isls. p. 24 (Hawaii); Dole, Proc. Boston Soc. N. H. xii. p. 300 (1869) (Kona, 6000 feet, Kealakeakua, Hawaii : " by no means abundant "); id. Hawaiian Almanac, p. 48 (1879) ; Scl. Ibis, 1871, pp. 359 & 360 (" Whether this is the same as Gmelin's *Corvus tropicus* is very doubtful "); id. Ibis, 1879, p. 92; Sharpe, Cat. B. iii. p. 13 (footnote) (1877); Gray, Hand-l. ii. p. 14 (1870) (subgenus *Physocorax*).

Peale, Cassin, Sclater, and others were fully justified in rejecting Gmelin's name *C. tropicus* (Syst. Nat. i. p. 372), which was based on Latham's *Tropic Crow* (Gen. Syn. i. p. 384, 1781), being described from a specimen in Sir Joseph Banks's collection. The bill is described as "pretty broad at base, tips of both mandibles notched." Plumage " glossy black above, dull on the under parts." The wings are said to have "a gloss of green" and the "vent and side-feathers tipped with dusky white; " length 12½ inches." *Most of these parts* are quite different, the length nearly twice as large. The name of *C. tropicus* is therefore inadmissible without a query, as the description does not suit the Hawaiian bird. Possibly Vieillot (Nouv. Dict. v. p. 356) was not quite wrong in referring Gmelin's (or rather Latham's) bird to some kind of *Cractieus*; but such a bird is not likely to occur on Hawaii, and therefore we must suppose that Latham described some entirely different bird with a wrong locality attached to it.

Adult male. Plumage sooty black, the tail of a deeper black and slightly glossy. Wings sooty brown, shafts of the quills black. Feathers of the throat stiff, with hair-like webs and lighter greyish shafts. In the wing the fifth primary is longest, the first being only as long as the secondaries. The tail is slightly graduated, the two central rectrices being about an inch longer than the outer ones. Bill and feet black; soles greyish flesh-colour. Iris dark brown. Palmer records a female specimen (no. 1463 of his collection) as having the iris "dark blue." This is clearly exceptional.

Five specimens marked as males measure as follows:—

	Wing. in.	Tail. in.	Metatarsus. in.	Culmen. in.
	11·4	7·4	2·4	2·15
	12·0	7·5	2·5	2·4
	12·4	7·9	2·5	2·3
	12·7	7·8	2·4	2·5
	12·9	7·9	2·5	2·4

Five marked as females measure:—

	11·5	7·5	2·45	2·2
	12·3	7·6	2·4	2·3
	12·3	8·0	2 5	2·2
	12·3	8·0	2·5	2·6
	12·5	8·0	2·4	2·25

Mr. Wilson says that immature specimens have the whole plumage of a more rusty shade, and the primaries light ochreous.

This Raven is known only from the island of Hawaii, where, according to Mr. Wilson, it is confined to elevations of 1100 to 6000 feet in the districts of Kona and Kau. In the former it was met with, not uncommonly, both by Palmer and Wilson. The former sent me 16 specimens. The cry is apparently truly corvine, for Peale compared it with that of the North-American Fish-Crow (*Corvus ossifragus*), and Palmer with that of the English Rook.

In November Palmer saw these Crows collecting nesting-material, and in June Mr. Wilson states that the broods had already left the nest. The eggs are unknown; the nest is said to be a loose structure of dead twigs and placed on trees.

The principal food consists of the fruit and flowers of the Ié (*Freycinetia arborea*), but Palmer says that it also eats meat, insects, and various fruits.

This Crow is not shy, and can easily be shot, when attracted by an imitation of its cry.

52. ASIO ACCIPITRINUS SANDWICHENSIS (*Bloxam*).

PUEO.

Strix sandwichensis, Bloxam, Voy. ' Blonde,' p. 250 (1826).

Otus brachyotus, Peale, U.S. Expl. Exped., Birds, p. 75 (1848).

Asio brachyotus, Sclater, Voy. ' Challenger,' Birds, p. 96 (1881).

Brachyotus galapagoensis (non Gould!), Cassin, U.S. Expl. Exp., Mamm. & Orn. p. 107 (1858) (It is remarkable that Cassin says the Hawaiian examples are *larger* and darker than North-American examples, while in fact they are smaller and not at all like *A. galapagoensis*); Dole, Proc. Boston Soc. Nat. Hist. 1869, p. 296; id. Hawaiian Alman. 1879, p. 43.

Asio sandvicensis, Blyth, Ibis, 1863, p. 27 (footnote).

Asio sandwichensis, Sharpe, Cat. B. Brit. Mus. ii. p. 238, p. 239 (note) (1875).

Asio accipitrinus, Stejneger, Proc. U.S. Nat. Mus. x. p. 85 (1888) ; Wilson & Evans, Aves Hawaiiens. pt. ii. (text) (1891).

As will be seen by the name I apply to the Short-eared Owl from the Sandwich Islands, this bird differs from typical *Asio accipitrinus*. Dr. Sharpe (*l. c.*) noticed the remarkably smaller size of the Hawaiian form, but the blackish frontal patch he mentioned is evidently due to the make of his two skins. Not much importance can be attached to the vague statement of Mr. Gurney in Newton's edition of Yarrell (vol. i. p. 167) that he had seen "typical examples from the Sandwich Islands"; but it is certain that Stejneger denied (*l. c.*) the smaller size of the Sandwich Island form, while his measurements rather confirm it. He says that the largest individual is about equal to the "average of the species, while the length of the wing, if it had grown to its full length, would not have fallen far behind the largest." He then gives the measurements of tail and wings of 4 skins, one of which is in moult and one "very much abraded." The measurements he quotes are thus: wing 290 to 304 mm. (= 11·4 to 12 inches), tail 138 to 144 mm. (= 5·4 to 5·7 inches), but he quotes no measurements of typical *A. accipitrinus*! I find that my eight adult skins from the Sandwich Islands, collected by Palmer on Kauai, Oahu, and Lanai, measure as follows :—

	Wing.	Tail.
	in.	in.
1.	11·4	5·5
2.	11·5	5·5
3.	11·5	5·5
4.	11·5	5·5
5.	11·6	5·5
6.	11·7	5·4
7.	11·7	5·6
8.	11·8	5·6

Fifteen adult examples of C. L. Brehm's collection, taken out of the drawer at random, are as follows:—

	Wing. in.	Tail. in.
Wüstenwetzdorf, Orlthal	12·5	6·0
Wüstenwetzdorf, Orlthal	12·5	5·9
Orlthal	12·3	5·9
Sandersleben	12·8	6·1
Sandersleben	12·3	5·9
Renthendorf	12·5	5·9
Renthendorf	12·6	6·0
Renthendorf	12·2	5·7
Renthendorf	12·4	4·9
Madrid	12·5	6·0
Wolga	12·6	5·6
Gothenburg	12·8	6·3
Mexico	12·5	6·2
Blue Nile	12·1	5·9

The other skins in my collection measure:—

	Wing. in.	Tail. in.
Bournemouth (C. Rothschild)	12·3	6·2
Dorset	12·4	6·0
Tring	13·0	6·3
Kamtschatka (Guillemard coll.)	12·8	6·3
Enzeli, Persia (Mocquerys coll.)	12·4	6·2
Japan (Owston)	12·2	5·9
Sambhar Lake, India (Marshall)	12·8	6·0
Corsica (Whitehead coll.)	12·2	6·0
Jalisco, Mexico (Dr. Buller)	12·5	6·3
Jalisco, Mexico (Dr. Buller)	12·2	6·3
Prov. Rio Janeiro (Paul Neumann)	13·0	6·3

Thus we see that *Asio accipitrinus* from the Old and New World have the wings from 12·1 to 13 inches, the tails 5·7 to 6·3 inches long; while my Sandwich Islands series has the wings 11·4 to 11·8 inches, the tails 5·4 to 5·6 inches long, and this makes about an inch or more difference in tails and wings.

Cassin referred the Hawaiian Short-eared Owl to *Asio galapagoensis* (Gould), but the latter is a perfectly distinct species [1], and Hawaiian examples do not in any way approach *A. galapagoensis*.

THE Short-eared Owl is not at all rare in the Sandwich Islands, where it seems to be sedentary and breeds. It is known to breed on Hawaii, Kauai, and Lanai, and is said to be common on Oahu and seen on Maui. Mr. R. W. Meyer, a gentleman who assisted Palmer very much with kind hospitality and advice, found it also breeding on Molokai. They probably breed on all the islands.

[1] See Ridgway in Proc. U.S. Nat. Mus. xix. p. 585, and Nov. Zool. vi. p. 175.

Palmer sent clutches of 4 eggs which were found by some of his friends in an open place in a small hollow patch in the grass in the middle of November on Lanai. He says it lays sometimes as many as six eggs, and that the female commences to sit as soon as the first egg is laid, so that young of very different size may be found in one nest. The eggs measure 41 : 31·5, 41 : 32, 43·5 : 31·5, 40 : 32·5, 40·5 : 32, 44·5 : 30·6, 41·3 : 32, and 41 : 30 mm., thus reaching and surpassing the average measurements of European and North American eggs of *Asio accipitrinus*.

Young in down were taken on Kauai in January. They are like specimens from Europe in C. L. Brehm's collection, but perhaps a shade paler.

The natives are afraid of the Owl, as they are of night-birds in many countries.

53. CIRCUS HUDSONIUS (*L.*).

— -

Falco hudsonius, Linnæus, Syst. Nat. ed. xii., i. p. 128 (1766).
Strix delicatula (sic), Dole, Proc. Boston Soc. N. H. 1869, p. 295 (teste Dole, 1879!).
Accipiter hawaii, Dole, Hawaiian Alman. 1879, p. 43 (reprint in Ibis, 1880, p. 241).
Circus hudsonius, Wilson & Evans, Aves Hawaiiens. (text) part ii. (1891).

THIS Harrier is a winter straggler to the Sandwich Islands. Mr. Wilson (*l. c.*) is inclined to think that it only visits Oahu, but Dole (1879) says :—"*Hab.* Hawaii, rare on the rest of the group." This ornithologist has not distinguished himself in connection with this bird, first referring to it under the name of *Strix delicatula*, then describing it as a new species of Sparrow-hawk in the following terms :—

"*Accipiter hawaii.* 'Io.' 14 in. long. Dark brown above; throat dull white; breast mottled brown and white; dull white feathers on legs and abdomen. Legs feathered below tarsi. Strong back claw. Legs and feet light and scaly. Never before described. Confounded with *Strix delicatula* of Samoa and Fiji Islands in previous lists. Sparrow-hawk. Similar to young *Accipiter rufitorques* of Fiji Islands.

" *Hab.* Hawaii, rare on the rest of the group.

" Preys on small birds, chickens and mice, and probably larger animals, as the following incident would suggest. . . . . ."

Here follows an incredible story, describing how one such bird " circled over a dog, flew to a pile of stones and took one in its claws, flew back to its old position over the dog as if intending to drop it on to the dog, but finally carried it back and placed it on the pile where it had taken it from." Mr. Dole believes that " this manœuvre would seem to show that the bird was accustomed to use stones in this way for killing its prey or its enemies."

The author then adds : " I suggest the above name for this bird. Specimen in Mills's collection, Hilo Hawaii," Mr. Wilson is inclined to refer the story of the stone to *Buteo solitarius*. He has examined two specimens and was the first to correct the mistake of Dole.

Palmer did not procure a specimen, but he says that he saw it twice.

54. BUTEO SOLITARIUS, *Peale.*

IO.

(?) *Brown Hawks or Kites,* Cook, Voy. Pacif. Oc. ii. p. 227 (Atooi) (1784).

Buteo solitarius, Peale, U.S. Expl. Exp., B. p. 62, pl. xvi. (1848); Hartl. Arch. f. Naturg. 1852, i. p. 131 ; Scl. P. Z. S. 1878, p. 348 (Owhyhee) ; id. Ibis, 1879, p. 92 ; id. Rep. Voy. 'Challenger,' B. p. 96, pl. xxi. (1881) ; Gurney, List Diurn. B. Prey, p. 64 (1884) ; Wilson & Evans, Aves Hawaiiens. pt. ii., 3 plates & text (1891).

Pandion solitarius, Cass. U.S. Expl. Exp., Mamm. & Orn. p. 97, Atlas, pl. iv. (Hawaii, type in Mus. Acad. Philadelphia) (1858) ; Dole, Proc. Bost. Soc. N. H. 1869, p. 295 (Hawaii, Niihau & Molokai); id. Hawaiian Almanac, 1879, p. 42.

Pandion (Polionëtus) solitarius, Gray, Cat. B. Trop. Is. p. 1 (1859) ; id. Hand-l. i. p. 15 (1869).

Onychotis gruberi, Ridgw. Proc. Ac. Philad. 1870, p. 140 ; id. Rep. U.S. Geol. & Geogr. Surv. 1876, p. 135 ; Baird, Brewer, & Ridgw. Hist. N. Amer. B. iii. p. 254 (1874) (woodcut & descr. specimen supposed to be from California?) ; Sharpe, Cat. B. i. p. 158 (1874) (footnote); Gurney, Ibis, 1876, p. 476; id. Ibis, 1881. p. 396, pl. xii. (spelt *O. grueberi*) ; id. List Diurn. B. Prey, p. 71 (1884).

Polionëtus solitarius, Sharpe, Cat. B. i. p. 452 (1874).

Onychotes solitarius, Ridgw. Proc. U.S. Nat. Mus. 1885, p. 38 (identity with *O. gruberi* demonstrated).

Buteo (Onychotes) solitarius, Gurney, Ibis, 1891, p. 21.

Size small; bill long and pointed, strongly curved. Nostril rounded. Feet strong, claws very powerful. Metatarsus feathered on its upper part for about one inch. Plumage either of a deep blackish brown or chocolate-brown all over, or brown above and whitish barred or striped with brown on the underside, or below white with brown patches along the sides, or uniform buff or whitish. In all these plumages the quills have the inner webs white towards the base and show indistinct black cross-bars ; the tail is narrowly barred with blackish brown, these bars being less distinct in the darkest individuals.

The late Mr. Gurney sen. (*l. c.*) believed that the birds with uniform rusty-buff undersides were in their first plumage, while those which were more or less barred or striped were adult or nearly adult, and the black ones were "melanistic specimens." The fact that he adds "normal" in brackets to his descriptions indicates that he considered "melanistic specimens" as probably occurring in young birds as well. I am unable to see why the birds with more or less uniform undersides should be young, as the specimens of this Hawk show the same individual variations which are known to occur in *Buteo buteo, Buteo galapagensis,* and in other species of the genus, in which they clearly represent different permanent " phases," but not stages of age. Palmer sent me

2 K 2

both dark and variegated examples, all evidently adult, but none with uniform rusty-buff under surface.

The female is larger. My five females measure:—

Wing.	Tail.	Metatarsus.
in.	in.	in.
11·4	6·7	2·7
12·1	7·1	2·7
11·1	6·5	2·7
11·6	6·8	2·8
11·7	6·6	2·8

My four males measure:—

10·3	6·0	2·6
10·25	5·9	2·4
10·4	6·6	2·5
10·0	6·2	2·6

Palmer records the colours of the bare parts as follows:—

♂, no. 1366. "Irides streaky grey. Upper mandible black; lower mandible black at tip, light grey at base. Cere pale greenish yellow. Tarsi dirty flesh-colour with a yellow tinge, toes greenish yellow, soles of feet yellow, claws black."

♀, no. 1487. "Iris dark hazel. Legs and feet yellow, with a bit of a greenish tinge."

The iris of other specimens is described as grey, mottled with brown, or brown clouded with grey.

WE have at present no certain evidence of the occurrence of this Hawk anywhere else but on the island of Hawaii. Dole tells a story of a Hawk on Kauai, which might have been this species, and the natives told Wilson that a large Hawk was to be found in the mountains of Maui, but all specimens known are from Hawaii, where Palmer met with it frequently.

It is generally seen soaring high in the air or perched motionless on a tree or rock. Sometimes Palmer found it very difficult to approach, at other times it was rather fearless.

The food of this Buzzard is evidently varied. Wilson found small birds in its stomach; Palmer mice, small birds, moths, and spiders.

Wilson writes:—"On the 23rd of June I was so fortunate as to find a nest of this species, containing a single young bird in the down; it was placed in a Koa tree (*Acacia koa*) about 50 feet from the ground, in a fork between two thick branches, and was a large structure of nearly circular form, being a foot and a half deep, and a foot in diameter, composed of dead koa branches and twigs."

THE GENUS PENNULA.

THIS genus has been somewhat vaguely described by Judge Dole as follows :—

"*Pennula millei* [1]. Moho. Not previously described. 6½ in. long. Bill ¾ in. long, black, straight, sides compressed, curved at tip. Tail not visible. Wings rudimentary, hidden in the long, loose, hairy feathers. Plumage dark, dull brown, ashy under the throat; feathers loose, hairy, long. Lower part of tibia naked. Legs long, set far back. Toes : 3 front, 1 back. Habitat, uplands of Hawaii. Nearly extinct. Specimen in Mills's Coll."

All ornithologists have recognized the genus *Pennula*. Dr. Sharpe, in vol. xxiii. of the 'Catalogue of Birds,' has placed the genus between *Porzana* and *Aphanolimnas*. It is thus separated from *Porzanula* by the genera *Corethrura*, *Rallicula*, *Thyrorhina*, *Ortygops*, and *Poliolimnas*. I believe, however, that the genera *Porzanula* and *Pennula* should be placed close together in a natural arrangement of the *Rallidæ*. These two genera are probably closely allied. *Pennula*, however, has the wings still softer, the rectrices about 13 mm. long, with very stiff shafts, concealed by the soft coverts, but easily felt. Bill much like that of *Poliolimnas* and *Porzanula*. Tibia bare for about 7 mm. Metatarsus covered in front with nearly a dozen transverse very distinct scales, and very distinctly reticulated behind.

Two species can be recognized, both being extinct :—

Upperside uniform	. *Pennula millei*.
Upperside distinctly spotted .	. . . . . *Pennula sandwichensis*.

THE history of the Hawaiian Rails is rather confused, and some recent authors have even added to the confusion. The first notes about a Rail on those islands are doubtless those of Latham, and of King in the 'Voyage to the Pacific Ocean,' iii. p. 119, where, in fact, a scientific name is added to the note on such a bird. Unfortunately, however, the description of King is quite insufficient. All King says is : "A Rail with very short wings and no tail, which, on that account, we named *Rallus ecaudotus*" (*sic*! evidently misprint for *ecaudatus*!). I fail to see how it is possible for any zoologist to accept a name which is founded on such a diagnosis. If there was only one small Rail with short wings and tail on the Sandwich Islands the name might be accepted; but as we are convinced that there are, or rather have been, at least two species of *Pennula* on the islands, now both extinct, this name is quite inadmissible. Wilson, Sharpe, Hartlaub, and other authors can therefore not be followed in

---

[1] Doubtless misprint for *millei*, as the letterpress shows.

accepting the name of King ("*Rallus ecaudolus*") for the little Rail from Hawaii with unspotted notæum, in opposition to "*Rallus sandwichensis*" with a spotted upper surface.

From all we know at present of the specimens of Sandwich Island Rails (namely, one in Leyden with a spotted back, and five with a uniform upper surface, of which three are now in England and two in Honolulu), I must, together with Wilson, Hartlaub, and Sharpe, recognize two distinct species. For one of these (the Leyden specimen) I accept the name of Gmelin : *Rallus sandwichensis*, based upon Latham's "Sandwich Rail." It is distinctly said that " the feathers on the upper parts are darkest in the middle"; and in the Latin diagnosis (Index Ornith. *l. c.*), " *Rallus* pallide ferrugineus, *supra maculis obscuris.*" Altogether Latham's description suits the Leyden bird very well, in fact about as well as any of Latham's descriptions suit a bird known to us. This has already been recognized by Dr. Hartlaub. It is also known from the sale-catalogue of the Bullock collection that Temminck bought a Sandwich Island Rail from that collection in 1819 (Hartlaub, *l. c.*). Although the *Pennula* now in Leyden is not labelled to that effect (Finsch, *l. c.*), we can hardly doubt that it is the same specimen ; and it is—notwithstanding the doubts of Dr. Finsch —quite *possible* that this is one of the birds described by Latham, many of which had passed into the Bullock collection. Finsch, it is true, has cleverly proved that the Leyden Rail is not necessarily and beyond doubt the type of Latham's Sandwich or Dusky Rail ; but after all there is a possibility that it is so. It is certain that only some, not all, of Cook's specimens were merely dried and *not* skinned, as I have been told by Mr. Robinson of Liverpool. But, accepting Dr. Finsch's assertion that his bird is not a " type," I fail entirely to see the reason for giving it a new name. Dr. Finsch does not at all show that his bird disagrees with Latham's diagnosis. It is true that Mr. Wilson (*l. c.*) says that " the specimen does not correspond with the Sandwich Rail of Latham"; but I agree with Dr. Hartlaub that it does. There is therefore no reason for giving a new name to the Leyden Museum bird, because nobody will follow Dr. Finsch in renaming a specimen *only* because it is not " the type "—for this is exactly Dr. Finsch's case.

The best proof, however, for the identity of " *Rallus sandwichensis* " with " *Pennula wilsoni* " is the drawing of Ellis, now published by Mr. Scott Wilson, which agrees exactly with Latham's description. Comparing this with Mr. Frohawk's and Mr. Keulemans's figures taken from the Leyden specimen, one must notice the most striking resemblance, the differences being nothing but a paler colour of the old drawing and a want of artistic skill ; but then it must be remembered that the date of these plates is about 120 years apart, and that our present men are more or less accomplished ornithological artists, while Ellis was certainly (*cf.* plate) not an artist in bird-drawing. I can easily show plates done quite recently from the same specimens by different draughtsmen which differ as much and even more than the plate of Ellis and those of Keulemans or Frohawk.

If Latham's " Dusky Rail " was really from the Sandwich Islands, it was probably the same as his " Sandwich Rail," as it cannot refer to *Pennula millsi*, having a streaked upper surface.

55. PENNULA MILLSI, *Dole*.

MOHO.

——— —

? *Rallus ecaudatus*, King in Voy. Pacif. Ocean (Cook's last,, iii. p. 119 (1784).

Wingless bird on Hawaii, which the natives call Moho, now nearly extinct, Pease (not Peale!), Proc. Zool. Soc. London, 1862, p. 145.

Pennula millsi (errore typogr. "*millei*"), Dole, Hawaiian Alman. 1879, p. 54 (reprint in Ibis, 1890, p. 241) (Uplands of Hawaii : named in honour of Mr. Mills, spec. in Mills's coll. : nearly extinct) ; A. Newton, P. Z. S. 1889, p. 5.

Pennula ecaudata (? King !), Wilson & Evans, Aves Hawaiiens. pt. v. text & plate (from specimen in Cambridge) (1891) ; Hartl. Abh. naturw. Ver. Bremen, xii. p. 396 (1892) ; Sharpe, Cat. B. Brit. Mus. xxiii. p. 114 (1894) (deser. spec. in Mus. Rothschild—synonymia partim !) ; Hartl. Abh. naturw. Ver. Bremen, xiv. p. 31 (1895) ; Hartl. Beitr. z. Gesch. ausgest. Vög. 2te Ausg. ("als MS. gedruckt ") p. 10 (1896).

Description of specimen in skin in my collection :—Upperside dark chocolate-brown, a little paler on the head, darkest and slightly mottled by the faint appearance of the almost black centres of the feathers on the rump and upper tail-coverts. Quills blackish, the edges chocolate-brown. Rectrices blackish, with chocolate brown tips. Sides of head with an ashy tinge. Underside lighter. Chin palest, whitish brown. Throat and chest vinous-brown, shading off into dusky chocolate-brown on the abdomen and under tail-coverts. Some of the feathers on the sides of the rump with indications of rufous-buff cross-bars. Under wing-coverts dark chocolate-brown. Total length about 5·5 inches, wing 2·7, tail 0·55, metatarsus 1·1, middle toe with claw 1·5, hind toe with claw 0·55, culmen 0·75.

The second specimen in my collection is perfectly similar, except that the lateral rump-feathers are all more distinctly barred with buff, and that also two of the rectrices have buff variegations. None of my specimens is marked with reddish buff on the outer web of the first primary, as is the Cambridge specimen.

This bird is figured in the ' Aves Hawaiienses,' but that figure is much too reddish and too light, and the barring on the lateral rump-feathers is too wide and coarse. My figure is taken from the skin, and I consider it very good, except that the feathers of the upper-side have a little too distinct lighter edges, thus giving the back a less uniform appearance than it has in the specimens in my collection and in that in the Cambridge Museum No other specimens are known to exist in Europe. These three form part of a series of five specimens procured by an old native bird-catcher named Hawelu, and were in the late Mr. Mills's collection. The other two are in the Bishop-Pauahi Museum, in Honolulu.

THERE can be no doubt that this Rail is now quite extinct. Neither Mr. Wilson nor Mr. Perkins found it. Moreover, I sent out to Palmer a dog which was specially trained for the purpose of getting Rails, and he stayed a good time at Olaa without seeing a sign of Rails. Not the fact that the old Hawaiian kings were fond of the Moho on the table, but that the foolishly introduced mungooses swarm in the scrub-covered lava-flats south of the Volcano House, halfway between Hilo and the Volcano of Kilauea, where Mr. Mills's birds were caught, accounts for its extinction. These pests are so common there, that Palmer could see them almost continually crossing the path when travelling through the bush. The man in charge of the lower volcano house told Palmer that he shot "twenty-nine mungoose last month, and three yesterday." A number of natives, encouraged by the promise of large reward, searched the country for Mr. Wilson, and afterwards again for Palmer, who went out shooting with his dog as long as he stayed at Olaa, but all without success. Mr. Wilson was told by Hawelu, that the mail-carrier had seen the Moho cross his path between 1884 and 1887, and that this bird outruns any dog. It could be discovered by the cry it utters—"a whirring sound resembling the rising of a bevy of Quail," and its nest stood on the ground. One of the Kanakas told Palmer that he used to catch many in former times, but that now they were all "pau," *i. e.* gone. The cry, as imitated by the natives to Palmer, "sounded much like that of the Laysan Rail." The natives also told Palmer that the last Moho was seen in 1884. It lived in the grass, not in swamp or wood, and, besides the obnoxious mungoose, the frequent fires must have been destructive to these birds.

Wilson describes the locality in which the Moho lived as follows:—

"The aspect of the region where the Moho was found much resembles a Scotch moor, with a short densely-growing *Vaccinium* in the place of heather; this is intermingled with a species of *Carex* and Ukiuki (*Dianella ensifolia*), a bright silver-leaved plant bearing a blue berry—the whole forming the thickest of cover. The only trees in this region are scrubby stunted Ohias, though here and there are thickets of fern interspersed with small bushes."

56. PENNULA SANDWICHENSIS (Gm.).

—— ——

? *" Rail with very short wings and no tail, which, on that account, we named Rallus ecaudatus,"* King in Voy.
   Pacif. Ocean (Cook's last), iii. p. 119 (1781).
Sandwich Rail, Latham, Gen. Synops. B. iii. p. 237 (Sandwich Is., also Tanna, Sir Joseph Banks) (1785).
Rallus sandwichensis, Gmelin, Syst. Nat. i. p. 717 (1788); Tiedemann, Anat. Naturg. Vög. ii. p. 434 (1814);
   Vieill. Nouv. Dict. d'Hist. Nat. éd. 2, xxii. p. 564 (1817); id. Tabl. Enc. Méth., Orn. p. 1069 (1823).
Rallus sanduicensis, Latham, Ind. Orn. p. 759 (1790).
Zapornia sandwichensis, Reichenb. Handb. spec. Orn., Rasores, pl. cix. figs. 1184, 1185 (1846).
Corethrura sandwichensis, G. R. Gray, Gen. B. iii. p. 595 (1846).
Ortygometra? sandwichensis, Gray, Cat. B. Trop. Isls. p. 52 (1859).
Crex sandwichensis, Schleg. Mus. P.-B., Ralli, p. 25 (1865).
Ortygometra sandvicensis, Dole, Proc. Boston Soc. N. H. ii. p. 302 (1869); id. Hawaiian Almon. 1879, p. 53
   (partim—non descriptio !).
Rallus sandwichensis, Hartl. Abhandl. naturw. Ver. Bremen, xii. p. 397 (1892).
Pennula sandwichensis, Hartl. Abhandl. naturw. Ver. Bremen, xiv. p. 30 (1893); Hartl. Beitr. z. Gesch.
   ausgest. Vög. 2te Ausg. (" als MS. gedruckt ") p. 40 (1896); Stone, Proc. Acad. N. Sc. Philad. 1894,
   p. 147; Sharpe, Bull. Brit. Orn. Club, i. p. xx (1892) (*Pennula ecaudata* and *P. sandwichensis* believed
   to be identical); Hartl. t. c. p. xxiv (*P. ecaudata* with uniform notæum different from *P. sandwichensis*
   with largely spotted notæum !); Sharpe, Cat. B. Brit. Mus. xxiii. p. 336 (1894); Wilson & Evans, Aves
   Hawaiiens. pt. vii. text and two plates (one sub nomine *P. wilsoni*).
Pennula wilsoni, Finsch, Notes Leyden Mus. xx. p. 77 (1898); Wilson & Evans, Aves Hawaiiens. pt. vii.
   text & plate (1899).
? *Dusky Rail,* Latham, Gen. Synops. B. iii. p. 237 (1785) (Sandwich Islands, Mus. Lever.—Upperside streaked
   with black, therefore not to be identified with *P. millsi*).
? *Rallus obscurus,* Gmelin, Syst. Nat. i. p. 718 (1788); Latham, Ind. Orn. p. 759 (1790); Donndorff, Orn.
   Beytr. i. p. 1151 (1794); A. Newton, Proc. Zool. Soc. Lond. 1889, p. 5 (probably identical with *Pennula
   millsi*); Tiedemann, Anat. Naturg. Vög. ii. p. 434 (1814).
? *Ortygometra obscura,* Dole, Proc. Boston Soc. Nat. Hist. xii. p. 302 (1869); id. Hawaiian Almon. 1879, p. 53.

Description of the only known example in the Leyden Museum by Sharpe :—
" *Adult* (type of species).   General colour above ruddy brown with blackish centres to the
feathers, producing a broadly striped appearance; wing-coverts like the back and
very much elongated; quills blackish, with rusty-brown edges; tail-feathers blackish,
completely hidden by the long feathers of the rump; head more uniform brown, with a
ruddy tinge; sides of face like the head; throat and under surface of body dark vinous
red, a little paler on the latter.   Total length 5·3 inches, culmen 0·8, wing 2·8, tail 0·7,
tarsus 1·3, middle toe and claw 1·35.   (*Mus. Lugd.*) "

2 L

THERE is no evidence whatever about the home of this Rail, except Latham's statement that it inhabits the Sandwich Islands. If it was obtained there during Cook's voyage it was most likely from Hawaii, where most of his birds were procured.

The plate in the 'Aves Hawaiienses' is made by Frohawk, and the one in this work by Keulemans from the Leyden example. Both artists had specially to travel to Leyden for the purpose.

57. FULICA ALAI, *Peale.*

ALAI KEOKEO.

Fulica atra (non Linnæus!), Bloxam, Voy. 'Blonde,' p. 251 (1826).
Fulica alai, Peale, U.S. Expl. Exp., Birds, p. 224, pl. lxiii. fig. 2 (1848); Hartl. Arch. f. Naturg. xviii. p. 137
(1852); id. Journ. f. Orn. 1854, p. 176; Cass. U.S. Expl. Exped., Mamm. & Orn. p. 306, Atlas, pl. xxxvi.
(1858); id. Proc. Ac. Philad. 1862, p. 322 (Hilo, Hawaii); G. R. Gray, Cat. B. Trop. Is. Pac. Oc. p. 54
· (1859); Scl. Ibis, 1871, p. 361; id. P. Z. S. 1878, p. 351; id. Rep. Voy. 'Challenger,' Birds, p. 90;
Pelz. Verh. zool.-bot. Ges. Wien, 1873, p. 159; Streets, Bull. U.S. Nat. Mus. no. 7 (Contr. N. H. Hawaii
& Fanning Is.), p. 21 (1877, Oahu); Finsch, Ibis, 1880, p. 78; Ridgw. Proc. U.S. Nat. Mus. iv. p. 331
(1882); Stejn. op. cit. x. p. 80 (1887), xi. p. 95 (1888, Kauai); Wilson & Evans, Aves Hawai. pt. iv.
(1893); Sharpe, Cat. B. Brit. Mus. xxiii. p. 225 (1894).
Fulica alae, Dole, Proc. Boston Soc. N. H. xii. p. 302 (1869); id. Hawaiian Alman. 1879, p. 54.

Adult male. Bill and frontal shield ivory-white. Iris cherry-red. Plumage slate-colour, darker
above, almost black on the head and neck; abdomen and breast ashy grey. Under
tail-coverts white, the central ones blackish slate-colour. Rectrices slaty black.
Primaries dirty brown, outer webs darker; first primary with a narrow white outer
margin. All the secondaries with the exception of the innermost ones broadly tipped
with white. Entire margin of the wing white. Under wing-coverts ashy grey,
sometimes mixed with whitish. Legs and feet light slaty blue; claws blackish.
Wing 183 to 192 mm. (7·2 to 7·55 inches), bill from gape 32–35 mm. (1·25 to 1·4 in.),
metatarsus 53–63 mm. (6·05 to 6·13 in.), middle toe without claw 70–76 mm.
(2·77 to 3 in.).

Adult female. A little smaller, and the frontal shield smaller and narrower. Wing
166–174 mm. (6·55 to 6·85 inches), bill from gape 30–31 mm. (about 1·2 in.), metatarsus
50 mm. (1·98 in.), middle toe without claw 66–67 mm. (2·6 to 2·63 in.).

Young. Upperside of a paler slate-colour and with a more brownish tinge; chin, throat,
and fore-neck white; rest of under surface pale slate-grey; the feathers more or less
tipped with white, giving the underside a mottled appearance.

Nestling in down. Bill crimson, with a blackish tip. Iris dark hazel. Skin of crown red,
the bare spot representing the future frontal shield purple. Down of chin and face
orange-scarlet, of rest of body black; on throat, neck, and upper surface widely tipped
with golden yellow, on the under surface tipped with silvery grey. "Tarsi and toes
dark brown with a reddish tinge."

2 L 2

THIS species is peculiar to the Sandwich Islands, and is evidently most nearly related to *Fulica americana*, from which it differs only in the much greater development of the frontal 'shield, the generally smaller wing, and the much wider white tips to the secondaries. In the 'Catalogue of Birds' it is stated to be nearest to *Fulica leucoptera*, but to have a much larger bill; but on comparing my series with a bird named *F. leucoptera* by Dr. Sharpe, we find the latter to be very much larger and to have a very much larger bill than the Hawaiian birds.

Fulica alai is evidently distributed over all the Hawaiian Islands. Palmer sent specimens from Oahu, Kauai, Molokai, and Niihau. Cassin mentions it from Hawaii, and Finsch saw it in Maui.

Nests and hard-set eggs were found on Niihau on July 20th. The nests were built up some inches above the water, of rushes. They measured about 10 inches in diameter, and contained three and five eggs each.

58. GALLINULA SANDVICENSIS, *Streets.*

ALAI or ALAE.

" *Common Water or Darker Hen,*" King, in Cook's Last Voy. Pacif. Ocean, iii. p. 120 (1784).

Fulica chloropus (non Linnæus !), Bloxam, Voy. ' Blonde,' p. 250 (1826).

Gallinula chloropus, Peale, U.S. Expl. Exp. i. p. 220 (1848) ; Hartl. Arch. f. Naturg. xviii. p. 137 (1852) ; Dole, Proc. Boston Soc. N. H. xii. p. 302 (1869) ; id. Hawaiian Alman. 1879, p. 53.

Gallinula galeata (partim !), G. R. Gray, Hand-l. B. iii. p. 66 (1879).

Gallinula sandvicensis, Streets, Ibis, 1877, p. 25 (Oahu, figure of forehead) ; id. Bull. U.S. Nat. Mus. no. 7 (Contr. N. Hist. Hawaii & Fanning Is.), p. 19 (1877) ; Finsch, Ibis, 1880, p. 78 (Maui, Oahu) ; Ridgw. Proc. U.S. Nat. Mus. iv. p. 331 (1882) ; Wilson & Evans, Aves Hawaiienses, pt. iv. text & plate (1893).

Gallinula sandwichensis, Wallace, Island Life, p. 296 (1881) ; Sharpe, Cat. B. Brit. Mus. xxiii. p. 180 (1894).

Gallinula galeata sandvicensis, Stejn. Proc. U.S. Nat. Mus. x. p. 78 (1887) ; xii. p. 380 (1889).

The Moorhen inhabiting the Sandwich Islands differs somewhat from *Gallinula galeata* by the more swollen and extended frontal shield, which reaches beyond the eyes. The phases of plumage and coloration of the bare parts agree with those of *Gallinula chloropus,* from which the adult *G. sandvicensis* differs most conspicuously in the much greater development of the frontal shield, bill, and feet.

The figures will best show the differences of the frontal shields of *G. galeata, G. chloropus,* and *G. sandvicensis.*

For a discussion of the various forms allied to *Gallinula chloropus,* see Hartert's article in ' Novitates Zoologicæ,' v. pp. 62–64 (1898). He recognizes there six apparently distinct forms, namely :—

1. *Gallinula chloropus* (L.), inhabiting Europe, Africa, Asia generally, and Guam in the Marianne Islands.

2. *G. chloropus orientalis,* Horsf., which he considers as doubtfully separable, from the Malay Archipelago.

3. *G. chloropus pyrrhorhoa,* Newt., from Mauritius, Madagascar, and Réunion, with buff under tail-coverts.

4. *G. galeata,* Bp., America generally.

5. *G. galeata garmani,* Allen, from Lake Titicaca and Chili, of much larger dimensions.

6. *G. sandvicensis,* Streets, from the Sandwich Islands.

Prof. Oustalet (Nouv. Arch. Mus. Hist. Nat. Paris, ser. 3, viii. (1896) p. 34) unites the Moorhen from Guam with *G. sandvicensis,* but according to Mr. Hartert the Guam form is not separable from *G. chloropus.* Of *Gallinula sandvicensis,* Hartert says that it " cannot be put down as a subspecies of *galeata,* from which it differs considerably by the extent of the frontal shield, which reaches beyond the eyes, is more rounded on the hinder corners, and much more swollen."

Mr. Hartert considers *G. sandvicensis* much more distinct from *G. galeata* than *G. galeata* from *chloropus*. "*G. galeata* resembles very much our *G. chloropus*, from which it differs almost only in the form of the frontal shield, which is less rounded, but more truncated at the top."

Stejneger (*l. c.* 1887) came to the conclusion that there are no reliable differences between *G. galeata* and *G. sandvicensis*; but I think we must acknowledge that when a large series is examined the bulk of the specimens really exhibit slight differences. I, however, think that if a good series from every locality where the genus has been recorded can be compared, we shall find that, with the exception of *Gallinula tenebrosa, G. frontata*, and *G. angulata,* all the named forms will prove to be subspecies of *Gallinula chloropus* (Linn.).

The "Alae" is distributed all over the Archipelago. Mr. Scott Wilson has observed it frequently in the swampy taro-fields of Hawaii, Oahu, Maui, and Kauai. Palmer sent me examples from Kauai, Molokai, and Oahu.

Three eggs sent by Palmer resemble those of the European Moorhen, and measure 44 by 33, 44·5 by 32·5, and 43 by 33·2 mm.

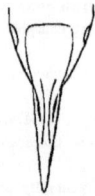

Ga!linula galeata.
(Brazil.)

Gallinula sandvicensis.
(Molokai.)

Gallinula chloropus.
(Guam.)

59. HIMANTOPUS KNUDSENI, *Stejn.*

AEO.

Himantopus nigricollis (non Vieillot !), Pelzeln, Verh. zool.-bot. Ges. Wien, xxiii. p. 159 (1873) (Honolulu) ;
  Finsch, Ibis, 1880, p. 79 (Maui).
Himantopus candidus (non Bonnaterre !), Dole, Hawaiian Alm. 1879, p. 52 ; Finsch, Ibis, 1880, p. 79.
Himantopus knudseni ("*knudseni*" (errore) in pl.), Stejneger, U.S. Nat. Mus. x. p. 81, pl. vi. fig. 2 (1887)
  (Kauai), xi. p. 91 (1888) (Niihau), xii. p. 381 (1889) ; Seebohm, Geogr. Distr. Charadr. p. 280 (1888) ;
  Wilson & Evans, Aves Hawaiienses, pt. iv. (1893) (plate) ; Sharpe, Cat. B. Brit. Mus. xxiv. p. 323
  (1896).

This Hawaiian species of Stilt differs both from *H. mexicanus* and *H. melanurus* (=*brasiliensis*, Brehm) in its much longer beak, the generally much more restricted amount of white on the forehead, but the most trenchant difference is the coloration of the rectrices, which are broadly tipped with greenish, with the exception of the central pair. The longer upper tail-coverts are much mixed with black, a character also not found in the American forms.

The *young* partially in down is white below, the small feathers of the head, neck, and upper part of back and wings blackish brown with golden-orange edges ; the down on lower back and rump is buffish grey, mixed with black, and with a distinct mesial black line.

The *adult* bird has the "inner ring of the iris brown, outer ring light red. Legs pink ; claws black ; bill dark brown, black at base."

The *nestling* has the "iris dark hazel ; bill bluish brown at base, black at the tip. Tarsi and feet greyish white."

THE "Aeo" is apparently distributed over all the islands, where it breeds in suitable places.

Palmer sent me skins from Oahu, Kauai, Niihau, and Molokai, but we have as yet no evidence of its occurrence in Hawaii, although Dole says that it is "common in ponds and swamps all over the group."

Pelzeln (*l. c.*) was the first who mentioned the occurrence of a Stilt on the Hawaiian Islands. He referred it doubtfully to *Himantopus nigricollis*. Finsch (*l. c.*) also believed it to be the same as the American species, while Dole called it "*H. candidus.*"

On Molokai Mr. Scott Wilson found young in down in June, and Palmer obtained the same on Niihau about the middle of July.

Like its allies, the "Aeo" is very fond of its young, and keeps crying and flying at close distances round the intruder, often swooping down and almost touching his head, or they imitate a sick or wounded bird on the ground, to attract the attention of the enemy.

On Kauai Palmer found the stomachs full of the larvæ of a large kind of dragonfly.

60. CRYMOPHILUS FULICARIUS (*Linn.*).

RED PHALAROPE.

Tringa fulicaria, Linn. Syst. Nat. ed. 10, i. p. 148 (1758).
Crymophilus fulicarius, Sharpe, Cat. B. Brit. Mus. xxiv. p. 693 (1896); Henshaw, Auk, xvii. p. 203 (1900).

" BROTHER MATTHIAS, of the Catholic Brotherhood, has a mounted specimen of this bird in winter dress which he shot, together with several others, on the island of Maui in December 1894. So far as I am aware, this is the first record of the bird's occurrence upon the Islands. Brother Matthias informs me that the Phalaropes frequent some small inland ponds at Kahului and are of not uncommon though irregular occurrence, two or three years often elapsing between their visits.

" At the same time and place Brother Matthias shot two American Curlews (not the Bristle-thighed), which I judge from his description to be probably *Numenius hudsonicus*. These specimens are still extant, and later I hope to be able to see and to identify positively the species.

" Since the above was written I shot another Red Phalarope from a flock of Akekeke (*Arenaria interpres*) on the Hawaiian coast, near Hilo, April 6th. The bird may yet be found to be an irregular winter visitor to Hawaii, coming down with the flocks of Plovers and Turnstones. That it should associate with the Turnstone, and with them feed in the upland cane-fields, is rather remarkable. The flock from which my specimen was shot was on its way from upland to its roosting-places on the coast." (*Henshaw*.)

2 M

61. GALLINAGO DELICATA (Ord) (?).

WILSON'S SNIPE.

Scolopax delicata, Ord; Wilson, Orn. ix. p. ccxviii (1825).
Gallinago delicata, Sharpe, Cat. B. Brit. Mus. xxiv. p. 642 (1896); Henshaw, Auk, xvii. p. 204 (1900).
Gallinago wilsoni of many authors.

"Mr. George C. Hewitt, Manager of the Naalcho Plantation, Kau, informs me that he killed a 'Jack Snipe' near Naalcho some years ago. Mr. Hewitt is a sportsman and is very sure that the bird was no other than *Gallinago*, with which he is well acquainted.

"I feel sure that Island records of the shore-birds of the north-west will multiply as time goes on. The immense flocks of Plover and Turnstone that each year wend their way from the American coast to the Islands must surely prove a magnet to attract other species hither, to say nothing of occasional individuals that mingle with these species in migration and unwittingly accompany them in their flight till all unwittingly they find themselves on foreign shores.

"The whole subject of the migration of the Plover and other species to and from the distant mainland is of exceeding interest. Especially interesting would be any book bearing upon the manner of the migrations and the time taken in the flight.

"As is well known, both the Plover (*Charadrius dominicus fulvus*) and the Akekeke (*Arenaria interpres*) leave the island early in May in immense numbers and return in August. My friend Mr. Patton, of Kakolau, Hawaii, has several times observed parties of Plover making the land, and always in a tired, if not an exhausted, condition. Once on land they seem to desire nothing but a chance to rest, but soon recuperate and go to feeding.

"Capt. Chas. Matson has captained ships for years between San Francisco and Hilo. He tells me that only twice has he seen migratory birds, once flocks of Ducks flying north from the islands, and once great numbers of Plovers taking the same course. It is worth noting that in both instances this vessel was about 2000 miles to the north and west of Hawaii, and the inference is that the birds were steering a straight course for the Aleutians. I hope to learn of other masters of vessels who can furnish notes upon this subject, and especially do I hope to find some one who has seen the migrating flocks of Plover resting upon the ocean; for it does not seem probable that such good swimmers as are the Plover and Turnstone attempt to make so long a flight without rest, even if their powers of wing are equal to a task of such magnitude, which may be doubted." (*Henshaw*.)

2 M 2

62. HETERACTITIS INCANUS (*Gm.*).

ULILI.

Totanus canus, antea, Part I. p. 15 (1893).
Heteractitis incanus, Sharpe, Cat. B. Brit. Mus. xxiv. pp. 454–456 (1896).

PALMER obtained specimens on Kauai from January to April 1891, on Molokai in January 1693, and on Niihau in July of the same year. It frequents the shores of the islands mostly in winter.

Professor Schauinsland sent me a series collected on Laysan in August and September.

---

63. HETEROPYGIA ACUMINATA (*Horsf.*).

SIBERIAN PECTORAL SANDPIPER.

Totanus acuminatus, Horsf. Trans. Linn. Soc. xiii. p. 192 (1821).
Heteropygia acuminata, Sharpe, Cat. B. Brit. Mus. xxiv. pp. 566–570 (1896).

PALMER procured three females on the island of Maui on the 6th of October, 1892.

Dr. Sharpe gives the distribution of this species as follows :—" Eastern Siberia, Kamtchatka, the Commander Islands and Alaska, breeding in the latter localities, and passing on migration by Japan and China to the Malayan Archipelago, Australia, and New Zealand."

Two specimens are recorded from Great Britain. The first was obtained near Yarmouth in September 1848, the second on August 29th, 1892, at Breydon in Norfolk.

64. NUMENIUS TAHITIENSIS (*Gm.*).

(Synonymy, description, etc. : see Part I. p. 17.)
Sharpe, Cat. B. Brit. Mus. xxiv. p. 368 (1896).

PALMER shot two males on Molokai in February 1893, and one on Niihau in July of the same year. It is also recorded from Oahu, Hawaii, Maui, and Kauai, and was seen on Lanai. Since I wrote about this bird on Laysan, I have received from that island a fine series of eight specimens collected by Professor Schauinsland. It is evident from the series now before me that the female has, as a rule, a much longer bill than the male, and the smallest males have much shorter wings than any female before me.

Schauinsland's series measure as follows :—

		Wing.	Bill from end of feathering in straight line.
		mm.	mm.
a.	" ♀ "	255	83
b.	" ♀ "	262	94
c.	" ♀ "	253	88
d.	" ♀ "	255	103
e.	" ♂ "	236	68
f.	" ♂ "	260	87
g.	" ♂ "	259	85
h.	" ♂ "	250	86

In three of Professor Schauinsland's females, and in the one bird marked " ♀ " by Palmer, the fore-neck and chest are very heavily streaked with deep brown, while in all the other specimens the striping is much narrower and fainter. It is probable, from what is known of other species of *Numenius*, that the heavily striped birds are the perfectly adult ones.

According to Schauinsland, it is quite evident that this bird, though very frequent on Laysan, *never breeds there* !

65. CALIDRIS ARENARIA (*Linn.*).

SANDERLING.

Tringa arenaria, Linnæus, Syst. Nat. ed. 12, i. p. 251 (1766).
Calidris arenaria, Wilson & Evans, Aves Hawaiiens. pt. iii. p. 153 (May 1892) ; Sharpe, Cat. B. Brit. Mus.
  xxiv. pp. 526-535 (1896) ; Henshaw, Auk, xvii. p. 204 (1900).

PALMER collected seven examples of this species on Kauai in April 1891.   Both Stejneger
and Wilson have also recorded the Sanderling from Kauai.   According to Mr. Henshaw,
" it appears to visit the Kau and Kona coasts of Hawaii annually in small numbers."   In
October 1899 this ornithologist shot two at Kaalualu, Kau, and Mr. Sam Kauani assured
him that it was by no means uncommon.
      Professor Schauinsland shot one specimen on Laysan.

---

66. ARENARIA INTERPRES (*L.*).

THE TURNSTONE.

Strepsilas interpres, antea, Part 1. p. 13 (1893) ; Wilson & Evans, Aves Hawaiiens. pt. iii. (1892).
Arenaria interpres, Sharpe, Cat. B. Brit. Mus. xxiv. pp. 91-103 (1896).
Tringa oahuensis, Bloxam, Voy. ' Blonde,' p. 251 (1826).

PALMER sent a large series from Kauai, collected in the months of February, March, April,
and July, and also one from Niihau shot in July, and some from Oahu, which were killed
in May 1893.   Mr. Wilson records it from Molokai, and Palmer saw it on Hawaii.   It will
doubtless occur on every island.
      Professor Schauinsland found the Turnstone common on Laysan in August and
September.

2 N

67. CHARADRIUS DOMINICUS FULVUS, *Gm.*

KOLEA. (LESSER GOLDEN PLOVER.)

Charadrius fulvus, antèa, Part I. p. 11 (1893).
Charadrius dominicus (partim), Sharpe, Cat. B. Brit. Mus. xxiv. pp. 195–208 (1896).
Charadrius dominicus fulvus of Ridgway, Coues, Stejneger, A. O. U. Check-list, Hartert, and others.

PALMER sent one specimen, shot on Niihau on July 27th, 1893, and a large series from Kauai, shot in December, January, February, March, and April. He writes that these birds leave the islands by the end of April and beginning of May.

Henshaw ('Auk,' xvii. p. 205) says that this Plover is well known to leave Hawaii in immense numbers, and to return in August. Henshaw further writes :—" It is of interest to note that by no means all the Kolea, Akekeke (*Arenaria interpres*), and Ulili (*Heteractitis incanus*) leave the islands in spring. Thousands of the two former species remain all the summer in the uplands, and the Ulili is by no means uncommon along the shore. I have examined numbers of such loiterers and find them, without exception, to be young birds, apparently birds of the year, probably too immature to feel the mating impulse."

Professor Schauinsland collected many specimens on Laysan from the middle of July to September.

68. PLEGADIS GUARAUNA (*L.*).

Scolopax guarauna, Linnæus, Syst. Nat. ed. 12, i. p. 242 (1766).
Plegadis guarauna, Ridgw. Proc. U.S. Nat. Mus. 1878, p. 163 (Kauai!); Baird, Brewer, & Ridgw. Water-B.
N. Am. i. p. 97 (1884); Stejneger, Proc. U.S. Nat. Mus. 1887, p. 84 (Kauai); Sharpe, Cat. B. Brit.
Mus. xxvi. p. 34 (1898); Scott Wilson & Evans, Aves Hawaii. pt. vii. (1899).

ONE single immature example, which doubtless belongs to *P. guarauna* (not being *P. falci-nellus*), was sent to the U.S. Nat. Mus. by the late Mr. Knudson from Kauai in 1872.
Nothing else is known on the Hawaiian Islands about the occurrence of any Ibises.

69. NYCTICORAX NYCTICORAX NÆVIUS (Bodd.).

AUKU KOHILI, AUKU.

—

Ardea nævia, Boddaert, Tabl. Pl. Enl. p. 56 (1783).
Ardea exilis (non Gmelin !), Dole, Proc. Boston Soc. N. H. 1869, p. 303; Peale, U.S. Expl. Exp. p. 216
    (1848).
Nycticorax nycticorax nævius, Stejneger, Proc. U.S. Nat. Mus. x. p. 84 (1887), xi. p. 102 (1888).
Nycticorax griseus, Wilson, Aves Hawaiienses, pt. vii. (June 1899).
Nycticorax nycticorax (partim), Sharpe, Cat. B. Brit. Mus. xxvi. p. 146 (1898).

I PREFER to keep the American form of the Night-Heron subspecifically distinct, because in nearly every case the bill and wings are decidedly larger than in the Old World form. The Hawaiian birds certainly belong to the American form, their beaks especially being of very large size. Palmer describes the bill as black, shaded with green, the tarsi as brownish and pale yellow, the iris as pale red.

Palmer procured ten adult birds, six young, and two nestlings on the islands of Oahu, Kauai, Molokai. Wilson and Knudsen obtained it on Kauai, while Perkins reports it to be commonly breeding all over the islands. On March 20th, 1893, Palmer found a breeding-place on Oahu. The nests were built of small dead sticks, and varied much in size, some being almost as big as a crow's nest, others being so small that they hardly appeared to hold the eggs, which could easily be seen from below. The nests were either placed in forks on dead branches of trees or on the top of the green "Lankala" or "screw-pine" bushes. The majority of the nests contained two eggs, but one had three, and one only a single hard-set egg, while several had already small young ones. Most of the eggs were more or less hard-set. The young are, needless to say, not white, as described by Dole. They have, according to Palmer, the "iris yellow ; tarsi greenish with a yellow tinge, soles yellow ; upper mandible blackish brown, with a light grey ridge ; lower mandible yellowish at tip, dark brown at base."

The eggs sent from Kaalualu, Oahu, measure : 51 by 39·5, 50·5 by 38·8, 51 by 37, 52·6 by 39·1, 50·7 by 37·5, 50·6 by 37, 50 by 35, 52·5 by 35, 49 by 36·5 mm.

A number of eggs of the European *Nycticorax nycticorax*, taken by Hartert on the Hansag in Hungary in May 1891, measure : 49 by 35, 49·5 by 33·5, 51 by 36, 51 by 32·5, 48 by 36, 50 by 37·5, 49·5 by 35·2, 48·5 by 36, 50 by 36·5, 48·6 by 35 mm.

It would thus appear that the eggs of the American and Hawaiian forms are, on an average, slightly more bulky.

70. DEMIEGRETTA SACRA (*Gm.*) (?).

AUKUU.

Sacred *Heron*, Latham, Gen. Syn. B. iii. p. 92 (1785) (Otaheite).

Ardea sacra, Gmelin, Syst. Nat. i. p. 640 (1788) ; G. R. Gray, Cat. B. Trop. Is. p. 48 (1859) ; Dole, Proc. Boston Soc. N. H. 1869, p. 303 ; id. Hawaii. Alm. 1879, p. 52 ; Finsch, Ibis, 1880, p. 79 ; Wilson & Evans, Aves Hawaii. pt. vii. (1899).

Demiegretta sacra, Wiglesw. Aves Polynes. p. 67 (1891) ; Sharpe, Cat. B. Brit. Mus. xxvi. p. 137 (1898).

G. R. GRAY mentioned the Sandwich Islands as a locality for this Heron, but he does not say on what authority. Mr. Sandford Dole's evidence is rather incorrect, and certainly his statement that it is " common all over the group " is erroneous. Dr. Finsch, who was well acquainted with this bird from his travels in the Pacific Ocean, says that he " observed the white form once at Kahalui." As no specimen has ever been closely examined by a competent ornithologist, the occurrence of this bird wants confirmation. Neither Mr. Wilson nor Palmer and Perkins ever came across it.

It is widely distributed. Sharpe (*l. c.*) says :—" Coasts of Burma and the islands in the Bay of Bengal, by the Malay Peninsula and islands to Australia and the Pacific, extending north to the islands in the Bay of Corea."

71. MERGUS SERRATOR, *Linn.*

RED-BREASTED MERGANSER.

Mergus serrator, Linnæus, Syst. Nat. ed. 10, i. p. 129 (1758) ; Henshaw, Auk, xvii. p. 203 (1900) (Hilo !).
Merganser serrator, Salvadori, Cat. B. Brit. Mus. xxvii. p. 479 (1895) ; A. O. U. Check-list, ed. 2, p. 47 (1895).

"I AM not aware that this Duck has hitherto been recorded from the Islands, where it is a casual and possibly a rather regular winter visitor. Nov. 8, 1899, one of these Mergansers was shot by Mr. Otto Rose, of Hilo, near the town, being one of two seen. Nov. 28, I shot the surviving bird a mile or two further down the coast in a small salt-water pond. It was fat and in fine order, and had in its throat two of the common fresh and brackish-water fish known to the natives as ' Ospu.'

"The natives, to whom I showed this Duck, seemed in no wise surprised, claiming to have seen the species before, though rarely. They gave it the name Molú ; but as this name is applied also, according to Mr. Dole, to the Shoveller, its correct application is open to doubt. Present-day natives know extremely little of Hawaiian birds, and usually are either unable to name a bird at all or are in doubt.

"In time, no doubt, as stated of the Gulls, particularly all the species of our north-west Ducks will be noted from the Islands, nothing being more likely than that a few stragglers will accompany the flocks of Shovellers and Pin-tails which are regular winter visitors." (*Henshaw.*)

72. A N A S   W Y V I L L I A N A, *Scl.*

KOLOA MAOLI.

Duck (of the Sandwich Islands), Bloxam, in Byron's Voy. p. 251 (1826).
Anas boschas (?), Hartl. in Wiegmann's Arch. f. Naturg. 1852, i. p. 137 (Oahu, in Mus. Berol.) (this is
 really *A. wyvilliana*, as examined by Mr. Hartert and acknowledged by Reichenow in a letter to
 Salvadori) ; G. R. Gray, Cat. B. Trop. Isls. p. 54 (1859) (Oahu) ; Cass. Pr. Ac. Phil. 1862, p. 322 ;
 Finsch & Hartl. Orn. Centralpolyncs. p. xxxix, no. 138 (1867) (part., Oahu).
Anas boschas, var., Hartl. Journ. f. Orn. 1854, p. 170 (Sandwich Is.).
Anas superciliosa a. sandwichensis (nomen nudum !), Bp. Compt. Rend. xliii. p. 649 (1856) (descr. nulla).
Anas superciliosa, var. (partim !), G. R. Gray, Cat. B. Trop. Isls. p. 54 (1859) (Oahu).
Anas superciliosa (non Gmelin !), Dole, Proc. Boston Soc. N. H. xii. p. 305 (1869) ; id. Hawaiian Alm.
 1879, p. 55 ; Sclater, Ibis, 1871, p. 360.
Anas wyvilliana, Scl. P. Z. S. 1878, p. 350 (Sandwich Is., juv.) ; id. Ibis, 1879, p. 92 ; id. Voy. 'Challenger,'
 Birds, p. 98, pl. xxii. (juv.) (1880) ; id. P. Z. S. 1880, p. 517 ; Ridgw. Proc. U.S. Nat. Mus. 1878,
 p. 251 (♀) ; Finsch, Ibis, 1880, p. 79 ; Wallace, Island Life, p. 296 (1881) ; Rehnw. Orn. Centralbl.
 1882, p. 19 ; Salvad. Orn. Pap. e Mol. iii. p. 396 (note ; 1882) ; Stejn. Proc. U.S. Nat. Mus. 1888,
 p. 98 (Kauai) ; Wilson & Evans, Aves Hawaii. pt. iv. plate (1893) ; Salvad. Cat. B. Brit. Mus. xxvii.
 p. 196 (1895).

Adult male. (Count Salvadori (*l. c.*) expressed a doubt whether this is the final nuptial
plumage. I, however, have little doubt that the four birds described hereafter are in
perfectly complete nuptial dress.) Top of the head blackish, the feathers tipped with pale
brown. A dark metallic green stripe from the eyes to the nape. Feathers of the neck
blackish, mixed with light brown, those of the upper back and interscapular region
blackish brown with crescent-shaped and undulated rufous-brown bands. Lower back,
rump, and upper tail-coverts brownish black, with but a few brown feather-edges and
spots. Inner tertials and greater scapulars brown, greyish in the middle and narrowly
edged with pale brown. Primaries dark greyish brown. The secondaries form a large
and fine speculum, bordered behind with a subterminal black line, followed by a white
terminal line, and in front by a black line, with a less defined grey band before it.
Above the speculum is bordered with a broad velvety-black stripe, formed by the black
outer webs of some of the tertials. Rectrices deep brown and blackish, with whitish-
brown edges and irregular arrow-shaped markings. The two central rectrices are black,
soft, and curled up as in *A. boschas* in one old male in abraded plumage, the next one to
it shows an inclination to curl up (the other corresponding one being absent), and in two
other males there is a distinct beginning to curl up. Sides of head and neck and throat

mottled with blackish brown and pale buffy brown.   Throat blackish in the oldest male. Breast rufous brown, with U-shaped blackish markings, or more or less rounded spots, these standing before the tips of the feathers and being followed by another blackish mark; the upper breast and sides more rufous.   Abdomen brownish buff, distinctly shaded with greyish in the oldest specimens, and varied with greyish-brown spots, in the most matured specimen distinctly, though faintly, cross-barred with ashy brown.   Sides of body pale rufous brown, with longitudinal or V-shaped deep brown markings, but in the oldest male some white feathers finely undulated with blackish brown appear on the flanks.   Under tail-coverts blackish and brownish, varying much and strongly tinged with rufous in older specimens.   Under wing-coverts and axillaries white.   Total length about 19 to 20 inches, wing 9·3 to 9·8, tail about 3·5, culmen 1·9, breadth of bill at base 0·65, in the middle 0·85, tarsus 1·6, middle toe with claw 2·2.

In less aged males the bars on the abdomen are not developed, the abdomen is more spotted, and the middle tail-feathers are not curled up, and when still younger the rectrices are uniform blackish, only edged with pale brown.

Young males resemble the females, but there are intermediate plumages, which would seem to indicate that the final plumage is not obtained before the lapse of several years. The younger males can be distinguished from the adult females by their much blacker and less varied back and rump, and by their rectrices being uniform blackish, only bordered with buffy brown.

Adult female. Blackish brown above with a slight gloss on the head, all the feathers broadly margined with brownish buff and mostly with one or two zigzag bars across.   Primaries dark greyish brown.   The secondaries form a fine speculum of *deep metallic purplish blue,* which in some specimens and under certain lights passes into green.   This speculum is bordered in front with a velvety-black terminal band, preceded by a whitish-grey band, which occasionally is absent or indistinct in a few specimens, and this speculum is bordered behind by a velvety-black band, followed by a white one. Towards the back the speculum is also terminated by a velvety-black line.   White tips are in a few specimens indicated or even distinctly developed on the coverts forming the black border in front of the speculum.   Rectrices blackish brown, bordered and barred with irregular lines of brownish buff.   Underparts buffy brown. darker on the breast, spotted with blackish brown, and more so along the sides of the body and on the breast.   Chin mostly quite unspotted and more reddish.   Under wing-coverts white, sometimes those near the margin dark brown with pale borders, a variability which, like some others, I cannot account for.   Axillaries white, in two specimens, one from Oahu and one from Hawaii, with a few dark brown spots.   A very young bird from Kauai, marked female, has the under wing-coverts almost unspotted white, the rectrices with a few pale bars only on the outer ones, besides being bordered all round with pale brown.

The females vary very much.   Some few have a distinctly indicated pale superciliary line; some have the spots on the lower parts much less bold than others, and I take them to be younger individuals, as the already-mentioned very young bird (with the wing-feathers only half-grown) has them also less bold. Some variations in the gloss of the speculum, the borders of the latter, and the colour of the under wing-coverts are mentioned above. The total length of the females is about $16\frac{1}{2}$ to $17\frac{3}{4}$ inches, wing 8·5 to 9, tail 3 to 3·3, culmen 1·6 to 1·8, tarsus 1·5.

Young in down. Almost exactly like the young in down of *Anas boschas*, but the olive tint is apparently more variable and generally less intense.

THE Sandwich-Island Duck has only one somewhat near ally, i. e. *Anas laysanensis*, Rothsch., from Laysan. This latter, however, can generally be distinguished by the white ring round the eye, its smaller size, more regularly barred rectrices, always spotted axilliares, more reddish colour, and other characters. In its curled-up central tail-feathers it resembles *A. boschas*, and this singular character is also developed in *A. laysanensis*, but in no other Duck besides.

There can be absolutely no doubt that Dr. Sclater's name of *wyvilliana* must be accepted for the Sandwich-Island Duck, as Bonaparte's name of *sandwichensis* was not accompanied by any description.

The "Koloa Maoli," or "Koloa," is probably an inhabitant of all the Sandwich Islands, where it most likely breeds wherever it finds suitable localities.

After the breeding-time they keep in flocks. On the lake on Niihau, Palmer observed large flocks, sometimes probably not less than a hundred together.

In habits and nidification this Duck does not seem to differ from *Anas boschas*.

It is evidently distributed all over the islands. Palmer sent me a large series from Kauai, Hawaii, Niihau, and Oahu. The nestlings were taken on May the 6th at Waialua, Oahu.

73. DAFILA ACUTA (*L.*).

Anas acuta, Linnæus, Syst. Nat. ed. 10, i. p. 126 (1758).
Dafila acuta, Stejneger, Proc. U.S. Nat. Mus. 1888, p. 97 (Kauai) ; Scott Wilson, Aves Hawaiienses, pt. iv.
(1893) ; Salvadori, Cat. B. Brit. Mus. xxvii. p. 270 (1895).

PALMER procured two males on Hawaii in December 1891. Stejneger and Wilson recorded this species from Kauai and Hawaii.

For full synonomy and descriptions, see the 'Catalogue of Birds.'
This species is only a winter visitor to the Islands.
Prof Schauinsland procured it also on Laysan.

74. SPATULA CLYPEATA (*L.*).

Anas clypeata, Linnæus, Syst. Nat. ed. 10, i. p. 124 (1758) ; Dole, Proc. Boston Soc. N. H. xii. p. 305
(1869) ; id. Hawaiian Almanac, 1879, p. 55.
Spatula clypeata, Scott Wilson, Aves Hawaiienses, pt. iv. (1893) ; Salvadori, Cat. B. Brit. Mus. xxvii. p. 306
(1895).

PALMER obtained five specimens, two males and two females on Hawaii on December 17th, 1891, and one male on Molokai in February 1893. Wilson saw this Duck on Hawaii; Peale and Stejneger record it from Hawaii, Oahu, and Kauai.

This species is also a winter visitor to the Hawaiian Islands. The reader is referred to the 'Catalogue of Birds' for full synonymy and descriptions.

Professor Schauinsland obtained one specimen from Laysan.

2 P

75. ANSER ALBIFRONS GAMBELI, *Hartl.*

Anser gambeli, Hartl. Rev. et Mag. Zool. 1852, p. 7; Salvadori, Cat. B. Brit. Mus. xxvii. p. 95 (1895).
(*Anser albifrons gambeli* of most authors.)

PALMER shot a male of this Goose on December 18th, 1891, on the lake on Mr. Clark's estate at Honokaohau, on the island of Hawaii. He says that he shot this bird by moonlight, and that it was in company with another Goose of presumably the same species. The iris is recorded as " dark hazel : upper mandible grey at base, pink ridge, tip very light grey; lower mandible grey and orange; tarsi and feet orange-red, claws dark brown."

This Goose is barely separable from the White-fronted Goose, *Anser albifrons*, Scop.; in fact, except by the slightly larger dimensions, it is indistinguishable from it, and can certainly only be regarded as a subspecies.

76. CHEN HYPERBOREUS (*Pall.*).

Anser hyperboreus, Pallas, Spic. Zool. vi. p. 25 (1767); id. Zoogr. Rosso-Asiat. ii. p. 227, pl. 65 (1811).
Chen hyperboreus, Salvadori, Cat. B. Brit. Mus. xxvii. p. 84 (1895).

PALMER sent me a very well-mounted adult specimen of this interesting Goose, procured on the island of Maui by Brother Matthias.

For full synonymy, description, and distribution see Cat. B. Brit. Mus. xxvii. p. 84.

77. BRANTA NIGRICANS (*Lawr.*).

Anser nigricans, Lawrence, Ann. Lyc. N. York, iv. p. 171, pl. xii. (1846).
Branta nigricans, Salvadori, Cat. B. Brit. Mus. xxvii. p. 123 (1895).

PALMER obtained a specimen of this bird from Brother Matthias, who got it at Kahului, on the island of Maui, in 1891.

For description and literature on this Goose, see Cat. B. Brit. Mus. xxvii. p. 123.

78. BRANTA CANADENSIS MINIMA (*Ridgw.*).

Branta minima, Ridgway, Proc. U.S. Nat. Mus. viii. p. 22 (1885).
Branta canadensis minima of the American Orn. Union Check-lists.
Bernicla munroii, Rothsch. Ann. & Mag. Nat. Hist. (6) x. p. 108 (1892) (Kauai).

PALMER shot the one recorded specimen of this small Goose on the 16th of March, 1891, near Waimea on the island of Kauai, and made the following remarks in his diary:—
" ♂ of Hawaiian Goose. Length 22⅜ inches. Stomach contains small seeds; bill and legs black, iris dark brown."

When looking over Palmer's birds in 1892, I at once saw that this bird was not at all *Nesochen sandvicensis*, as Palmer imagined ; but, as no other Goose had been recorded from these islands, I erroneously concluded it must be new. However, even before I discovered the true name of this bird I had suspected my mistake, when Palmer sent me three other American species of Geese.

THE GENUS NESOCHEN.

Nesochen, Salvadori, Cat. B. Brit. Mus. xxvii. p. 81 (1895).

THE Sandwich Island Goose has been placed by most authors in the genus *Bernicla*, or, as it should be called, *Branta* (*cf.* Cat. B. Brit. Mus. xxvii. p. 111). Some other authors (Eyton, Reichenbach, Gray, and recently Heine & Reichenow) have placed it (in my opinion without much reason) with the South American genus *Chloëphaga* ("*Taenidiesthes*" of Heine l). Count Salvadori was the first to create a separate genus for this species. The only character he gives is the "deeply excised webs of the feet," these being "not deeply excised" in the genus *Branta*. This character is very conspicuous, and the legs and feet are altogether very strong and large, causing the birds to stand and walk very firmly and very high, so that there is perhaps sufficient excuse for the genus *Nesochen*.

79. NESOCHEN SANDVICENSIS (*Vig.*)

NENE.

"*Geese—not unlike the Chinese Geese,*" Ellis, Narrat. Voy. ii. p. 143 (1782).

Bernicla sandvicensis, Vig. P. Z. S. 1833, p. 65 (descr. nulla) ; id. op. cit. 1834, p. 43 (diagnosis) ; Stanley, tom. cit. p. 41 (hatched in England) ; Jard. & Selby, Ill. Orn. ser. 2, pl. viii. (1836) ; G. R. Gray, Gen. B. iii. p. 607 (1844) ; Reichenb. Syn. Av., Natat. pl. 99. figs. 2355, 2356 (1850) ; Hartl. Arch. f. Naturg. 1852, i. p. 137; Scl. P. Z. S. 1859, p. 206 (incubation) ; Dole, Proc. Bost. Soc. Nat. Hist. xii. p. 305 (1869) (Hawaii and Maui) ; id. Hawaiian Almanac, 1879, p. 54 ; Pelz. Verh. zool.-bot. Ges. Wien, 1873, p. 159 ; Finsch, Ibis, 1880, p. 81 ; Scl. P. Z. S. 1880, p. 504 ; Montluz. Bull. Soc. Accl. 1886, pp. 146, 147 (fig.) ; Tristram, Cat. Coll. B. p. 51 (1889) ("type") ; Wilson & Evans, Aves Hawaiiens. pt. iv. plate & text (1893).

Chloëphaga sandvicensis, Eyton, Mon. Anat. p. 81 (1838) ; id. Syn. Anat. p. 26 (1869).

Anser hawaiiensis, Eydoux & Souleyet, Voy. 'Bonite,' Zool. i. p. 104, pl. x. (1841).

Anser hawaiensis, Peale, U.S. Expl. Exp., B. p. 249, pl. lix. (*Anser hawaiensis*) (1848) ; Hartl. Arch. f. Naturg. 1852, i. p. 122.

Bernicla sandwichensis, Cass. U.S. Expl. Exp., Mamm. & Orn. p. 348 (1858) ("Apparently peculiar to Hawaii") ; Gray, Cat. B. Trop. Is. p. 54 (1859) (Hawaii).

Anser sandwicensis, Schleg. Dierent. p. 278 (1864) ; id. Mus. P.-B., Anseres, p. 106 (1866).

Branta (Leucopareia) sandwichensis, Gray, Hand-l. B. iii. p. 76 (1871).

Anser (Brenthus) sandwichensis, Reichen. Orn. Centralbl. 1882, p. 36 ; id. Vög. d. Zool. Gärt. p. 64 (1882).

Nesochen sandvicensis, Salvadori, Cat. B. Brit. Mus. xxvii. p. 126 (1895).

Adult male. Chin, throat, sides of head to about 1 cm. beyond the eye, and crown black. The black of the crown runs in a narrow band along the back of the neck and joins a narrow brownish-black collar which encircles the lower neck. Remainder of sides of head and neck brownish buff ; the neck-feathers narrow and pointed, and arranged in wavy lines so as to show the blackish bases of the feathers and producing an irregularly striped appearance. Upperside dark umber-brown, with brownish-white tips to the feathers (not bars, as described in several works). Rump dusky black ; upper tail-coverts white, sometimes slightly mottled with brown. Primaries dusky black. Secondaries blackish brown, outer webs paler and more hoary brown. , Rectrices brownish black. Entire chest, breast, and middle of abdomen pale greyish brown ; feathers on sides of breast and abdomen and flanks with wide brownish-white tips and a subterminal brown bar. Vent and under tail-coverts white. Iris dark hazel ; bill black ; feet blackish. Total length about 28 inches, wing 14 to 15, tail about 6, culmen 1·8 to 1·85, metatarsus 3, middle toe with claw 3·5.

2 Q

Adult female. Similar to the male, only very slightly smaller; wing about half an inch shorter. Mr. Wilson states that the female has the black "extending further down the throat, and occupying a greater space below the eye," that the "feathers on flanks are paler than in the adult male," and that the lower breast is not so pale. In my series of nine adult Geese of both sexes these differences are not visible. Wilson further says that the young male is like the adult female in colour. I am inclined to think that his supposed old female is not quite adult. Salvadori describes the young in down as follows: "Upper parts greyish brown; forehead, sides of head, throat, and middle of the underparts whitish; a white spot on each side of the back at the base of the wing."

PALMER sent nine adult examples, all collected on Hawaii in December 1891. He found the birds not uncommon on the slopes of Hualalei on Hawaii, on places where the Ohelo-berries were abundant. These berries, according to Palmer, form their principal food. We have no absolute proof that this bird breeds anywhere else than on the lava-flows of Hawaii, although specimens are said to have been observed on Kauai and Niihau, on neither of which islands, however, does it breed. Finsch and Wilson both say that they were informed of its breeding in the crater of Haleakala on Maui.

Already the early travellers, Ellis and Bloxam, have noticed this remarkable Goose, but Vigors, in 1833, was the first to give it a scientific name. It has been frequently imported alive to Europe, and has successfully bred in several public and private menageries and gardens, both in Great Britain and on the Continent.

On Hawaii most ornithologists have noticed the "Nene," which is its native name. It breeds at great altitudes, probably not below 5000 feet above the sea. The eggs are white, like those of other Geese. The food consists of berries and grass. According to Mr. Wilson, the "Ohelo" (*Vaccinium reticulatum*), the strawberry (*Fragaria chiliensis*), and a black berry called "Popolo," are the principal food of this Goose.

Mr. Wilson says:—"It is easy of approach, and I am told that when one of a flock is wounded the remainder will not leave their companion, so that the collector, if heartless enough, may kill the entire number." Palmer also mentions the fearlessness of the bird, as compared with other Geese, but at the same time found them often so wild that he did not succeed in procuring one or two seen, and that he had to use a rifle once or twice to kill a specimen he wanted.

The flesh of this Goose is highly esteemed as an excellent food. Mr. Wilson mentions for the first time a "peculiar sweet musky scent in the neck of the Nene," which is "a fact well known to Hawaiians."

LARIDÆ.

80. STERNA FULIGINOSA, *Gm.*

Haliplana fuliginosa, anteà, Pt. I. p. 39 (synonymy, literature, description) (1893).
Sterna fuliginosa, Saunders, Cat. B. Brit. Mus. xxv. p. 106 (1896) ; Wilson & Evans, Aves Hawaii. pt. vii. (text) (June 1899).

I HAVE only seen two examples from Oahu in the British Museum, but Dr. Stejneger has recorded it from Kauai, where "it appears to be common." Palmer sent it only from the Laysan group, but not from the real Sandwich Islands, where he did not pay sufficient attention to sea-birds.

81. STERNA LUNATA, *Peale.*

Haliplana lunata, anteà, Pt. I. p. 39 (synonymy, literature, description) (1893).
Sterna lunata, Saunders, Cat. B. Brit. Mus. xxv. p. 100 (1896); Wilson & Evans, Aves Hawaii. pt. vii. (text) (June 1899).
? *Sterna panaya* (non Latham, non *St. anætheta*, auct.), Dole, Proc. Boston Soc. N. II. 1869, p. 306; id. Hawaiian Almanac, 1879, p. 56.

It was probably this Tern which Dole erroneously recorded as *St. panaya*. Knudsen sent it to America from Kauai; Wilson "obtained specimens from the Sandwich Islands"; Palmer sent it from Hawaii and Niihau; he saw it near Oahu.

82. ANOUS HAWAIIENSIS, *Rothsch.*

NOIO.

Anous hawaiiensis, anteà, Pt. I. p. 43 (1893); Saunders, Cat. B. Brit. Mus. xxv. p. 148 (1896); Wilson & Evans, Aves Hawaii. pt. vii. (1899) (text & plate).
Anous stolidus, Dole (? partim !), Proc. Boston Soc. N. II. 1869, p. 307; id. Hawaii. Alm. 1879, p. 57 (1879).

BLOXAM had already collected this Tern in 1825, one of his specimens being now in the British Museum. Palmer only procured it on Kauai and in the Laysan group of islands; Mr. Knudsen obtained it from Niihau; and Mr. Perkins observed it "quite commonly throughout the group."

83. GYGIS ALBA (*Sparrm.*).

? "*White Pigeon*" (!), King, Voy. Pac. Ocean, iii. p. 120 (1784).
Gygis alba, anteà, p. 35 (1893); Wilson & Evans, Aves Hawaii. pt. vii. (1899).
Gygis candida, Saunders, Cat. B. Brit. Mus. xxv. p. 142 (1896).

WITH the exception of King's rather doubtful "White Pigeon" (!!) there is no authentic record of the occurrence of this bird on the Hawaiian Islands proper, though it is a common breeding species on Laysan.

2 Q 2

84. LARUS GLAUCESCENS, *Naum.*

Larus glaucescens, Naumann, Vög. Deutschl. x. p. 351 (1840); Saunders, Cat. B. Brit. Mus. xxv. p. 284 (1896); Henshaw, Auk, xvii. p. 201 (1900).

PALMER obtained two immature birds in first and second year's plumage on Hawaii on June 26th, 1892.

Henshaw (*l. c.*) writes :—" This Gull (*Larus glaucescens*) is becoming an irregular though a rare visitor to the island of Hawaii, following vessels from San Francisco to Hilo. I learn from the captains of several vessels sailing between the two ports that the numerous Gulls that frequently attend the course of outward bound vessels usually turn about when off shore a hundred miles or so. Occasionally, however, one or two Glaucous Gulls, for some reason or other, fail to join their fellows on their homeward course, and day after day steadily follow in the course of the Island-bound vessel. Such birds frequently, perhaps always, roost at night upon the yards.

" Recently two Glaucous Gulls followed one of the U.S. transports from San Francisco clear into Hilo harbour, where they lingered for many weeks and then disappeared, no one knows where. This particular transport happens to be painted white, which fact recalls the statement of an old mariner that Gulls are much more likely to follow in the wake of a white vessel than of any other, the simple explanation being that the birds are not so likely to lose track of a white vessel.

" I have examined two Glaucous Gulls, shot in Hilo harbour, during my five years' residence in Hilo, out of five or six that have been reported in this interval. One of them was in fine condition, but the other weak and much emaciated.

" I believe that none of these wanderers ever attempt to return to America, but their final fate is unknown. No hint of the Glaucous Gull establishing itself upon the Hawaiian Islands is recorded, so far as I know, and the Islands are but illy adapted to their habits. The bird islands to the north-west, Laysan and others, would seem to be in every way adapted to this bird, and there in time the Glaucous Gull may become established.

" That other species of American Gulls occasionally find their way to the Islands in the wake of vessels, especially to the harbour of Honolulu, is highly probable, and only the paucity of observers has prevented their detection and record."

85. LARUS PHILADELPHIA (*Ord*).

Sterna philadelphia, Ord, in Guthrie's Geogr., 2nd Amer. ed. ii. p. 319 (1815).
Larus philadelphia, Saunders, Cat. B. Brit. Mus. xxv. p. 185 (1896).

A YOUNG female specimen of this American species was obtained by Palmer at Poli-hule lake on Kauai on the 15th of March, 1891. Palmer found the " iris brown, feet and legs flesh-colour, bill black."

So far as I am aware, this is the first specimen of this Gull ever obtained in the Sandwich Islands. For synonymy and description, see Cat. B. Brit. Mus. xxv. p. 185.

PROCELLARIIDÆ.

86. OCEANODROMA CASTRO (*Harcourt*).

Thalassidroma castro, Harcourt, 'Sketch of Madeira,' pp. 123, 166 (1851); Grant, Ibis, 1898, p. 314; Saunders, Manual Br. B. ed. ii. p. 734 (1899).

Oceanodroma cryptoleucura, anteà, Pt. I. p. 53 (1893); Salvin, Cat. B. Brit. Mus. xxv. p. 350 (1896).

THIS interesting little Petrel was first described by the late Mr. E. Vernon Harcourt in 1851 under the name of *Thalassidroma castro*, but was entirely lost sight of till it was again described by Mr. Ridgway in 1882 as *Oceanodroma cryptoleucura*. It has an immense range, specimens being on record from Niihau, Sandwich Islands, Madeira and surrounding islets, Cape Verde Islands, St. Heleua, the Galapagos Islands, Washington City (picked up), Drogden lightship off Copenhagen, Kobbergrunden in the Kattegat, and lastly one picked up dead at Littlestone, Kent, in December 1895.

In Part I. of this book there is a printer's error, the wing measurement being given as 5·75 to 8·3 instead of 6·3 inches.

87. BULWERIA ANJINHO (*Heineken*).

Anteà, Pt. I. p. 51 (sub nomine *B. bulweri*); Salvin, Cat. B. Brit. Mus. xxv. p. 420 (1896) (*B bulweri*).

KNUDSEN obtained this bird on Kauai.

88. PUFFINUS CUNEATUS, *Salvin*.

Anteà, Pt. I. p. 47 (1893); Salvin, Cat. B. Brit. Mus. xxv. p. 371 (1896); Anthony, Auk, 1898, pp. 39, 313, 316 (San Benedicte and Socorro Islands).

THIS Petrel is only known in the Sandwich group from Kauai, where it was procured by Mr. Knudsen. Palmer only sent specimens from the Laysan group.

Professor Schauinsland's Laysan series consists of adult birds of the phase with white under surface, and young in down. The latter are pale grey, lighter and almost white on the neck, breast, and middle of abdomen. The feet, according to Schauinsland, are bluish white or whitish grey (not yellow or red!); the bill grey with black tip.

In Part I. of this work I stated that the underparts were brownish grey in the young. I find, however, from the observations of Mr. A. W. Anthony on Socorro and San Benedicte islands, that these entirely grey birds are not necessarily young, but represent a regular phase of plumage, found also in other species of Petrels and also in the Skuas.

Mr. Anthony found on the above-named islands that the dark birds outnumbered the light by two to one; while on Laysan, in a series of at least fifteen skins, not one dark bird was found, and the only specimens with dark underside I possess from this region were procured by Palmer on French Frigate shoals. I possess also two specimens with white under surface from the Bonin Islands.

89. PUFFINUS NEWELLI, *Hensh.*

Puffinus newelli, Henshaw, Auk, xvii. p. 246 (1900).

I DO not know this Petrel, and therefore quote Mr. Henshaw's original description and information:—

"Above, including upper surface of wings and tail, clear and somewhat glossy black. Border of under wing-coverts black. Beneath, including under tail-coverts, pure white. Maxilla and edge and tip of mandible black; rest of maxilla light brown. Tarsus and feet light yellow, but black along the outer posterior side of tarsus, the outer toe, and half the middle toe. Wing 8·65 inches, tail 3·75, bill 1·28, tarsus 1·80.

" The above is a description of a Shearwater obtained by Mr. M. Newell of Hilo (Brother Matthias of the Catholic Brotherhood) in Waihee Valley, Island of Ulani, in the spring of 1894, and by him recently presented to the author. The sex was not determined. The bird was taken from its burrow with several others by natives and brought to Mr. Newell alive. The latter saved two specimens. One, the type, is in my possession ; the other is probably still extant and in Honolulu.

" In 1894 the species was numerous enough in the above-mentioned locality, but its present status is doubtful, for the mongoose, which is rapidly exterminating the native Puffins elsewhere upon the islands, is an inhabitant also of Ulani.

" As this Puffin was quite unknown to me, and as no account of it appears in either Rothschild's or Wilson's works upon the Island birds, I sent the specimen to Mr. Ridgway, who kindly compared it with National Museum material. Mr. Ridgway's remarks upon the specimen are as follows:—'The Puffin which you sent for identification is without doubt a new species. It comes nearest to *P. auricularis,* Townsend, of Clarion Island (Revillagigedo group, N.W. Mexico), but differs in blacker colour of upper parts, wholly white malar region, more extensive, more uniform, and more abruptly white anterior and central under tail-coverts, more extensive and 'solid' blackish border to under wing-covert region, and especially in the very abrupt line of demarkation along sides of neck between the black of upper parts and white of underparts. *P. auricularis* also has the bill entirely black and also stouter.'

" The species is dedicated to Mr. Newell, who has paid considerable attention to Hawaiian birds and has made extensive collections."

Besides the type there are two or three examples in the Bernice Pauahi Bishop Museum in Honolulu.

90. ÆSTRELATA PHÆOPYGIA SANDWICHENSIS, Ridgw.

UUAU.

?? *Procellaria alba* (non Gmelin !), Bloxam in Byron's Voyage, p. 252 (1826) ; Dole, Proc. Boston Soc. Nat.
   Hist. xii. p. 308 (1869); id. Hawaii. Alman. 1879, p. 55.
Æstrelata sandwichensis, Ridgway in Baird, Brewer, & Ridgw. Water-B. N. America, ii. p. 395 (1884) ; id.
   Proc. U.S. Nat. Mus. ix. p. 95 (1886) ; id. op. cit. xi. p. 104 (1888) ; id. op. cit. xix. p. 649 (1896) ;
   Stejn. op. cit. x. p. 77 (1887).
Œstrelata phæopygia, Wilson & Evans, Aves Hawaii. pt. v. (text) (1894), pt. vii. (plate) (1899) (full
   synonymy) ; Salvin, Cat. B. Brit. Mus. xxv. p. 408 (partim, specimen *c* !) (literature imperfect, neither
   the first description of *Æ. sandwichensis* nor Wilson & Evans' work quoted, *Æ. sandwichensis* united
   with *Æ. phæopygia*).

THERE is nothing to indicate which species of Petrel was meant by Bloxam and Dole
(*l. c.*), but possibly they had seen the present species.

Ridgway (*l. c.*) was the first to describe the Hawaiian form. Salvin afterwards united it
with his *Æstrelata phæopygia* (*cf.* Proc. U.S. Nat. Mus. xix. p. 618; Nov. Zool. vi. p. 198),
in which he was followed for a time by Ridgway (1887), while this same author afterwards
(1896) was very doubtful as to the identity of *Æ. sandwichensis* and *phæopygia*. It is
certainly a mistake to unite these two forms unhesitatingly, as Salvin did, thus stopping
future researches about a question which he makes us believe to be finally settled.
Mr. Hartert and I have examined the skin from Kauai (collected by Knudsen) in the British
Museum, and, comparing it with a series of Galapagos skins, we find the following
differences:—In *Æ. sandwichensis* the bill is a little slenderer, the culminicorn, measured in

Nasal tube of *Æ. phæopygia*.   Nasal tube of *Æ. sandwichensis*.

a straight line from the end of the hard covering of the curved anterior half to the tip,
measures 0·7 in.; there is no dark patch in front of the base of the axillaries; the nasal
tubes are smaller and their covering is less horny; the dimensions are generally smaller.
In *Æ. phæopygia* the bill is a little larger and bulkier, the anterior culminicorn measures
about 0·75 to 0·8 in.; there is always a large slate-coloured patch in front of the base of the
axillaries, and the latter are often mixed with slate-colour; the nasal tubes are longer, and

their covering is more horny and is nearer to the end of the feathering on the forehead; the dimensions are generally larger.

The Kauai bird in London has the wing 11·5 inches, tail 5·7, metatarsus 1·35.

The Galapagos birds have the wings from 11·6 to 12·5 inches, averaging from 11·8 to 12 inches, tails 5·8 to 6·3 and 6·4, metatarsus about 1·4 to 1·5 inches.

Ridgway makes the following statements:—

" *A. sandwichensis*.—Bill smaller (culmen from base of nasal tube 1·20, from anterior end of same 0·90) and nasal tube shorter (0·30); hind-neck and sides of neck light sooty-slate, like back; feathers of back and scapulars without paler tips; inner webs of primaries without any definite white space, though basal portion is whitish.

" *A. phœopygia*.—Bill larger (culmen from base of nasal tube 1·30–1·36, from anterior end of same 0·92–1) and nasal tubes longer (0·33–0·38); hind-neck and sides of neck black, like top of head; feathers of back and scapulars (especially the latter) with distinct narrow greyish-white tips; inner webs of primaries with an extensive definite space of white, occupying (except on the first) at least the basal half."

It will be seen from this that our observations made on the specimen of *A. sandwichensis* in the British Museum agree in some important points with those made independently (before) on the one in the U.S. National Museum by Mr. Ridgway. I cannot, however, corroborate the differences in the coloration supposed to exist by Mr. Ridgway, but it seems to me that the dark patch on the sides of the breast in *A. phœopygia* is quite remarkable.

Under the circumstances it is necessary to regard the Hawaiian Petrel as a subspecies of *Æ. phœopygia*, though further researches on Hawaiian examples should be made.

I know only of the existence of two adult birds in collections—namely, the two above-mentioned ones collected on Kauai, but there are probably some in the Bernice Pauahi Bishop Museum at Honolulu. Mr. Wilson writes:—" I obtained a young bird—said to be of this species—in the down from a native, whilst staying at Kilauea in the month of September 1887, and was told that a considerable number had their nests in holes in the ground in the vicinity, more particularly on the slopes of Mauna Loa. At Kilauea we used to hear at evening-time the peculiarly harsh cry of a bird flying over our heads, and the natives told me it was the Uuau. The flesh is esteemed a great delicacy by the Hawaiians."

91. DIOMEDEA IMMUTABILIS, *Rothsch.* (*anteà*, p. 57).

Professor Schauinsland has sent me a series of twelve skins in all stages of growth and plumage. The *nestling as hatched is covered with down of a greyish-white colour*, the basal half of the down being dark sooty-brown. When we examine the next stage, in which the young is covered with dark brown down, it becomes evident how the change from the downy stage to the first plumage takes place without a real moult. The dark brown down when it displaces the first nestling-down appears to be composed of integral downy plumes. As the birds grow the down is pushed out further and further till we perceive that it is not composed of integral plumes, but that it is attached to the end of the webs of the feathers. These

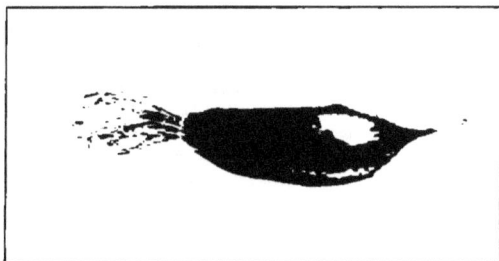

Feather with down attached.

downy filaments are then gradually worn off until we see the final feathers of the first white plumage, which induced me to name the species *D. immutabilis*, to distinguish it from most of the other Albatrosses, which had a first plumage different from that of the adult. I have since, however, learnt that by far the greater number of the seventeen species of Albatrosses have a similar metamorphosis of plumages to the present species.

Professor Schauinsland describes the feet of the adult bird as pale bluish grey, the bill yellowish with a greenish-grey tip; and of a young bird in brown down the feet as dark brownish grey, bill dark leaden grey.

I have also received a series of eight eggs of *D. immutabilis*, which vary very much both in shape and coloration. The two extremes are as follows :—

1. Very elongate, length 111·5 mm., width 62·5 mm.; ground-colour dirty white, marked with numerous large and small blotches of a brownish-maroon colour, which are principally massed at the two ends, though there are also a few in the central zone.

2. Very thick and short, length 100 mm., width 70; colour uniform brownish buff without any markings whatever.

The majority of the specimens before me are dirty white, with irregular patches and

2 n

spots of brownish maroon at the larger end. They measure: 106·5 by 70 mm., 102 by 66, 104·5 by 71, 106 by 69, 101 by 66, 103 by 65.

I have also received two more albinistic specimens of this Albatross from Laysan, one of which is entirely white, while the other has the wings dull grey.

Professor Schauinsland sent me also for identification a most curious Albatross which is doubtless a hybrid between *Diomedea immutabilis* and *D. nigripes*. The description of this bird is as follows :—Forehead white, merging into the dark ashy-grey colour of the crown. Large patch in front of the eyes black, under the eye backwards a white line, sharply separated from the blackish ashy grey of the sides of the hinder part of the crown. Chin and sides of the head pale grey; remainder of the head and neck ashy grey, the feathers white at base. Upper parts deep ashy grey, wing-coverts almost blackish. Upper tail-coverts white. Rectrices blackish slate-colour, white at base. Underside from the fore-neck to the tail white. Maxilla 3·7 inches, mandible 3·5, wing 19·75, metatarsus 3·7, middle toe with claw 4·5.

D. immutabilis is, as a migrant, widely spread. Mr. Alan Owston sent me a specimen killed on Myiakejima, Japan, in October 1893 (Bull. B. O. Club, iii. p. xlvii, June 1894). In the Muséum d'Histoire Naturelle in Paris I have seen a specimen killed near Hawaii by Mons. Baillou. Mr. A. W. Anthony found this species near San Gerónimo and Guadalupe Islands on the coast of Lower California, and it is to be suspected that several reports of Albatrosses observed on the western coast of North America refer to this species, and perhaps also some of the specimens mentioned by Cassin (U.S. Expl. Exp. p. 399) might have been *D. immutabilis*. Certainly the birds mentioned by Pickering (*l. c.* p. 401) as being observed between Oahu and the north-west coast of America, and as being *all* " of a blackish or dark dove-color, with a white frontlet or a circle around the base of bill," were all *D. nigripes* and not the young of a white species; but the white birds described on p. 399 could only have been *D. immutabilis* or *D. albatrus*.

The literature referring to *D. immutabilis* is so far thus :—

Diomedea (an *exulans*?), Kittlitz, Mus. Senckenb. i. p. 120 (1834).
? *Diomedea brachyura* (partim—old white ones!), Cassin, U.S. Expl. Exp., Orn. p. 399 (1858).
? *Diomedea melanophrys*, Bean, Proc. U.S. Nat. Mus. v. pp. 170, 173 (1882).
Diomedea immutabilis, Rothsch. Bull. B. O. Club, i. p. xlviii (June 1893) (reprint in Ibis, 1893, p. 448) ;
    id. anteà, Pt. I. p. 57 and plates (1893) ; id. Bull. B. O. Club, iii. p. xlvii (June 1894) (reprint in Ibis, 1894, p. 548) ; Salvin, Cat. B. Brit. Mus. xxv. p. 416 (1896) ; Anthony, Auk, xv. p. 38 (1898) ; id. Auk, xvi. p. 99 (1899); Schauinsland, ' Drei Monate auf einer Koralleninsel,' pp. 46, 52, 101 ; Scott Wilson, Aves Haw. pt. vii. text (1899).

Mr. Scott Wilson in part vii. of his 'Aves Hawaiienses ' has recorded in the introduction and the text the two Albatrosses known to breed on Laysan under three names, viz. : " *Diomedea albatrus* (chinensis)," *D. nigripes*, and *D. immutabilis*. This is partly due to my having erroneously identified Temminck's name of *D. chinensis* with *D. nigripes*; but if the authors of 'Aves Hawaiienses' had consulted page 446 as well as page 444 of vol. xxv. of the 'Catalogue of Birds,' they would have found Mr. Salvin's correction of my mistake. I may also say that if the authors had carefully read my article on what I called *D. chinensis* in Part 1. of this work, they would have found out my mistake just as easily as Mr. Salvin did.

To make the matter quite clear, I herewith quote the two headings of the 'Catalogue of Birds ' :—

1. DIOMEDEA NIGRIPES, *Aud.*

D. nigripes, Cat. B. Brit. Mus. xxv. p. 445 (1896).
D. chinensis, Rothsch. Avif. Laysan, pt. i. p. 55 (1893) (non *D. chinensis*, Temminck !).

2. DIOMEDEA IMMUTABILIS, *Rothsch.*

D. immutabilis, Cat. B. Brit. Mus. xxv. p. 446 (1896).

It is thus evident that only two species of *Diomedea* are known to occur in the seas of the Hawaiian possessions, and that the true *D. albatrus* (the adult of *D. chinensis*, Temminck, 1820) has never been taken there.

STEGANOPODES.

92. PHAËTHON RUBRICAUDA (*Bodd.*).

Phaëthon rubricauda, Grant, Cat. B. Brit. Mus. xxvi. p. 451 (1898) (*antea,* Pt. I. pp. 33, 34) ; Wilson & Evans, Av. Hawaii. pt. vii. (1899).

Since writing the first paragraph of the present work, I have received a magnificent series of the Red-tailed Tropic-bird from Professor Schauinsland, collected on Laysan in July, August, and September. This series contains young in various stages. The downy chick is not at all pure white nor uniform grey. The upperside is brownish grey, the down being white at the base ; face, breast, and abdomen almost pure white. The iris is light blue ; bill blackish blue ; lores bare and blackish blue. The feathers of the first plumage appear first on the wings and scapulars. The upperside of the young in first plumage is broadly barred with black ; the primaries white, with black shafts and black longitudinal spots along the shafts near the tips, of varying sizes. The bill is blackish.

Palmer sent me two skins, male and female, which he obtained on Niihau 27.7.1893. According to Wilson, " It breeds in several places in the group, especially on Kauai and Niihau, and chooses holes in almost inaccessible cliffs wherein to deposit its eggs." Mr. Perkins " considers this species much more uncommon " than the white-tailed species.

Comparing my series of Red-tailed Phaëthons from various localities, I find them to belong to two well-marked forms. While my series from Laysan and Niihau and two caught in the Pacific Ocean at lat. 21° 10′ N., long. 115°, belong to a smaller form, with narrower and slenderer beak, shorter wings, and with a very slight rosy tinge, the birds from the Kermadec group, Norfolk and Lord Howe's Islands—to the north of New Zealand—are larger, with a thicker, stronger and longer bill, and have a very pronounced rosy-red tinge in their plumage, especially on the wings. This beautiful red tint is so strong that even in skins which are about four or five years older than those recently collected by Schauinsland this colour is much stronger.

Mr. Grant (*l. c.*) attributes the " pinkish tint " to freshly-killed specimens. This is true in so far as that tint fades more or less in skins, especially when not kept in the dark, but it does not in the least account for the differences in the birds from various localities, as described above.

I took the following measurements from my series in the Tring Museum :—

KERMADEC ISLANDS.		PACIFIC. Lat. 21° 10' N.		LAYSAN.		NIIHAU.	
Bill from swollen protuberance on upper jaw at gape.	Wing.	Bill as before.	Wing.	Bill as before.	Wing.	Bill as before.	Wing.
mm.	mm.	mm.	mm.	mm.	mm.	mm.	mm.
75	355	78	330	70	310	70	320
82	345	87	330	72	320	80	318
80	338			74	310		
75	330			75	308		
83	345			77	312		
82	352			72	305		
82	346			75	320		
78	337			73	308		
80	342			77	325		
80	338			76	315		
81	340			76½	323		
82	360						
85	345			70–77	310–325		
80	330						
87	330						
80	340						
82	340						
86	347						
83	340						
82	340						
85	360						
81	344						
84	355						
81	340						
81	343						
81	346						
82	347						
81	335						
75–87	330–360						

Mr. Hartert has examined the series of *Phaëthon rubricauda* in the British Museum, which, however, is poor in fresh skins. The specimens from Mauritius and Round Island, near Mauritius, are not nearly so reddish as those from the Kermadec Islands, and their measurements are :—

Bill, measured as before.	Wing.
mm.	mm.
77	320
84	330
76	325
78·5	336
80	330
76–80	320–336

These birds belong, therefore, clearly to the typical *P. rubricauda*. The specimens from Norfolk Island agree with those from the Kermadec Islands, but the skins in the

British Museum are not good. Those from Christmas Island are rather reddish and measure :—

Wing.	Bill as above.
mm.	mm.
320	73·5
341	73
328	75·5

They seem thus to be somewhat intermediate between the smaller and whiter and the larger and more reddish form, but a larger series must be studied to confirm this impression. In any case there are two distinct subspecies—one the typical *P. rubricauda* ; the other

PHAËTHON RUBRICAUDA ERUBESCENS, subsp. nov.

[*Ex* Banks's Icon. ined. pl. 31, and Gray, List of Birds in B. M. pt. iii. p. 184 (1844), and Ibis, 1862, p. 250 —nomen nudum !!]

P. rubricauda erubescens differs from *P. rubricauda* in being larger, with stronger bill, and in having a more reddish plumage.

Hab. Kermadec, Norfolk, Lord Howe's Islands.

In the synonymy of *P. rubricauda* in the 'Catalogue of Birds' are two quotations which cannot without hesitation be referred to that species, namely :—

1. *P. melanorhynchus*, Gmelin, *ex* Latham, Black-billed Tropic Bird, Gen. Syn. iii. pt. ii. p. 619.

   The description of this supposed form does not suit *P. rubricauda*. It is described as smaller than *P. æthereus*, *P. lepturus*, and *P. fulvus*, and as having a black bill, while the description of the plumage indicates a young bird. I should consider it rather to be the young of *P. æthereus*, under which species it might be quoted with a query.

2. *P. novæ-hollandiæ*, Brandt, Mém. Acad. Pétersb. (6) v. pt. ii. p. 272 (1840), *ex* Latham's New Holland Tropic Bird, Gen. Hist. B. x. p. 448 (1824).

   This refers also to a young *Phaëthon*, and can only be quoted with a big query under the head of one of the known species.

93. PHAËTHON LEPTURUS, *Lacép. & Daudin.*

Phaëton lepturus, Lacép. & Daudin in Buffon's Hist. Nat. ed. 18mo, par Lacépède, chez Didot, Tabl. d. Ois. in Quadr. xiv. p. 319 (1799).

Phaëton candidus, Temm., } of most authors.
Phaëton flavirostris, Brandt, }

Phaëthon lepturus, Grant, Cat. B. Brit. Mus. xxvi. p. 453 (1898).

Phaëthon æthereus (non Linnæus), Scott Wilson & Evans, Aves Hawaii. pt. vii. (1899).

PALMER sent me skins of birds shot on Kauai in March, Oahu in April, and on Maui in October. He describes the bill as " orange-yellow, streaked with grey ; tarsi and upper portion of foot pearl-white with a blue tinge, the webs black."

Dole (Proc. Boston Soc. Nat. Hist. 1869, p. 308, and Hawaii. Alm. 1879, p. 58) records *Phaëthon æthereus* from the Islands, but he gives no details nor exact locality. Unfortunately, by an error, writing from memory, I had written to Mr. Scott Wilson that Palmer obtained *P. æthereus*, while in fact he only got *P. lepturus*. I need not say that I am very sorry for my mistake. According to Mr. Wilson, Mr. Perkins states that he met with *Ph. æthereus* on the cliffs round Honolulu and elsewhere " breeding not uncommonly on the rocky ledges." As no specimen of *Ph. æthereus*, but only *P. lepturus* (=*candidus*), has been so far produced from the Sandwich Islands, I have no doubt that Dole and Perkins were mistaken in their identifications, and that the white-tailed Tropic-birds which breed on the Hawaiian Islands are all *P. lepturus* and not *P. æthereus*.

94. FREGATA AQUILA (*L.*).

IWA.

Fregata aquila, Stejneger, Proc. U.S. Nat. Mus. 1888, p. 102 (Kauai) ; *anteà*, p. 21 (1893) ; Grant, Cat. B. Brit. Mus. xxvi. p. 443 (1898) ; Wilson & Evans, Aves Hawaii. pt. vii. (text) (1899).
Tachypetes palmerstoni, Dole, Hawaii. Alm. 1879, p. 58.

On the islands of the Hawaiian possessions at least the first plumage shows the head, neck, and chest cinnamon-rufous, and, so far as we can ascertain, this is not proved to be the case elsewhere.

According to Dole, Kittlitz states that this bird breeds on " Nihoa," but so far I have been unable to find where he made this statement. Mr. Perkins saw it on Oahu, and Mr. Knudsen obtained a female on Kauai.

The photograph of a young bird opposite to p. 21 is wrongly lettered as that of a young Frigate Bird ; it is, however, the young of the Red-footed Booby (*Sula piscatrix*)!

95. SULA SULA (*L.*).

Anteà, p. 29 ; Cat. B. Brit. Mus. xxvi. p. 436 (1898).

This species is recorded from off the island of Niihau.

Palmer never saw a single specimen of this bird on Laysan, and it remained to Professor Schauinsland to procure it there. He sent me an adult male which he captured on August 29th, 1897. " Bill pale greenish blue ; face and gular sac azure-blue ; eyelids of a darker blue. Feet light bluish green. Iris pale grey."

96. SULA PISCATRIX (*L.*).

Anteà, p. 27 ; Cat. B. Brit. Mus. xxvi. p. 432 (1898).

This Gannet, as might be expected, has been recorded as an occasional visitor to the seas round the northern Hawaiian Islands.

Also of this species Professor Schauinsland sent a fine series, showing every gradation from the downy young of a few days' old to the fully-adult white breeding-plumage. The specimens in the latter plumage show no trace of brown colour on the tail, so conspicuous in *Sula piscatrix websteri* from the Eastern Pacific (*cf.* Nov. Zool. vi. p. 177, 1899). Even in birds in the second year's plumage, where the back and wings are still practically all brown in colour, the tail shows a strong admixture of white.

The nestling is covered with uniform white down. The bare parts of the pullus are: " Feet of a dark ivory-colour, bill black, lores and face greyish blue."

On p. 28 I have said that the photograph showing a nest of this bird in a palm-tree is a picture of the trees on Laysan mentioned by Kittlitz. This, I am sorry to say, is an error. The plate having been taken by Mr. Williams on Lehua, and not being separately labelled, I thought it formed part of the Laysan series with which it was sent to me.

As a matter of fact no living palm-trees exist at present on Laysan, though Prof. Schauinsland informs us that the remains of stumps and roots prove them to have been at one time rather numerous on that island. The palm on Mr. Williams's photograph is *Pritchardia gaudichaudi,* and presumably those formerly growing on Laysan belonged to this species also.

LIST OF BIRDS INTRODUCED INTO THE
SANDWICH ISLANDS.

ALAUDA ARVENSIS. The Skylark has been met with by Palmer on Oahu, where it had been introduced recently.

PASSER DOMESTICUS (*teste* Finsch, Ibis, 1880, p. 78, Schauinsland and Wilson). Oahu, Molokai, and probably other islands.

CARPODACUS FRONTALIS. Palmer sent a series of skins from Kauai and Hawaii. Schauinsland observed it on Molokai. He seems to think that it immigrated from America, but this is most improbable. Palmer also says that it is introduced (Abh. nat. Ver. Bremen, xvi. 3). (Home: North America.)

MUNIA TOPELA. Palmer sent me a dozen from Kauai, where it is very common. (Home: China.)

ACRIDOTHERES TRISTIS. Palmer found it very numerous and very harmful to the native birds. He sent me skins from Kauai. (Home: India.) *Cf.* Finsch, Ibis, 1880, p. 77. Prof. Schauinsland saw it on Molokai, but he says it is most frequent on Oahu, where it was introduced by the botanist Dr. Hillebrand.

TURTUR CHINENSIS (*teste* Finsch, Ibis, 1880, p. 78, and Wilson). Also Schauinsland, Abh. nat. Ver. Bremen, xvi. Heft 3 (1900) [1]. Prof. Schauinsland says "*Turtur chinensis* und andere."

ORTYX VIRGINIANUS and LOPHORTYX CALIFORNICUS. Plentiful on several islands. Palmer sent some chicks from Hawaii. (Home: North America.)

PORPHYRIO MELANOTUS. Palmer shot two on Oahu, where they were said to be imported. (Home: Australia, Tasmania, New Zealand, and (?) parts of New Guinea.)

PEACOCKS. These birds have become feral on the island of Maui, where Palmer found them in large numbers. They were probably *Pavo cristatus.*

---

[1] I cannot quote the page, as I have hitherto only seen a separate copy, for which I am much obliged to its author, but this copy is separately paged.

DOMESTIC FOWL. In addition to those kept by the inhabitants, a number of these birds have become feral and have developed tails somewhat approaching the long tails of the famous Japanese breed. Palmer sent seven specimens in various plumages and colours from the island of Kauai.

PHEASANTS (probably *Phasianus torquatus*). Common on several islands. Very numerous on Molokai (Schauinsland, *l. c.*).

PARROTS have also been introduced into some of the islands, but I am not aware to which species they belong, except that Palmer shot two *Platycercus palliceps* on Maui.

Several of the introduced birds have become a nuisance and are a danger to the native birds, taking away their nesting opportunities and food, and the *Acridotheres* kills and eats the young and eggs of small birds.

BIRDS ERRONEOUSLY REPORTED TO OCCUR
ON THE SANDWICH ISLANDS.

MYZOMELA NIGRIVENTRIS, *Peale.*

Myzomela nigriventris, Dole, Proc. Boston Soc. N. H. xii. p. 298 (1869) ; id. Hawaii. Alm. 1879, p. 46.

This is a bird inhabiting Samoa. Mr. Dole says it is found in the dense forests in Hawaii and Samoa, and that a (? Hawaiian) specimen is preserved in the Smithsonian Institution. Needless to say that now we know better, and that no true *Myzomela* inhabits Hawaii.

TATARE OTAITIENSIS, *Less.*

Tatare otaitiensis, Dole, *l. c.* p. 299 (1869) ; id. Hawaii. Alm. 1879, p. 47.

The above name is a synonym of what is now generally called (Cat. B. Brit. Mus. vii. p. 525) *Tatare longirostris,* and only known from the Society and Paumotou groups of islands.

Dole says that it inhabits reedy marshes in Kauai and Hawaii. I am not aware how this mistake could have arisen—can such a bird as a *Tatare (Acrocephalus)* have existed on the Sandwich Islands? It is not at all an impossibility, as Laysan is inhabited by *Acrocephalus familiaris*!

EMBERIZA SANDVICENSIS, *Gm.*

Emberiza sandvicensis, Dole, *l. c.* p. 301 (1869); id. Hawaii. Alm. 1879, p. 49.

Originally stated by Latham (Syn. ii. p. 202) to have come from "Sandwich Sound," and afterwards by mistake transferred to the Sandwich Islands.

EMBERIZA ATRICAPILLA, *Gm.*

Emberiza atricapilla, Dole, *l. c.* p. 301 (1869); id. Hawaii. Alm. 1879, p. 49.

Erroneously stated by Latham (*l. c.* p. 203) to come from the Sandwich Islands.

Brotogerys pyrrhopterus, Coriphilus kuhli, and *C. taitianus* have been erroneously said to occur on the Sandwich Islands.

STERNA BERGII, *Licht.*

Sterna bergii, Dole, *l. c.* p. 306 (1869); id. Hawaii. Alm. 1879, p. 55.

Dole gives no exact locality, but enumerates this Tern among the Hawaiian birds without any comment.

DOUBTFUL:—

Sandwich Thrush, Latham, Gen. Syn. ii. p. 39 (1783).

Turdus sandwichensis, Gmelin, Syst. Nat. i. p. 313 ; Wilson & Evans, Av. Haw., Introduct. p. xiii.

It is doubtful what is meant by the "Sandwich Thrush." Perhaps a *Phæornis,* but the white forehead is not found in *Phæornis.* The mensure is too large to suit *Oreomyza bairdi.*

2 s 2

ADDITIONS, SUPPLEMENTS, AND CORRECTIONS

TO

PARTS I. & II. OF THIS WORK.

———————•••••———————

PART I.

ABOUT seven years have elapsed since the appearance of the first two Parts of this work. It is therefore not surprising that additions have become necessary. The most important event during these years has doubtless been the expedition to Laysan of Professor Schauinsland, of Bremen. His attention was called to Laysan as a perfect egg-paradise, through my book, and he stayed there three months, principally to collect materials for his embryonic studies; but he was wise enough not to neglect other branches, and he brought home a large collection of well-prepared bird-skins, of which I obtained a magnificent set.

Professor Schauinsland's accounts of the bird-life on Laysan, in his little book ' Drei Monate auf einer Koralleninsel' (Bremen, 1899), are among the most interesting and fascinating ones I ever read. I quote here a few of his notes, rather freely translated :—

"The effect on those who for the first time visit the island of the tameness and absolute confidence of most of the birds is simply stupendous. We always took our meals in company with some of the pretty yellow *Telespiza*. When we took our seats, some of these impertinent little chaps arrived at once and pecked at the bread in front of us, and even sat down on our plates and took part of our rice and bacon, so that we had to remove them with our hands, like so many troublesome flies, if we wished to eat our food in peace. When in the hot hours of the day we sat in the shade of our hut and enjoyed the cool breeze, usually one or the other of the neat grey little *Acrocephalus* came and seated himself on our knee or the back of the chair, looked at us fearlessly, and sang its lovely little song. Often the *Telespiza*, the best songsters of Laysan, sang away when we had caught them in our hands, though I would not like to say if that was really sheer mirth or the expression of a sort of queer feeling and doubt. Our constant companions when at work were the funny little Rails. Immediately when we opened the door of our laboratory, some of these little fellows entered with us and searched busily among our collections for flies, which they caught with great alacrity. It was too droll to see them stop from time to time, and to hear them put forth their most peculiar song, which has some resemblance to the noise of a high-sounding alarum-clock. They also tried to hop on to our table in order to snatch a piece of flesh or fat which we had put aside when skinning.

"The sea-birds were equally tame. When we took our way through one of the Albatross-rookeries the birds did not give way at all, so that we had to be careful if we did not wish to hurt them. Often enough we could not help coming so near, that they most indignantly pinched our legs, which was by no means pleasant. This in any case was the habit of the *young* Albatrosses ; but also the *old* ones did not fly away until they found out that we really meant to harm them. It was thus possible for us to take all the birds we wanted without the help of a gun, except a few, namely the Duck, the *Himatione*, and such species which visit the island only occasionally. The tameness of the birds sometimes became troublesome. A Frigate-bird once took the cap off a Japanese workman, and carried it high up into the air before it dropped it again, and this play was repeated for several days. A Japanese, returning with two full baskets from an egg-collecting tour, was hit on the neck by a flying Albatross, so that he tumbled down and rolled among the eggs.

"Laysan is an actual bird-paradise, such as may be found again on the globe's surface. While land-birds are of lesser importance, and must be content if they are left unmolested, the sea-birds are the dominating and domineering occupants. They characterize the whole island. From an extended portion of the North Pacific Ocean they flock to this island, which, on account of its sandy soil, is more convenient to them than many others which, though being uninhabited, lack the sand and consist merely of rock, in which the Petrels, which deposit their eggs in long tunnels, cannot burrow.

"The quantities of birds nesting on Laysan are prodigious. When approaching Laysan we saw clouds of birds marking the situation of the island, and the Terns (*Sterna fuliginosa*), which were seeking their nesting-places, appeared from far off like swarms of bees. It is most difficult to guess at the number of these masses of birds; but those we saw flying numbered tens of thousands, if not hundreds of thousands. In many places almost every square foot of ground is literally occupied by breeding birds, so that the wanderer finds it— specially at night—hardly possible to put down his feet without hurting the birds. But the birds are not only distributed horizontally over the island but also vertically, so that they dwell one above the other as well as side by side. Wide portions of the island, especially those with soft sand and but little vegetation, are honeycombed by the Petrels. Nothing is more tedious than to walk over such ground. The thin ceiling of the birds' burrows breaks through continually, and one sinks in to above the knee at every step. On places where thick scrub grows, especially the *Chenopodium sandvicheum*, it happens that not only two parties, but even *four*, live above each other ! On the top of the bushes the Frigate-birds have built their nests ; deeper below, in the same bushes, the little land-birds (mostly *Acrocephalus*, sometimes also *Himatione*) are breeding ; below, on the ground, overshaded by the branches, breed the beautiful Tropic-birds ; and, down in the ground, the black Petrel hatches out its young in subterranean passages. Thus the birds live in four stories, and a comparison with the series of flats in large towns is opportune.

"In spite of this excellent use of all the available space, the birds which have chosen Laysan for their breeding-home would not be able to breed if they arrived there all *at the same time !* They are therefore obliged to make room for others, so that some species of sea-birds leave the place as soon as their young are strong enough to fly ; and while the former occupant is leaving, the new comers already begin to arrive. Thus we see a constant coming and going ; and it follows that breeding-species can be found at almost every season of the year, a fact which is remarkable even in the tropics, where the breeding-season is generally less regular. In this way a most definite succession, which probably dates back thousands of years, takes place year after year in the arrival and departure of certain species.

For several years the observation has been made that *Œstrelata hypoleuca*, which has undermined nearly the whole island, arrives on Laysan between the 15th and 18th of August. I remember most vividly the evening of the 17th of August, 1896. It was less noisy on the island than before, for the clamorous Terns had reared their young, and thousands of Albatrosses had left their ancestral home for the boundless ocean, which would in future be their dwelling-place. We were just leaving the little hill from where we had been looking for the sail which should take us back again to civilized countries. The golden glow of the sunset was fading away, and the slender sickle of the new moon began to shine; then our eyes, which had become well acquainted with every one of the characteristic motions of our feathered companions of the island from week-long observations, were struck by a new phenomenon. Against the dissolving evening glow was sharply traced the silhouette of a magnificent flier, which cut through the air with the keenest and at the same time most elegant movements, inaudible and almost without movement of its wings. The manner in which it dashed along was unknown to us, and we saw that a new arrival had reached our island. The next evening there were more, and on the third thousands filled the air. The new guests were pretty birds, barely of the size of a domestic Pigeon; but they began to domineer all over the island in such a way that the few pairs of Tropic-birds, Terns, and others which were still breeding made way before them, as if they could not stand these noisy neighbours. They are, on land, entirely nocturnal, and at once took possession of their innumerable subterranean burrows. In the bright moonshine one could see how they were busily engaged in removing the loose sand from the holes, most of which had more or less collapsed since they had left them. Loving couples selected their nests and fought hard for them against later intruders. Quarrels, fights, and clamour became unceasing; in a few days there was no spot with sandy soil where the horrid 'song' of these Petrels could not be heard! Under every bush, between our luggage and cases, and, alas! also under our bedroom, their tune was raised, which stood about in the middle between that which 'drives, men to madness' and the cries of new-born babies, which are only harmonious to their devoted parents. The face of the island was entirely changed!

"A few months later the aspect is again altered by an invasion of a still more imposing kind. During the last days of October the first vanguard of the mighty Albatrosses appeared and a few days afterwards the island looked, from an elevated point, as if it was densely covered with large snowflakes. There was hardly a spot of ground on which the dazzling white plumage of an Albatross was not apparent; and the number of these birds is often so large, that many are obliged to deposit their eggs on rather unsuitable spots, and many others fail altogether to find a nesting-place, and must leave the overcrowded area.

"Of the other sea-bird invasions I will only mention that of the Terns, which is so enormous that during the first days, when the birds have not yet selected a spot for breeding, the island appears from a distance as if a heavy cloud of smoke were hanging over it."

It is to be regretted that Professor Schauinsland has not yet given us the more detailed accounts about the birds of Laysan which he announced he was going to publish (see 'Drei Monate auf einer Koralleninsel,' p. 100, note 17).

HIMATIONE FRAITHI, *Rothsch.*

Anteà, Pt. I. p. 3 (spelt "*freethi*") ; Schauinsland, Drei Monate Korallenins. pp. 43, 101 (1899).

Professor Schauinsland tells us that the nearest ally of *H. fraithi*, the well-known *H. sanguinea*, is still one of the commonest birds in the higher portions of the Hawaiian Group, where it usually takes its food from the bright red flowers of the Metrosideros-trees, which bear a striking resemblance to its red plumage. " In Laysan," he continues, " this plant does not exist ; but here, too, the *Himatione fraithi* flits from bush to bush and visits the flowers for food, preferring the large flowers of *Capparis sandwichoana*. This bird offers a good example of how a new species has evolved by isolation. In spite of its similarity to *H. sanguinea*, it differs from the latter in the different shade of its red plumage, in having some brownish feathers on the underside of the tail, and by its shorter bill."

The nests and eggs were found by Dr. Schauinsland, and Dr. Studer of Bern has also received nest and eggs from one of his correspondents. These eggs are of a glossless white, and marked with deeper-lying mauve-coloured patches, and above these with rufous-brown patches and spots. The markings are more or less confined to the thick end, and often form a circle. The eggs resemble large eggs of *Certhia brachydactyla* and also those of *Hirundo rustica*, which, however, are much more glossy. They vary in size : one clutch of three (which is the regular number) from Professor Schauinsland measuring 19·5 by 14, 20·2 by 14, and 20 by 14 mm; three eggs, kindly lent me by Dr. Studer, 17·8 by 13·4, 17·8 by 13, and 17·9 by 13·2 mm.; while Schauinsland gives 20·5 by 14·5 and 19·75 by 14 mm. The nests are well built and have a fairly deep cup of about 35 to 40 mm. depth. The material consists of fine rootlets, with a few dry pieces of grass. Sometimes the nests are interwoven with feathers of Petrels or the down of Albatrosses. I have examined four nests.

TELESPIZA CANTANS, *Wilson.*

Anteà, p. 5 and p. 7 (sub nomine *T. flavissima*) ; Rothschild, Bull. B. O. Club, viii. p. lvi (1899) ; Schauinsland, Drei Mon. Korallenins. p. 42 (habits) (1899).

As I have already stated (*l. c.*), the series collected and observations made by Schauinsland have proved beyond doubt that *T. flavissima* is only the perfectly adult *T. cantans*. Schauinsland speaks of this bird as follows :—" This bird is now entirely carnivorous. Among other things it has found that the eggs of sea-birds, which breed here at almost all seasons, are as nutritious as they are tasty : with a few knocks of its strong, sharp bill it breaks them open, and completely sucks out the contents. It is so saucy, that the parents of the eggs are, on his account, most unwilling to leave their clutches even for one short moment ! When the parents take their duties over from each other, the new comer stands close by and pushes the former occupant from the eggs, so that they are not left free for the shortest space of time. Nevertheless the little robber is often able to steal an egg."

Four eggs belonging to Dr. Studer agree with those formerly described, and measure :— 22·4 by 17·4, 22·7 by 16·4, 22·7 by 16·7, and 22·4 by 16·4 mm.

PORZANULA PALMERI, *Froh.*

Antea, Pt. I. p. 9; Schauinsland, Drei Mon. Korallenins. p. 42 (1899).

Eggs in Dr. Studer's possession agree with those found by Palmer. Schauinsland says :—"This funny little Rail has become accustomed to a totally new life; it lost its power of flight completely, and hardly uses the rudiments of its wings to help it when running like a shadow across the sand with mouse-like speed. Originally more a swamp-bird and dependent on worms, it has here become omnivorous, and the sea-birds must furnish its principal food. Although it cannot open their eggs with its thin beak, I have often seen it partake of the tasty inside of an egg when a *Telespiza* had broken one. It does not even despise corpses of birds, which are so frequent here, but it tears the flesh off in pieces and devours it; it feeds also chiefly on flies and the numerous beetles (*Dermestes*)."

ANAS LAYSANENSIS, *Rothsch.*

Anteà, p. 19; Salvad. Cat. B. Brit. Mus. xxvii. p. 190 (1895).

The series collected by Professor Schauinsland reveals the following facts :—

The Plate in Part I. of this work is too pale, owing to Palmer's specimens being all in rather worn plumage. The crown of the head of the very old male in fresh plumage is quite black.

The adult female differs from the male in being much smaller, and in having the upper-side somewhat more distinctly spotted with pale rufous; but the speculum is not less developed in the quite adult females. The specimens with the speculum only indicated are younger birds.

The pullus in down is rufous olive-brown on the back, with the buff spots obsolete and irregular in number. Forehead and sides of head and neck buffy rufous, a narrow blackish line through the eye; top of head a little darker than the back. Underside buff; chest more rufous.

The wing in five males varies from 8·1 to 8·7 inches; in five females from 7·5 to 8·2 inches.

SULA CYANOPS (*Sundev.*).

Anteà, p. 25; Grant, Cat. B. Brit. Mus. xxvi. p. 430 (1898).

Professor Schauinsland's series shows that the youngest birds described by Mr. Grant in the 'Catalogue of Birds' are really in the first plumage, one of the series still having the feathers mixed with down, though otherwise agreeing with Mr. Grant's description.

The nestling with the first feathers just showing through the down has, according to Schauinsland, " the iris dark blue; feet mouse-grey with a yellow tinge; the beak dark grey, with a yellow point ; face and gular sac leaden-grey."

PROFESSOR SCHAUINSLAND'S explorations have further added the following birds to the Laysan list, the majority of which are only more or less irregular winter visitors :—

1. LIMOSA LAPPONICA NOVÆ-ZEALANDIÆ, *Gray.*
    ♂. Laysan, 5. 11. 1896, in the Tring Museum.   Several more were procured, which are in Bremen.

2. HETEROPYGIA ACUMINATA (*Horsfield*).
    One specimen. (Bremen Museum.)

3. CALIDRIS ARENARIA (*Linn.*).
    One specimen. (Bremen Museum.)

4. TRINGA AMERICANA, *Cass.*
    (*Teste* Schauinsland, ' Drei Monate auf einer Koralleninsel,' p. 101, 1899.)

5. CRYMOPHILUS FULICARIUS (*Linn.*).
    (*Teste* Schauinsland, *l. c.*)

6. ANAS BOSCHAS, *Linn.*
    ♂ ad. Laysan, 7. 11. 1896.   (Bremen Museum.)

7. SPATULA CLYPEATA (*Linn.*).
    Laysan, October, November, January. (Bremen Museum.) (*Anteà*, p. 275.)

8. DAFILA ACUTA (*Linn.*).
    Laysan, three specimens. (Bremen Museum.) (*Anteà*, p. 275.)

9. NETTION CAROLINENSE (*Gm.*).
    1 ♀. 27. 10. 1896, Laysan.   (Bremen Museum.)   The females of *N. crecca* and *N. carolinense* are apparently indistinguishable ; but *N. crecca* is not recorded from the Pacific coasts of America.

10. QUERQUEDULA QUERQUEDULA (*Linn.*).
    (*Teste* Schauinsland ; but I am inclined to consider it more likely to be *Q. discors.*)

11. MARECA AMERICANA (*Gm.*).
    ♀ juv. Laysan, 15. 10. 1896.   (Bremen Museum.)

12. CLANGULA ALBEOLA (*Linn.*).
    ♂ juv. 15. 1. 1897. (Bremen Museum.)

13. LARUS GLAUCESCENS, *Naum.* (or *L. nelsoni*, Hensh. ?).
    I have examined one young bird, now in the Bremen Museum, procured on Laysan by one of Professor Schauinsland's correspondents.   The outermost primary is nearly pure white on the outer web, while the inner web shows a distinctly separated darker and a lighter half.   Under tail-coverts mixed irregularly white and ashy brown.   Wing 16·2 inches !   The small size points rather to *L. nelsoni* than to *L. glaucescens*; but probably the specimen is a young female of the latter.   Schauinsland (*l. c.* p. 101) mentions this Gull as *L. glaucus*, but it does not belong to the latter form. (*Anteà*, p. 286.)

2 T

14. OCEANODROMA FULIGINOSA (*Gm.*).

Sooty Petrel, Latham, Gen. Syn. B. iii. pt. 2, p. 409 (1785) ("Otaheite"!).

Procellaria fuliginosa, Gmelin, Syst. Nat. i. p. 562 (1788) (*ex* Latham!).

Oceanodroma fuliginosa, Stejneger, Proc. U.S. Nat. Mus. xvi. p. 620 (1893); Salvin, Cat. Birds Brit. Mus. xxv. p. 352 (1896); Schauinsland, Drei Monate Korallenins. p. 101, no. 19 (1899) (enumerated among the birds breeding on Laysan).

Professor Schauinsland did not meet with this Petrel alive when on Laysan, but from the examination of some fragments of skeletons which he had picked up, he concluded that another, hitherto unrecorded species of Petrel, with one nostril-opening, must occur there. This is, in fact, the case, for months after he had left a new species arrived in small numbers, and bred, like the others, in holes. One skin was sent to the Bremen Museum and determined by Professor Reichenow as *Oceanodroma fuliginosa*. I am obliged to Prof. Schauinsland for having lent me this example for description. There seems to be doubt that this bird is the same as that called *O. fuliginosa* by Dr. Stejneger and also in the 'Catalogue of Birds,' vol. xxv. Another question is, whether there is sufficient reason to refer it to Latham's "Sooty Petrel," and consequently to adopt Gmelin's name. The original description would suit many deep-brown species of approximately the same size, and is evidently equally well applicable to *Bulweria anjinho*, or more still to *B. macgillivrayi* from Fiji, and *Cymodroma mœstissima* from Samoa! As Latham's bird is from "Otaheite" it is necessary for us to receive so-called *O. fuliginosa* from there, to prove beyond doubt that the birds from the Japanese seas and Laysan are really rightly called *O. fuliginosa*. In Latham's description the slaty colour of the upperside is not mentioned, the brown rump of Latham is not marked in the modern bird, and the underside is brown (not slaty), but not paler than the upperside! The example from Laysan measures:—Wing 7·3 inches, tail 4·1, forked for 1·25, culmen (over curve) 0·8, bill from gape 1·1, metatarsus 1·15, middle toe with claw 1·15. Nasal tube rather prominent. (Sex not determined on label.)

15. PHALACROCORAX PELAGICUS, *Pall.*

One female, Laysan, 22. 10. 1896. Head brown, with whitish edges to the feathers; rest of plumage blackish brown, with some few white filamentous feathers scattered over the neck and back. Bill from end of feathering on forehead to tip 51·5 mm., wing 282 mm. (=11·1 inches), tail 185 mm. (=7·3 inches).

I have examined specimens, mostly single ones, of all these fifteen species, with the exception of *Tringa americana* and *Crymophilus fulicarius*.

PART II.

Add :—

PHÆORNIS OAHENSIS, *Wilson & Evans.*

Turdus sandwichensis, var., Bloxam, Voy. ' Blonde,' Appendix, p. 250 (1826) (Oahu).
Phæornis oahensis, Wilson & Evans, Aves Hawaii., Introduct. p. xiii (1899).
Ph. oahuensis, iid. t. c. p. xxiv.

WE know nothing of this evidently extinct bird, except the short description of Bloxam of what he called " *Turdus sandwichensis* var. from Oahu."

The description is as follows :—

" Length 7¼ inches. Upper parts olive-brown, extremities of the feathers much lighter colour; tail and wings brown ; bill bristled at the base."

PHÆORNIS PALMERI, *Rothsch.*

Anteà, Pt. II. p. 67 : Wilson & Evans, Aves Hawaii. pt. vi. (text & plate) (1896).

MR. PERKINS procured several specimens of this species, from which I find that my type specimen is not fully adult. Messrs. Wilson and Evans state that Mr. Gay reported this bird from a skin subsequently destroyed by rats. The truth is that, though very badly injured, the specimen forming the type of the species and described in Part II. of this work is the one seen by Mr. Gay.

Palmer only succeeded afterwards in getting two young birds barely fledged.

Messrs. Wilson and Evans give a very fair figure of the adult and semi-adult birds in Part vi. of their ' Aves Hawaiienses ' (1896).

The Genus CHASIEMPIS (*anteà*, p. 69).

MR. SCOTT WILSON has given in Part vi. of the ' Aves Hawaiienses ' (1896) a full account and plates of the three recognized species of this genus. I am glad to see that Mr. Wilson's conclusions fully agree with mine. I have, however, no doubt as to the " Spotted-winged Flycatcher " of Latham (*Muscicapa maculata,* Gmelin) belonging as a synonym to *Chasiempis sandwichensis.* In Part vii. Mr. Wilson figures the eggs of a *Chasiempis* spec.? and a photograph of the nest. Nest and eggs agree with those of *Ch. gayi* described *anteà* p. 76, and figured in this number. It is curious how Mr. Wilson could know that these eggs belonged to a *Chasiempis,* as he apparently does not even know the island they came from, or else he would have known the species.

2 T 2

HEMIGNATHUS ELLISIANUS, *Gray*.

Anteà, Pt. II. p. 87 (1893) ; Hartl. Abh. wiss. Ver. Bremen, xiv. p. 29 (1895).
Hemignathus lichtensteini, Scott Wilson & Evans, Aves Hawaii. pt. v. (text, and plate from the type in Berlin)

I HAD hoped that my acceptance of Gray's name *ellisianus* for the extinct Oahu species would be followed by everybody, as it was by Hartlaub (*l. c.*). Mr. Wilson, however, attempts to show that *Drepanis (Hemignathus) ellisianus*, Gray, should be quoted as a synonym of *Hemignathus obscurus*, Gm. This is clearly erroneous. Gray, in his Cat. B. Trop. Isl. Pacific Ocean, p. 9, gives as a first quotation in the synonymy of his *ellisianus* "Vieillot, Ois. Dor. tab. 53," which Mr. Wilson correctly places as a synonym of *H. obscurus*, but Mr. Wilson omits the fact that Gray quotes it with a query. This being the case, that quotation must be cast aside in determining the priority and meaning of the name *ellisianus*. The first quotation without a query is Gray's second one, "Lichtenstein, Abh. k. Akad. Berlin, 1838. p. 449, tab. v. fig. 1."

Gray must have clearly perceived that the description and figure of Lichtenstein referred to a bird different from the one described by Latham which formed the basis of Gmelin's "*Certhia obscura*." Lichtenstein's description and figure being fairly good, Gray considered Lichtenstein's bird as diagnosed and named it (*i. e.* the *H. obscurus*, Lichtenstein, non Linnæus) in honour of Ellis. Gray then quotes also Ellis's unpublished drawing of a *Hemignathus* which, as Mr. Wilson correctly says, is unquestionably *H. obscurus*, but it does not affect a name, if subsequently (in another line) wrong quotations are added to the synonymy. If this were the case there would be many names to be altered in zoology. Gray also quotes Cassin in his synonymy of *D. ellisianus*, but this also after referring to Lichtenstein, so that this cannot affect the name *ellisianus* either. It is quite true, as Mr. Wilson says, that Gray never saw a specimen of his *D. ellisianus*; but the fact remains that this name was bestowed on the bird of Lichtenstein, which was sufficiently characterized, and which therefore required no further diagnosis by Gray, who could not examine it. The erroneous quotations of Gray added to his name cannot be used as a reason for rejecting his name. Mr. Wilson is of opinion that the name of Ellis being chosen by Gray must mean that Ellis's unpublished drawing formed the type of his name, but this is an unfounded assumption. It frequently happens in ornithology that names of persons are given to birds which they incorrectly figured under another name. Besides, in determining the priority of names, the supposed meaning of the names should never be considered at all. I therefore believe that those who care for strict priority of names in zoology must accept the name *Hemignathus ellisianus* for the Oahu bird.

Mr. Hartert, during his visit to Berlin in 1893, made the following description of the type of *Hem. ellisianus* :—

"Above greenish olive-brown, more greenish on the back and rump and more greyish on the head and hind-neck ; the dark bases of the feathers on the head showing through. Lores deep brown. A distinct yellow superciliary stripe. Chin, throat, and middle of abdomen dull brownish white (apparently somewhat faded). Upper breast olive-greenish, sides of breast and flanks dull olive-greenish, more olive-brown on the flanks. Wings and tail deep brown, bordered with yellowish green. Under wing-coverts dull white. The bill is brown, somewhat horn-brown, but *not blackish* as in all the other species of *Hemignathus*.

" It is not probable that the bill and feet are faded, as in specimens of *Heterorhynchus lucidus* collected and stuffed at the same time and kept side by side with *H. ellisianus*, the bill and feet are still blackish and not brown.

"Wing 3·3 inches, tail 2·1, culmen 2·24, bill from gape to tip in a straight line 1·87, mandible from mental apex to tip 1·57 inches."

We are not aware of any other specimens in Europe or elsewhere! Mr. Wilson was told by a native that this bird still existed on Oahu. But, although the dense forests of parts of Oahu make collecting most difficult, the failure of such collectors as Wilson, Palmer, Perkins, and an employé of the Honolulu Museum to come across it, does not speak in favour of the theory that this bird is still living.

My Plate is delineated from a drawing made of the type in Berlin.

Key to the Adult Males of the Genus Hemignathus.

A. Bill extremely long, culmen more than 2·5 inches long : habitat Kauai . . . . . *H. procerus.*
B. Bill of medium length, culmen less than 2·5 inches, but more than 2.
    a. A conspicuous yellow superciliary line, bill brown : habitat Oahu . . . . . . *H. ellisianus.*
    b. No distinct yellow superciliary line, bill black : habitat Lanai . . . . *H. lanaiensis.*
C. Bill much smaller, culmen less than 2 inches long, altogether smaller : habitat Hawaii . *H. obscurus.*

HETERORHYNCHUS LUCIDUS (*Licht.*).

Anteà, p. 105 ; Wilson & Evans, Aves Hawaii. pt. v. plate & text (1894).

I AM now giving a Plate of this evidently extinct species. Fig. 1 represents the adult male in the Paris Museum, fig. 2 the one in Leyden, and fig. 3 the Frankfort Museum specimen.

In the synonymy given by Mr. Wilson he left out *Heterorhynchus olivaceus*, which I had shown belonged here. I said (*anteà*, p. 106) that I had examined the type of Lafresnaye's species ; but it was *not for that reason* that I placed it as a synonym of *H. lucidus*, but because the plates, both in the 'Magasin de Zoologie' and in the 'Voyage of the Venus' showed most distinctly the curved (not straight) mandible and the superciliary line. The plates are quite recognizable. Mr. Wilson has finally (Introduction to 'Aves Hawaii.' p. xxii, in part vii. 1899) admitted the correctness of my statements and accepted the name *H. wilsoni* for the Hawaiian form. He has seen the actual type out of Lafresnaye's collection, now in Boston. However, Lafresnaye's specimen was only a duplicate of the collection made by Neboux during the 'Venus' expedition. This is not mentioned in Lafresnaye's original description, but positively stated by Neboux in Rev. Zool. 1840, p. 289. Thus the specimens in the Paris Museum are apparently cotypes, and marked " types " according to the unfortunate habit of marking all original specimens as " types," which is still in use among some continental zoologists.

VIRIDONIA SAGITTIROSTRIS, *Rothsch.*

Antea, Pt. II. p. 109 (1893) ; Wilson & Evans, Aves Hawaii., pt. vii. plate & text (1899).

SINCE the discovery of this bird Mr. Perkins and Mr. Henshaw have collected specimens. Some of the latter are in my collection.

Palmer evidently made a mistake in saying that he obtained it at altitudes of from 500 to 1500 feet, for Perkins found it only at heights of about 2000 feet, and Mr. Henshaw tells me (*in litt.*) that he did not see it below 1800 feet. This ornithologist considers it to be a rare bird, and only to be found in a very limited section of the woods. Mr. Henshaw found the feet and tarsi of a *beautiful bluish grey* (not blackish grey); the base of the mandible and the cutting-edges of the maxilla are bluish grey. "The bird is an inhabitant of the densest forest, but nevertheless appears occasionally on the edge of the forest and even in clearings. Were it not for its curiosity it would be next to impossible to find the bird, so quiet is it and so retiring." *Henshaw in litt.*)

COMPLETE LIST OF THE BIRDS KNOWN FROM THE HAWAIIAN POSSESSIONS.

[The species which are peculiar to the group are marked with a circle. Among the sea-birds the distribution is not in every case certain, but those presumably nesting only in the neighbourhood are considered as peculiar to the group.]

○ 1. ACROCEPHALUS FAMILIARIS (Rothsch.).
○ 2. PHÆORNIS OBSCURA (Gm.).
○ 3. ,, MYIADESTINA Stejn.
○ 4. ,, OAHENSIS Wils. & Evans.
○ 5. ,, PALMERI Rothsch.
○ 6. ,, LANAIENSIS Wils.
○ 7. CHASIEMPIS SANDWICHENSIS (Gm.).
○ 8. ,, GAYI Wils.
○ 9. ,, SCLATERI Ridgw.
○ 10. LOXOPS COCCINEA (Gm.).
○ 11. ,, RUFA (Bloxam).
○ 12. ,, OCHRACEA Rothsch.
○ 13. ,, CÆRULEIROSTRIS (Wils.)
○ 14. CIRIDOPS ANNA (Dole).
○ 15. PALMERIA DOLEI (Wils.).
○ 16. VIRIDONIA SAGITTIROSTRIS Rothsch.
○ 17. OREOMYZA MACULATA (Cab.).
○ 18. ,, BAIRDI Stejn.
○ 19. ,, MANA (Wils.).
○ 20. ,, PERKINSI Rothsch.
○ 21. ,, NEWTONI (Rothsch.)
○ 22. ,, MONTANA (Wils.).
○ 23. ,, FLAMMEA (Wils.).
○ 24. ,, PARVA (Wils.).
○ 25. CHLORODREPANIS VIRENS (Gm.).
○ 26. ,, CHLORIS (Cab.).
○ 27. ,, WILSONI (Rothsch.).
○ 28. ,, STEJNEGERI (Wils.).
○ 29. HIMATIONE SANGUINEA (Gm.).
○ 30. ,, FRAITHI Rothsch.
○ 31. VESTIARIA COCCINEA (Forst.).
○ 32. DREPANIS PACIFICA (Gm.).
○ 33. DREPANORHAMPHUS FUNEREUS (Newt.).
○ 34. HEMIGNATHUS OBSCURUS (Gm.).
○ 35. ,, LANAIENSIS Rothsch.
○ 36. ,, ELLISIANUS Gray.
○ 37. ,, PROCERUS Cab.
○ 38. HETERORHYNCHUS LUCIDUS (Licht.)
○ 39. ,, HANAPEPE (Wils.)
○ 40. ,, AFFINIS (Rothsch.).
○ 41. ,, WILSONI Rothsch.
○ 42. PSEUDONESTOR XANTHOPHRYS Rothsch.

○ 43. PSITTIROSTRA PSITTACEA (Gm.).
○ 44. ,, OLIVACEA Rothsch.
○ 45. LOXIOIDES BAILLEUI Oust.
○ 46. TELESPIZA CANTANS Wils.
○ 47. RHODACANTHIS FLAVICEPS Rothsch.
○ 48. ,, PALMERI Rothsch.
○ 49. CHLORIDOPS KONA Wils.
○ 50. MOHO NOBILIS (Merrem).
○ 51. ,, APICALIS Gould.
○ 52. ,, BISHOPI (Rothsch.).
○ 53. ,, BRACCATUS Cass.
○ 54. CHÆTOPTILA (Peale).
○ 55. CORVUS HAWAIIENSIS Peale.
○ 56. ASIO ACCIPITRINUS SANDWICHENSIS (Bloxam).
57. CIRCUS HUDSONIUS (L.).
○ 58. BUTEO SOLITARIUS Peale.
○ 59. PENNULA MILLSI Dole.
○ 60. ,, SANDWICHENSIS (Gm.).
○ 61. PORZANULA PALMERI Frob.
○ 62. FULICA ALAI Peale.
○ 63. GALLINULA SANDWICENSIS Streets.
○ 64. HIMANTOPUS KNUDSENI Stejn.
65. LIMOSA LAPPONICA NOV.E-ZEALANDIÆ Gray.
66. CRYMOPHILUS FULICARIUS (L.).
67. GALLINAGO DELICATA (Ord) (?).
68. HETERACTITIS INCANUS (Gm.).
69. HETEROPYGIA ACUMINATA (Horsf.).
70. TRINGA AMERICANA Cass.
71. NUMENIUS TAHITIENSIS (Gm.).
72. CALIDRIS ARENARIA (L.).
73. ARENARIA INTERPRES (L.).
74. CHARADRIUS DOMINICUS FULVUS Gm.
75. PLEGADIS GUARAUNA (L.).
76. NYCTICORAX NYCTICORAX N.ÆVIUS (Bodd.).
77. DEMIEGRETTA SACRA (Gm.) (?).
78. MERGUS SERRATOR L.
79. ANAS BOSCHAS L.
○ 80. ,, WYVILLIANA Scl.
○ 81. ,, LAYSANENSIS Rothsch.
82. DAFILA ACUTA (L.).
83. SPATULA CLYPEATA (L.).
84. MARECA AMERICANA (Gm.).

85. NETTION CAROLINENSE (Gm.).
86. QUERQUEDULA QUERQUEDULA (L.).
87. CLANGULA ALBEOLA (L.).
88. ANSER ALBIFRONS GAMBELI Hartl.
89. CHEN HYPERBOREUS (Pall.).
90. BRANTA NIGRICANS (Lawr.).
91. „ CANADENSIS MINIMA (Ridgw.).
o 92. NESOCHEN SANDVICENSIS (Vig.).
93. GYGIS ALBA (Sparrm.).
94. STERNA FULIGINOSA Gm.
95. „ LUNATA Peale.
96. ANOUS STOLIDUS (L.).
o 97. „ HAWAIIENSIS Rothsch.
98. LARUS GLAUCESCENS Naum.
99. „ PHILADELPHIA (Ord).
100. OCEANODROMA CASTRO (Harcourt).

101. OCEANODROMA FULIGINOSA (Gmelin).
102. BULWERIA ANJINHO (Heineken).
103. PUFFINUS NATIVITATIS Streets.
104. „ CUNEATUS Salvin.
o 105. „ NEWELLI Hensh.
106. ÆSTRELATA HYPOLEUCA Salvin.
o 107. „ PHÆOPYGIA SANDWICHENSIS Ridgw.
o 108. DIOMEDEA IMMUTABILIS Rothsch.
109. „ NIGRIPES Aud.
110. PHAËTHON RUBRICAUDA (Bodd.).
111. „ LEPTURUS Lacép. et Daud.
112. FREGATA AQUILA (L.).
113. SULA SULA (L.).
114. „ PISCATRIX (L.).
115. „ CYANOPS (Sundev.).
116. PHALACROCORAX PELAGICUS Pall.

LISTS OF THE BIRDS KNOWN FROM THE VARIOUS ISLANDS.

[The species which breed in the group are marked with an asterisk.]

LAYSAN.

*ACROCEPHALUS FAMILIARIS.
*HIMATIONE FRAITHI.
*TELESPIZA CANTANS.
*PORZANULA PALMERI.
*ANAS LAYSANENSIS.
CHARADRIUS DOMINICUS FULVUS.
STREPSILAS INTERPRES.
HETEROPYGIA ACUMINATA.
THINGA AMERICANA.
CALIDRIS ARENARIA.
HETERACTITIS INCANUS.
LIMOSA LAPPONICA NOVÆ-ZEALANDIÆ.
NUMENIUS TAHITIENSIS.
CRYMOPHILUS FULICARIUS.
LARUS GLAUCESCENS.
*GYGIS ALBA.
*ANOUS STOLIDUS.
* „ HAWAIIENSIS.
*STERNA LUNATA.
* „ FULIGINOSA.

*PUFFINUS NATIVITATIS.
* „ CUNEATUS.
*ÆSTRELATA HYPOLEUCA.
*BULWERIA ANJINHO.
*OCEANODROMA FULIGINOSA.
*DIOMEDEA IMMUTABILIS.
* „ NIGRIPES.
ANAS BOSCHAS.
SPATULA CLYPEATA.
DAFILA ACUTA.
NETTION CAROLINENSE.
QUERQUEDULA QUERQUEDULA.
MARECA AMERICANA.
CLANGULA ALBEOLA.
PHALACROCORAX PELAGICUS.
*FREGATA AQUILA.
*SULA CYANOPS.
* „ PISCATRIX.
* „ SULA.
PHAËTHON RUBRICAUDA.

KAUAI (WITH NIIHAU).

*PHÆORNIS MYIADESTINA.
* „ PALMERI.
*CHASIEMPIS SCLATERI.
*HEMIGNATHUS PROCERUS.
*HETERORHYNCHUS HANAPEPE.
*OREOMYZA PARVA.
* „ BAIRDI.
*CHLORODREPANIS STEJNEGERI.
*HIMATIONE SANGUINEA.
*VESTIARIA COCCINEA.
*LOXOPS CÆRULEIROSTRIS.
*PSITTIROSTRA PSITTACEA.
*MOHO BRACCATUS.
*ASIO ACCIPITRINUS SANDWICHENSIS
*FULICA ALAI.
*GALLINULA SANDVICENSIS.
*HIMANTOPUS KNUDSENI.
HETERACTITIS INCANUS.
NUMENIUS TAHITIENSIS.
CALIDRIS ARENARIA.
ARENARIA INTERPRES.

CHARADRIUS DOMINICUS FULVUS.
PLEGADIS GUARAUNA.
*NYCTICORAX NYCTICORAX NÆVIUS.
*ANAS WYVILLIANA.
DAFILA ACUTA.
SPATULA CLYPEATA.
BRANTA CANADENSIS MINIMA.
*STERNA LUNATA.
* „ FULIGINOSA.
LARUS PHILADELPHIA.
*ANOUS HAWAIIENSIS.
*BULWERIA ANJINHO.
*PUFFINUS CUNEATUS.
* „ NEWELLI.
(?*)ÆSTRELATA PHÆOPYGIA SANDWICHENSIS.
*PHAËTHON RUBRICAUDA.
* „ LEPTURUS.
*FREGATA AQUILA.
*SULA SULA.
* „ PISCATRIX.

2 U

OAHU.

*PHÆORNIS OAHENSIS.
*CHASIEMPIS GAYI.
*HEMIGNATHUS ELLISIANUS.
*HETERORHYNCHUS LUCIDUS.
*OREOMYZA MACULATA.
*CHLORODREPANIS CHLORIS.
*HIMATIONE SANGUINEA.
*VESTIARIA COCCINEA.
*LOXOPS RUFA.
*PSITTIROSTRA OLIVACEA.
*MOHO APICALIS.
*ASIO ACCIPITRINUS SANDWICHENSIS.
CIRCUS HUDSONICUS.
*FULICA ALAI.

*GALLINULA SANDVICENSIS.
*HIMANTOPUS KNUDSENI.
HETERACTITIS INCANUS.
NUMENIUS TAHITIENSIS.
ARENARIA INTERPRES.
CHARADRIUS DOMINICUS FULVUS.
*NYCTICORAX NYCTICORAX NÆVIUS.
*ANAS WYVILLIANA.
SPATULA CLYPEATA.
*STERNA LUNATA.
" „ FULIGINOSA.
*ANOUS HAWAIIENSIS.
*PHAËTHON LEPTURUS.
*FREGATA AQUILA.

MOLOKAI.

*PHÆORNIS LANAIENSIS.
*OREOMYZA FLAMMEA.
*CHLORODREPANIS WILSONI.
*HIMATIONE SANGUINEA.
*PALMERIA DOLEI.
*VESTIARIA COCCINEA.
*DREPANORHAMPHUS FUNEREUS.
*PSITTIROSTRA PSITTACEA.
*MOHO BISHOPI.
*ASIO ACCIPITRINUS SANDWICHENSIS.
*FULICA ALAI.

*GALLINULA SANDVICENSIS.
*HIMANTOPUS KNUDSENI.
HETERACTITIS INCANUS.
NUMENIUS TAHITIENSIS.
ARENARIA INTERPRES.
CHARADRIUS DOMINICUS FULVUS.
*NYCTICORAX NYCTICORAX NÆVIUS.
*ANAS WYVILLIANA.
SPATULA CLYPEATA.
*ANOUS HAWAIIENSIS.

MAUI.

*HETERORHYNCHUS AFFINIS.
*OREOMYZA NEWTONI.
*CHLORODREPANIS WILSONI.
*HIMATIONE SANGUINEA.
*PALMERIA DOLEI.
*VESTIARIA COCCINEA.
*LOXOPS OCHRACEA.
*PSEUDONESTOR XANTHOPHRYS.
*PSITTIROSTRA PSITTACEA.
*ASIO ACCIPITRINUS SANDWICHENSIS.
*FULICA ALAI.
*GALLINULA SANDVICENSIS.
*HIMANTOPUS KNUDSENI.

HETERACTITIS INCANUS.
HETEROPYGIA ACUMINATA.
NUMENIUS TAHITIENSIS.
ARENARIA INTERPRES.
CHARADRIUS DOMINICUS FULVUS.
*NYCTICORAX NYCTICORAX GRISEUS.
DEMIEGRETTA SACRA.
*ANAS WYVILLIANA.
CHEN HYPERBOREUS.
BRANTA NIGRICANS.
*NESOCHEN SANDVICENSIS (?).
*ANOUS HAWAIIENSIS.
*PHAËTHON LEPTURUS.

LANAI.

*PHÆORNIS LANAIENSIS.
*HEMIGNATHUS LANAIENSIS.
*OREOMYZA MONTANA.
*CHLORODREPANIS WILSONI.
*HIMATIONE SANGUINEA.
*VESTIARIA COCCINEA.
*PSITTIROSTRA PSITTACEA.
*ASIO ACCIPITRINUS SANDWICHENSIS.
*FULICA ALAI.

*GALLINULA SANDVICENSIS.
*HIMANTOPUS KNUDSENI.
HETERACTITIS INCANUS.
NUMENIUS TAHITIENSIS.
ARENARIA INTERPRES.
CHARADRIUS DOMINICUS FULVUS.
*NYCTICORAX NYCTICORAX NÆVIUS.
*ANAS WYVILLIANA.
*ANOUS HAWAIIENSIS.

HAWAII.

* PHLEORNIS OBSCURA.
* CHASIEMPIS SANDWICHENSIS.
* HEMIGNATHUS OBSCURUS.
* HETERORHYNCHUS WILSONI.
* VIRIDONIA SAGITTIROSTRIS.
* OREOMYZA MANA.
*   „   PERKINSI.
* CHLORODREPANIS VIRENS.
* HIMATIONE SANGUINEA.
* VESTIARIA COCCINEA.
* DREPANIS PACIFICA.
* LOXOPS COCCINEA.
* CIRIDOPS ANNA.
* PSITTIROSTRA PSITTACEA.
* LOXIOIDES BAILLEUI.
* RHODACANTHIS PALMERI.
*   „   FLAVICEPS.
* CHLORIDOPS KONA.
* CHÆTOPTILA ANGUSTIPLUMA.
* MOHO NOBILIS.
* CORVUS HAWAIIENSIS.
* ASIO ACCIPITRINUS SANDWICHENSIS.
  CIRCUS HUDSONIUS.
* BUTEO SOLITARIUS.

* PENNULA MILLSI.[1]
* FULICA ALAI.
* GALLINULA SANDVICENSIS.
* HIMANTOPUS KNUDSENI (?).
  CRYMOPHILUS FULICARIUS.
  GALLINAGO DELICATA (?).
  HETERACTITIS INCANUS.
  NUMENIUS TAHITIENSIS.
  CALIDRIS ARENARIA.
  ARENARIA INTERPRES.
  CHARADRIUS DOMINICUS FULVUS.
* NYCTICORAX NYCTICORAX NÆVIUS.
  MERGUS SERRATOR.
* ANAS WYVILLIANA.
  DAFILA ACUTA.
  SPATULA CLYPEATA.
  ANSER ALBIFRONS GAMBELI.
* NESOCHEN SANDVICENSIS.
* STERNA LUNATA.
  LARUS GLAUCESCENS.
* ANOUS HAWAIIENSIS.
* DIOMEDEA IMMUTABILIS.
* PHAËTHON LEPTURUS.

[1] The home of *P. sandwichensis* is unknown!

www.ingramcontent.com/pod-product-compliance
Lightning Source LLC
Chambersburg PA
CBHW021355210326
41599CB00011B/880